THE ALKALOIDS

Chemistry and Biology

VOLUME **66**

THE ALKALOIDS
Chemistry and Biology

VOLUME **66**

Edited by

Geoffrey A. Cordell
Evanston, Illinois

ELSEVIER

Amsterdam • Boston • Heidelberg • London • New York • Oxford
Paris • San Diego • San Francisco • Sydney • Tokyo
Academic Press is an imprint of Elsevier

ACADEMIC
PRESS

Academic Press is an imprint of Elsevier
84 Theobald's Road, London WC1X 8RR, UK
Radarweg 29, PO Box 211, 1000 AE Amsterdam, The Netherlands
Linacre House, Jordan Hill, Oxford OX2 8DP, UK
30 Corporate Drive, Suite 400, Burlington, MA 01803, USA
525 B Street, Suite 1900, San Diego, CA 92101-4495, USA

First edition 2008

ISBN: 978-0-12-374520-0
ISSN: 1099-4831

For information on all Academic Press publications
visit our website at books.elsevier.com

Printed and bound in USA

08 09 10 11 12   10 9 8 7 6 5 4 3 2 1

CONTENTS

Numbers in parentheses indicate the pages on which the authors' contributions begin.

Brad J. Andersh (191), Department of Chemistry and Biochemistry, Bradley University, Peoria, IL 61625-0208, USA

Athar Ata (191), Department of Chemistry and Biochemistry, Bradley University, Peoria, IL 61625-0208, USA

Jaume Bastida (113), Departament de Productes Naturals, Biologia Vegetal i Edafologia, Facultat de Farmàcia, Universitat de Barcelona, 08028 Barcelona, Spain

Mary J. Garson (215), School of Molecular and Microbial Sciences, The University of Queensland, Brisbane, Qld 4072, Australia

Toh-Seok Kam (1), Department of Chemistry, University of Malaya, 50603 Kuala Lumpur, Malaysia

Kuan-Hon Lim (1), Department of Chemistry, University of Malaya, 50603 Kuala Lumpur, Malaysia

Maribel G. Nonato (215), Research Center for the Natural Sciences, College of Science, Graduate School, University of Santo Tomas, España, Manila 1015, Philippines

Edison J. Osorio (113), Grupo de Investigación en Sustancias Bioactivas, Facultad de Química-Farmacéutica, Universidad de Antioquia, A. A. 1226, Medellín, Colombia

Sara M. Robledo (113), Programa de Estudio y Control de Enfermedades Tropicales, Facultad de Medicina, Universidad de Antioquia, Medellín, Colombia

Hiromitsu Takayama (215), Graduate School of Pharmaceutical Sciences, Chiba University, Chiba 263-8522, Japan

The four chapters in this volume reflect some very interesting aspects of the diversity of global alkaloid research in its various chemical and biological applications with contributions from several different countries.

Kam and Lim review the structural diversity and biological activities represented by the monomeric and bis-monoterpenoid indole alkaloids isolated in recent studies of the alkaloids of *Kopsia*. An important aspect of this work is the continuing evolution of the structural diversity of the indole alkaloids, which represent some significant challenges in developing biogenetic pathways for their formation.

Osorio, Robledo, and Bastida summarize another important aspect of the biological applications of alkaloids, their antiprotozoal activity. This activity was founded in the 17th century discovery of the antimalarial activity of *Cinchona*, and extends today to a diverse array of alkaloid structures and protozoa.

The genus *Buxus* is an important source of a select type of steroidal alkaloids. As Ata and Andersh have summarized, in the recent past, this group has expanded rapidly to yield a series of interesting structures and biological activities.

Pandanus species are an important economic product in several countries of Southeast Asia, yet they have received limited chemical and biological study. A collaboration between groups in the Republic of the Philippines (Nonato), Japan (Takayama), and Australia (Garson) describes their studies on the alkaloids of the genus *Pandanus* and those of other research groups.

Geoffrey A. Cordell
Evanston, Illinois

## CHAPTER 1

Alkaloids of *Kopsia*

Toh-Seok Kam* and **Kuan-Hon Lim**

I. INTRODUCTION

Plants of the genus *Kopsia* (Apocynaceae) are distributed from southern China and Burma to northern Australia and Vanuatu. The genus is, however, most diverse in Peninsular Malaysia and Sarawak (Malaysian Borneo) (1). All species are shrubs

Department of Chemistry, University of Malaya, 50603 Kuala Lumpur, Malaysia

* Corresponding author.
E-mail address: tskam@um.edu.my (T.S. Kam).

The Alkaloids, Volume 66
ISSN: 1099-4831, DOI 10.1016/S1099-4831(08)00201-0

or small trees, and due to their attractive appearance (the most distinguishing feature being the showy white flowers with red, pink, or yellow "eyes"), a number have become widely cultivated as garden or ornamental plants. The genus was first published in 1823 by Blume in honor of the Dutch botanist J. Kops, with one species *K. arborea* (2). Later botanical studies include a preliminary partial revision by Markgraf (3) and a chemotaxonomic study by Sévenet *et al.* (4). The most recent and comprehensive revision of the genus, however, is that of Middleton in which 24 species are recognized and four new species are described (1,5). In this review, we shall follow the classification according to Middleton (1), with the species attributed in the original reports cited in parenthesis.

The first *Kopsia* alkaloid isolated was kopsine (**1**) (6). The structure was, however, only solved in the 1960s after considerable classical degradation studies coupled with the introduction of high-resolution mass spectrometry (7–17). Additional confirmation was later provided by chemical correlation of kopsine with minovincine (18), as well as by X-ray crystallographic analysis of the methyl iodide salt of (−)-kopsanone (19,20). Other notable examples of early *Kopsia* alkaloids include fruticosine (**2**) and fruticosamine (**3**) from *K. fruticosa* (14,21–25), and kopsingine (**4**) from *K. singapurensis* (26). These alkaloids have also been discussed in previous volumes of this, as well as other, series (27–31). In more recent times, plants of this genus have proven to be fertile sources of many alkaloids with unusual and fascinating molecular structures, as well as interesting biological activities, and a review chapter devoted exclusively to the *Kopsia* alkaloids appears timely. The present review shall therefore focus on the chemistry and pharmacology of these more recent *Kopsia* alkaloids. The organization of the chapter will be based on the alkaloid structure type, in order of increasing complexity, and approximately along the lines of a progressing biosynthetic pathway. Under each section, aspects of structure elucidation, chemistry, synthesis, and biological activity of the alkaloids concerned will be addressed. Finally, the occurrence of alkaloids in *Kopsia* species which have been chemically investigated will be summarized.

1 kopsine   **2** R^1 = OH, R^2 = H   **4** R = CO$_2$Me kopsingine
                      3 R^1 = H, R^2 = OH

II. MONOTERPENE ALKALOIDS

The monoterpene alkaloids constitute a relatively small group of compounds and occur in several species, including *K. pauciflora*, *K. profunda* (*K. macrophylla*), and *K. dasyrachis*, from which several new monoterpene alkaloids (**5–13**) related to skytanthine have been recently isolated.

The North Borneo species *K. pauciflora* provided six such monoterpene alkaloids, namely, kinabalurines A–F (**5–10**), which are hydroxyskytanthine derivatives (32,33). The first alkaloid isolated was kinabalurine A (**5**), which was obtained as colorless plates. The mass spectrum showed a molecular ion at m/z 183 ($C_{11}H_{21}NO$) accompanied by fragments due to loss of H, Me, and OH, and other fragments at m/z 84, 58, and 44, characteristic of skytanthine-type alkaloids. The IR spectrum indicated the presence of a hydroxyl group ($3357\,cm^{-1}$), and this was supported by the presence of an OH resonance ca. δ 3.27 in the ^{1}H NMR spectrum. The ^{13}C NMR spectrum accounted for all 11 carbon atoms and the presence of an oxymethine was confirmed by the resonance at δ 80.0. Other significant peaks in the ^{1}H NMR spectrum included a pair of three-H doublets at δ 0.97 and 1.06, corresponding to two CH_3CH– groups, and an *N*-methyl singlet at δ 2.25. The spectral data thus suggested that kinabalurine A is a hydroxyskytanthine derivative, and COSY and HETCOR experiments confirmed that hydroxy substitution was at C(7) and allowed the full assignments of the NMR spectral data. In addition, the observed J_{1-9} value of 10 Hz required a *trans* ring junction. The NMR data, however, were insufficient to establish the stereochemistry completely and unequivocally and for this purpose X-ray diffraction analysis was undertaken, which established the structure of kinabalurine A. Kinabalurine A was the second 7-hydroxyskytanthine reported, the first being incarvilline (**14**) isolated from the Chinese plant *Incarvillea sinensis*. The structure of incarvilline was also established by X-ray analysis (34). Kinabalurine A differs from incarvilline in having a *trans* ring junction, a 7β-OH substituent, and a 4α-methyl group. Kinabalurine B (**6**) is the 7-oxo derivative of kinabalurine A as shown by the spectral data, as well as by its ready formation via oxidation of kinabalurine A. Similarly, kinabalurine C (**7**) was readily shown to be the *N*-demethyl derivative of kinabalurine B from the spectral data (loss of the *N*-methyl signal in the ^{1}H and ^{13}C NMR, and the presence of a secondary amine absorption in the IR at $3400\,cm^{-1}$). The *trans* ring junction in kinabalurine C was clearly shown in the 600 MHz ^{1}H NMR spectrum, which showed the H(9) signal as a quartet of doublets ($J_{5\beta-9\alpha} = J_{1\beta-9\alpha} = J_{8\beta-9\alpha} = 12\,Hz$, $J_{1\alpha-9\alpha} = 4\,Hz$).

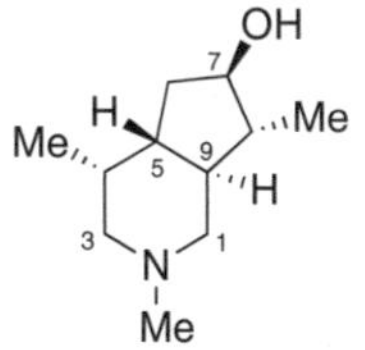

5 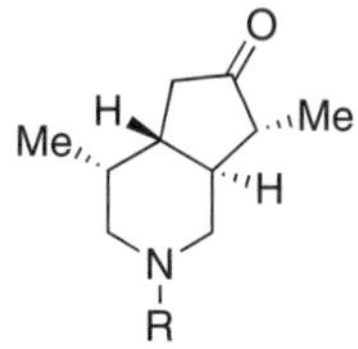   6 R = Me; 7 R = H 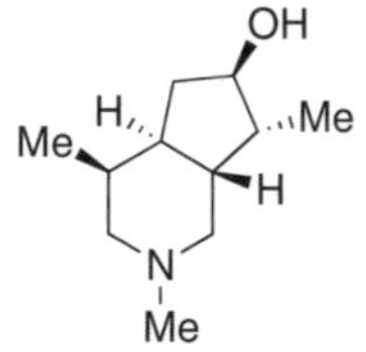   8 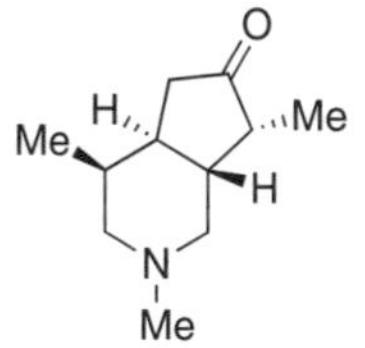   9

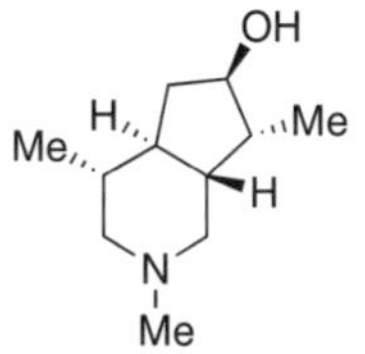

10 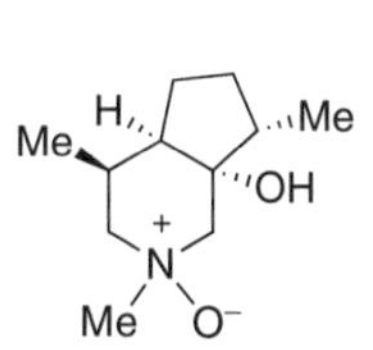   11 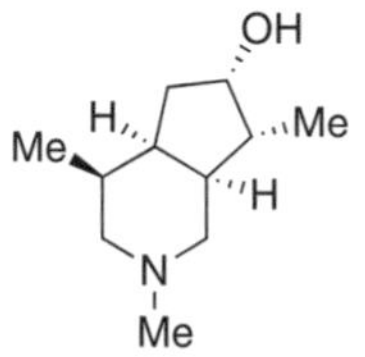   14 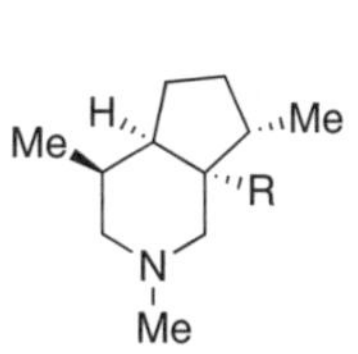   15 R = OH; 16 R = H

The spectral data for kinabalurine D (**8**) showed it to be yet another 7-hydroxyskytanthine diastereomer, but proved inadequate for definitive assignment of stereochemistry. To this end, kinabalurine D was converted to the quaternary ammonium salt, which provided suitable crystals for X-ray analysis. Kinabalurine D differs from kinabalurines A–C in having a 4β-methyl group and a *trans* ring junction in which the stereochemistry of H(5) and H(9) are now reversed. Kinabalurine E (**9**) is the 7-oxo derivative of kinabalurine D, as shown by the spectral data and by chemical correlation (PCC oxidation) with **8**.

Kinabalurine F (**10**) was obtained in minute amounts, and its structure elucidation relied mainly on analysis of the 600 MHz NMR data and by comparison with **5**, **8**, and incarvilline (**14**). The orientation of the 7-hydroxy group of kinabalurine F was deduced to be β based on comparison of the observed C(7) shift (δ 81) with those of **5** (δ 80) and **8** (δ 81), which also have a 7β-OH. The C(7) shift in incarvilline, which has a 7α-OH, was shifted upfield to about δ 73. The observed NOE interactions from H(7α) to the 8-methyl and from H(6α) to H(5) fixed their respective stereochemistry. Likewise, the observed H(1β)/H(8β) NOE interaction allowed the assignment of H(1α), which appeared as a triplet with $J = 10.5$ Hz, requiring H(9) and H(1α) to be *trans*-diaxial, which is possible only if H(9) is β. The observed H$_2$(3) signals as a triplet with $J = 11$ Hz and a doublet of doublets ($J = 11, 2$ Hz) are only consistent with H(4β), resulting in H(4β) and H(3α) being *trans*-diaxial to each other. The 4-methyl of kinabalurine F therefore has an α-orientation.

Kinabalurine G (**11**) was isolated from the leaf extract of *K. dasyrachis*, another *Kopsia* from Malaysian Borneo (35). It was the most polar alkaloid from the leaf extract. The mass spectrum showed fragments typical of skytanthine-type alkaloids, while the IR spectrum indicated the presence of a hydroxyl group. The ^{1}H NMR spectrum indicated the presence of two CH$_3$CH groups and an *N*-methyl group, which was rather deshielded at δ 3.17. This observation, coupled with the polar nature of this compound, and the observation of a strong M-16 fragment in the mass spectrum, suggested that compound **11** is an *N*-oxide. This was readily confirmed by FeSO$_4$ reduction of **11**, which yielded the parent monoterpene alkaloid **15**. The *N*-methyl signal was shifted upfield to δ 2.30, while the resonances of the two α-carbons, C(1) and C(3), were also shifted upfield from δ 67.3 and 67.2 to δ 62.5 and 57.7, respectively. The presence of a low-field, quaternary carbon signal at δ_C 79.1 indicated that the hydroxyl function was attached to a quaternary carbon, i.e., C(5) or C(9), based on a skytanthine-type carbon skeleton.

Detailed analysis of the ^{1}H and ^{13}C NMR spectral data (COSY, HMQC, HMBC, NOE) and comparison with δ-skytanthine (**16**) (36) enabled placement of the OH function on C(9) and allowed full assignment of the NMR spectral data, as well as elucidation of the stereochemistry. The parent monoterpene, 9-hydroxy-δ-skytanthine (**15**), is unknown, and was not detected in this study, although a 9-hydroxyskytanthine of unknown stereochemistry, as well as a β-skytanthine *N*-oxide, have been previously reported from *Tecoma stans* (37) and

Skytanthus acutus (38), respectively. The kinabalurines, together with incarvilline, provide a useful array of stereoisomers in this series with various ring junction, 7-hydroxy, and 4- and 8-methyl group stereochemistry.

K. *profunda* (K. *macrophylla*) (1) provided two more new monoterpene alkaloids, kopsilactone (12) and kopsone (13), in addition to the known alkaloids 5,22-dioxokopsane, dregamine, akuammiline, tabernaemontanine, deacetylakuammiline, norpleiomutine, and kopsoffine (39). The IR spectrum of kopsilactone (12) indicated the presence of a γ-lactone unit (1770 cm^{-1}), which was supported by the observation of a quaternary carbon resonance at δ 176. The observed J_{3-4} value of 11 Hz required a *trans*-diaxial arrangement between H(3α) and H(4β), while the estimated J_{4-5} value of ca. 4 Hz suggested a *cis* relationship between H(4) and H(5). An equatorial H(5) required a *cis* ring junction between the piperidine and the five-membered ring, which, in turn, fixed the stereochemistry of the lactone–piperidine ring junction.

The second monoterpene alkaloid, kopsone (13), gave a molecular ion, which analyzed for $C_{11}H_{19}NO$. The IR (1720 cm^{-1}) and ^{13}C NMR (δ 218) data indicated the presence of a ketone function. Other groups indicated by the NMR spectra were two CHMe, an *N*-methyl, three methylenes (one deshielded at δ_C 56), and four methines (one deshielded at δ_C 72). These, as well as a postulated common origin of 12 and 13 from the hypothetical 9-hydroxysky-tanthine precursor 17 (Scheme 1), led to the proposed structure for kopsone. The relative stereochemistry was deduced from analysis of the ^{1}H NMR spectrum.

The leaves of K. *dasyrachis* also gave kopsirachine (18), which is constituted from union of the flavonoid, catechin, and two units of skytanthine. The gross structure was deduced from spectroscopic and chemical

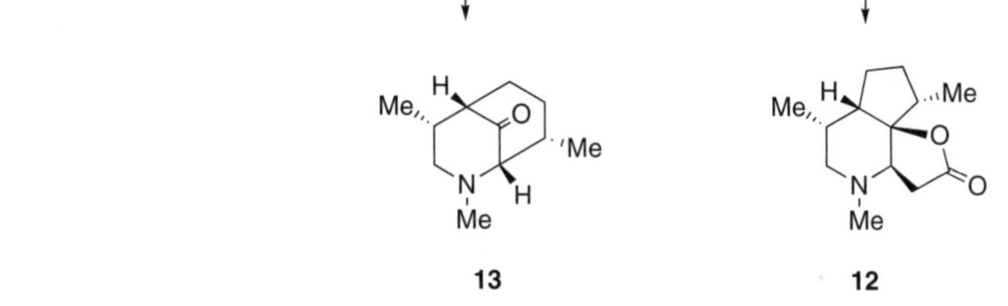

Scheme 1

evidence, but the stereochemistry of the skytanthine units in **18** remains to be firmly established (36).

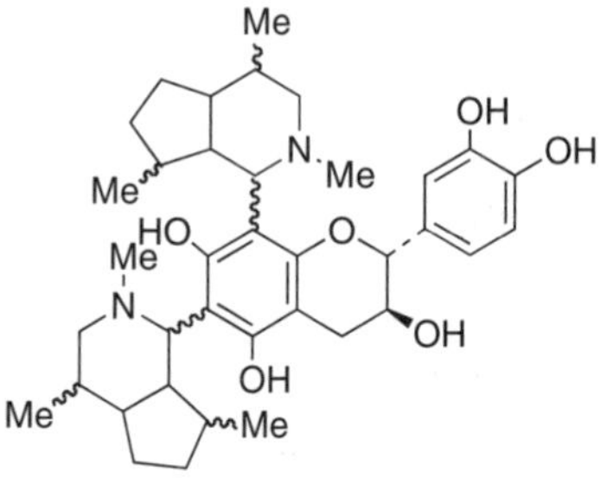

18 kopsirachine

III. SIMPLE INDOLE ALKALOIDS

The simple β-carboline alkaloid harmane (**19**), although widely distributed in several plant families, is rarely encountered in the Apocynaceae. It has been recently obtained for the first time from *Kopsia* from *K. griffithii* and was found to display moderate leishmanicidal activity against *Leishmania donovani* (40,41).

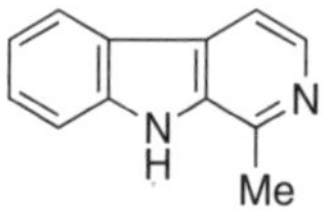

19 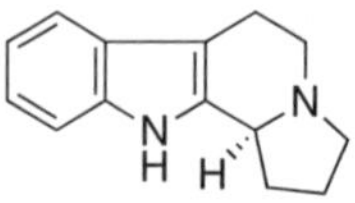   **20** 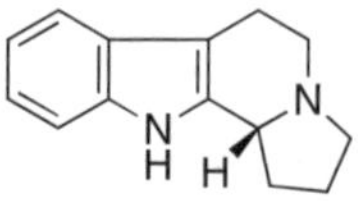   **21**

A new β-carboline, (+)-harmicine (2,3,5,6,11,11b-hexahydro-1*H*-indolizino[8,7-b] indole), which has been previously synthesized in racemic form (42–46) was also isolated for the first time as an optically active natural product from *K. griffithii*, and was assigned the structure **20** (40). Recent enantioselective syntheses of both (*S*)-(−)- and (*R*)-(+)-harmicine resulted in the correct assignment of the absolute configuration of naturally occurring (+)-harmicine as (*R*), as shown in **21** (47–49).

The synthesis of (*S*)-harmicine (Scheme 2) was based on the use of the (*S*)-1-allyl-1,2,3,4-tetrahydro-β-carboline (**23**) as the starting compound, which was in turn obtained by a diastereoselective C(1)-allylation of the appropriate β-carboline precursor **22**, incorporating a glutamic acid-derived chiral auxiliary. Subsequent hydroboration, followed in succession by removal of the trichloro-ethoxycarbonyl group, and cyclization through a Mitsunobu reaction, led eventually to (*S*)-harmicine (**20**) (47,48).

The synthesis of (*R*)-harmicine (**21**), on the other hand, was based on an asymmetric transfer hydrogenation of the appropriate iminium salt **24**, in the presence of the chiral Ru(II) catalyst (*S,S*)-**25**, as the key step (Scheme 3). The structure of the product was also confirmed by an X-ray crystallographic analysis (49).

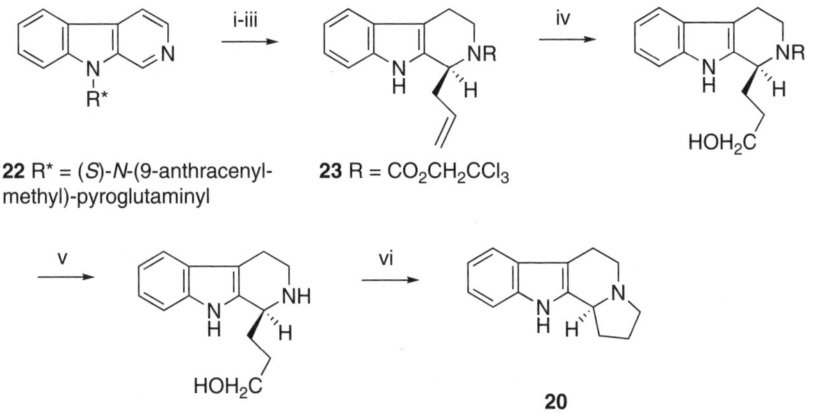

22 R* = (*S*)-*N*-(9-anthracenyl-methyl)-pyroglutaminyl　　**23** R = CO₂CH₂CCl₃

Scheme 2　Reagents: (i) Bu₃Sn–CH₂CH=CH₂, ClCO₂CH₂CCl₃; (ii) NaOH–THF/H₂O; (iii) Et₃SiH–TFA; (iv) BH₃; NaOH, H₂O₂; (v) Zn, AcOH, THF–H₂O; (vi) Ph₃P, DEAD, CH₂Cl₂.

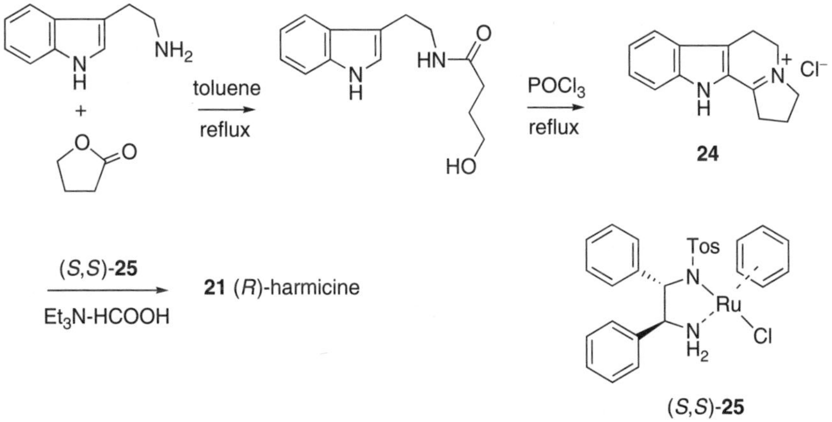

Scheme 3

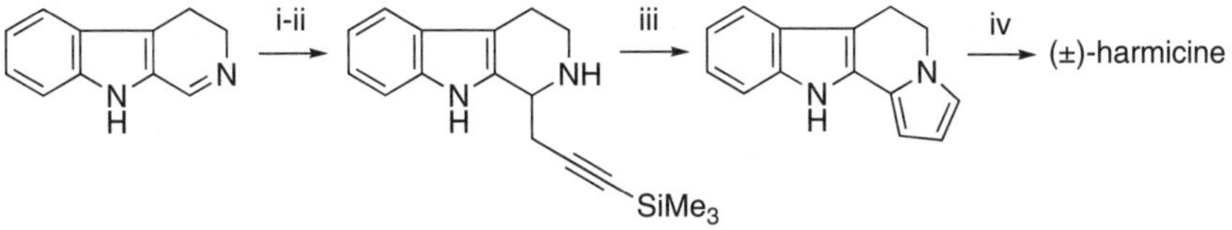

Scheme 4　Reagents: (i) BF₃·OEt₂, THF; (ii) Me₃SiCCCH₂MgBr, Et₂O; (iii) AgOAc, CH₂Cl₂; (iv) H₂, 5% Rh/C, MeOH–AcOH.

A recent synthesis of racemic harmicine is noteworthy for its economy and elegance (50). The synthesis commenced with 3,4-dihydro-β-carboline and involved three steps, namely, alkylation with 3-trimethylsilylpropargylmagne-sium bromide, Ag(I)-promoted cyclization to a pyrrole, and chemoselective hydrogenation of the pyrrole ring (Scheme 4).

IV. CORYNANTHE, AKUAMMILINE, VINCORINE, ASPIDODASYCARPINE, AND PLEIOCARPAMINE ALKALOIDS

Only a few alkaloids belonging to these groups occur in *Kopsia*. The most common include tetrahydroalstonine, akuammiline and its derivatives, aspido-dasycarpine, lonicerine, and pleiocarpamine. Recently, a pair of unusual indole regioisomers, arboricine (**26**) and arboricinine (**27**), were isolated from the stem-bark extract of *K. arborea* (51).

Aboricine (**26**) showed NMR spectral data consistent with that of a tetracyclic corynanthean alkaloid with an acetyl substituent at C(15) and an ethylidene side chain at C(20). The presence of the diagnostic Wenkert–Bohlmann bands in the IR spectrum was consistent with an axial bridgehead hydrogen within a *trans*-quinolizidine skeleton. Since on biogenetic grounds C(15) is an invariant stereogenic center in the corynanthean alkaloids with H(15α), the orientation of H(3) was assigned as β from the observed $J_{3-14\alpha}$ value of 10 Hz, requiring these hydrogens to be in a *trans*-diaxial arrangement, as well as from the observed NOEs for NH/H(14β), H(17)/H(14β), and H(15)/H(18), the latter NOE also confirming the *E*-geometry of the C(19)–C(20) double bond. The configuration at C(3) is therefore similar to that in deplancheine (**28**) (52,53), and aboricine (**26**) is thus the 15β-acetyl derivative of deplancheine. This in itself would not constitute an unusual finding, if not for the isolation of the regioisomeric alkaloid arboricinine (**27**) from the same plant.

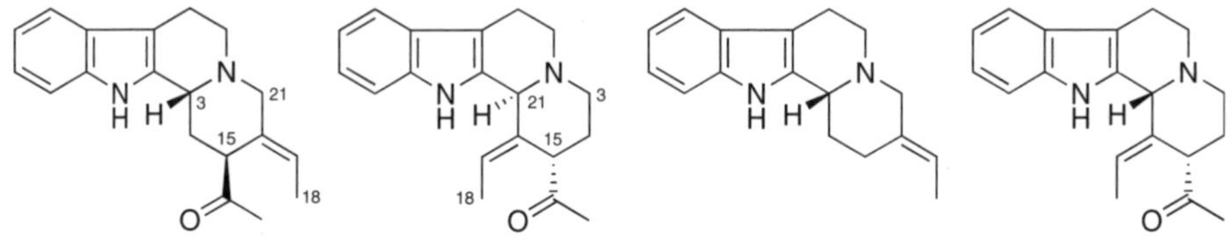

26 arboricine   27 arboricinine   28 deplancheine   29

Arboricinine (**27**) was an isomer of arboricine (**26**) as shown by the mass spectrum. The IR spectrum also showed similar Wenkert–Bohlmann bands indicating the presence of a *trans*-quinolizidine skeleton. Despite some similarities, there were however, significant differences in the MS fragmentation, ^{1}H, and ^{13}C NMR data. Furthermore, the COSY and HMQC data revealed the presence of different partial structures compared to those present in **26**. Thus, although the C(5)–C(6) unit appeared intact (with essentially similar chemical shifts), an isolated aminomethine was now present, as well as an NCH$_2$CH$_2$CH fragment. In addition, a strong NOE was observed in the case of **27** between the indolic NH and the vinylic H(19), which was not the case with **26**, although the H(18)/H(15) NOE seen for **26** was also observed for **27**. The HMBC data also revealed certain clear differences in **27** when compared with **26**. The key observations concerned the change in the location of the ethylidene function relative to the indole ring. The observed three-bond correlations from H(5) and

H(15) to the aminomethine and from the aminomethine-H to C(19) and C(2) (two-bond correlation) indicated that the aminomethine was linked to C(2), as well as to the quaternary C(20) from which the ethylidene side chain was branched. These observations sufficed to allow the assembly of the structure of arboricinine as shown in **27**, revealing a structure regioisomeric with **26** in which the principal changes have occurred in ring D. The structure was consistent with the extensive HMBC, as well as NOE/NOESY data, in particular, the observed strong NOE between the vinylic H(19) and the indolic NH, which becomes intelligible in the light of the rearranged ring D.

Arboricinine (**27**) is therefore a regioisomer of arboricine (**26**) and appears to have been derived from **26** following cleavage of the C(2)–C(3) bond, followed in succession by rotation about the C(5)–N(4) bond, isomerization to the C(21)–N(4) iminium ion, and finally, cyclization by bond formation between C(21) and C(2). A possible pathway based on these considerations is shown in Scheme 5, which leads to the numbering system adopted for **27**.

Based also on this presumed origin, as well as the *trans*-quinolizidine configuration required by the Wenkert–Bohlmann bands, the relative configuration of arboricinine is as shown in **27**, in which the configurations of C(21) (cf. C(3) in **26**), C(15), and N(4) are inverted compared to those in arboricine (**26**). The orientations of H(21) and H(15) were determined to be α and β, respectively (thus ruling out the alternative stereoisomer **29**), based on the observed NOEs between H(21)/H(17) and H(21)/NH. Although from a biogenetic viewpoint, somewhat similar processes have been proposed to link some of the monoterpenoid indole alkaloid classes (e.g., stemmadenine or preakuammicine to condylocarpine), this appears to be the first instance where such a regioisomeric relationship, arising from a rotated ring D, has been found in a pair of alkaloids of the same structure class and from the same plant. Both **26** and **27** showed no appreciable cytotoxicity against drug-sensitive and vincristine-resistant KB cells,

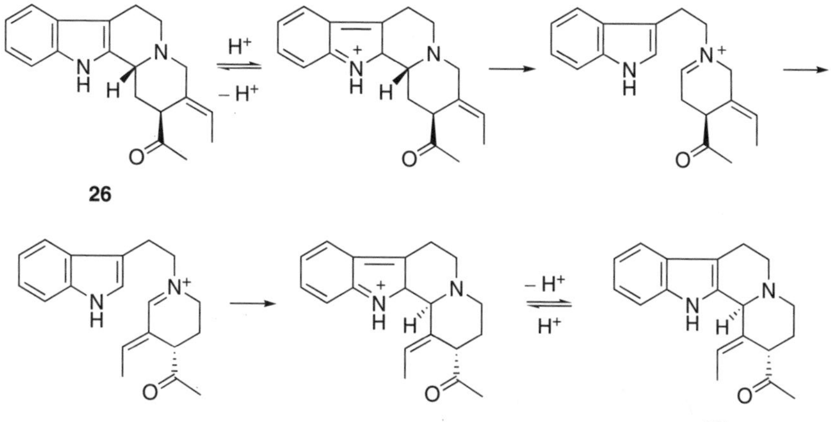

Scheme 5

as well as Jurkat cells ($IC_{50} > 25\,\mu g/mL$ in all cases), but showed moderate ability to reverse multidrug resistance in vincristine-resistant KB (VJ300) cells (IC_{50} 10.8 and $9.2\,\mu g/mL$ for **26** and **27**, respectively, in the presence of $0.1\,\mu g/mL$ of vincristine) (54).

Two recent new akuammiline alkaloids are the C(16) epimers of akuammiline (**30**) and deacetylakuammiline (**31**), namely, **32** (55) and **33** (56), respectively. The structures were readily assigned based on the NMR data, in particular the observed chemical shift changes involving the ester methyl, acetyl (in the case of **32**), and H-17 signals, when compared to akuammiline and deacetylakuammiline. The *N*-oxides of akuammiline and 16-*epi*-deacetylakuammiline (**34** and **35**, respectively) have also been recently isolated from *K. griffithii* (41).

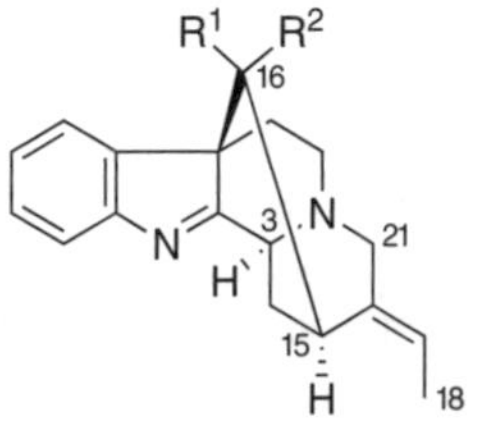

30 $R^1 = CH_2OAc$, $R^2 = CO_2Me$
31 $R^1 = CH_2OH$, $R^2 = CO_2Me$
32 $R^1 = CO_2Me$, $R^2 = CH_2OAc$
33 $R^1 = CO_2Me$, $R^2 = CH_2OH$
34 $R^1 = CH_2OAc$, $R^2 = CO_2Me$, N(4)→O
35 $R^1 = CO_2Me$, $R^2 = CH_2OH$, N(4)→O
36 $R^1 = CHO$, $R^2 = H$ rhazinoline

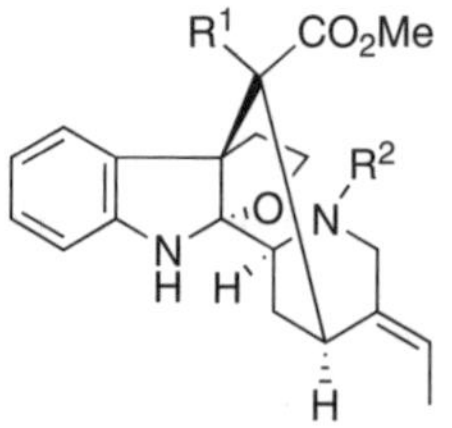

38 $R^1 = CH_2OH$, $R^2 = H$
39 $R^1 = H$, $R^2 = CHO$
40 $R^1 = CH_2OAc$, $R^2 = H$

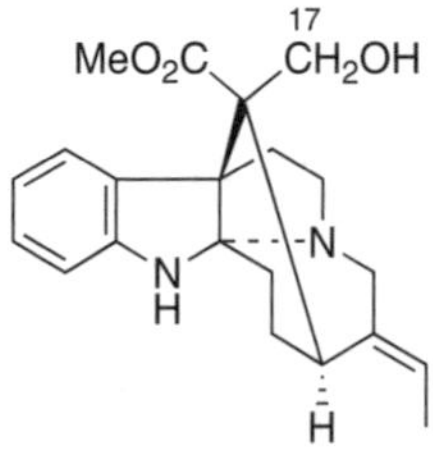

37 vincophylline

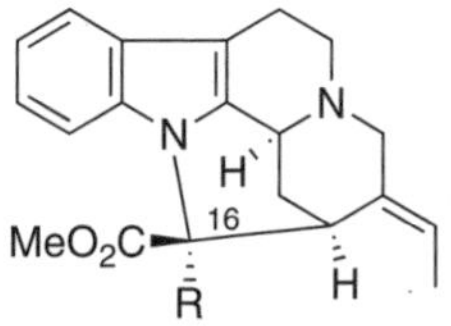

41 R = H
42 R = CH_2OH

Another new akuammiline alkaloid is rhazinoline (**36**), recently isolated from *K. arborea* (54). The UV spectrum showed absorption maxima characteristic of an indolenine chromophore, which was confirmed by the observed carbon resonance at δ 187.3 due to the imine carbon, while the IR and ^{13}C NMR spectra indicated the presence of an aldehyde. The signals observed in the ^{1}H NMR spectrum at δ 4.82 (d, $J = 5.1\,Hz$), 3.31 (br s), and 4.13 (br d, $J = 16.4\,Hz$), corresponding to H(3), H(15), and H(21β), respectively, are characteristic of an akuammiline-type alkaloid, while the COSY and HMQC data also disclosed partial structures that are reminiscent of an akuammiline skeleton, except that the NCHCH$_2$CH partial structure, corresponding to C(3)–C(14)–C(15), has been extended to include a CHCHO fragment, corresponding to C(16)–C(17). These observations revealed the similarity of rhazinoline to the akuammiline-type alkaloid, strictalamine, previously reported from *Rhazya stricta* (57). However, the *W*-coupling observed between H(14α) and H(16) requires the configuration at C(16) to be assigned as *S*, thus indicating that rhazinoline (**36**) is the C(16)-epimer of strictalamine. This was further supported by the NOEs observed between H(6β)/H(16), H(16)/H(18), and H(14β)/H(17). The latter NOE also indicated that the formyl H(17) was located within the shielding zone of the imine double bond which accounted for the unusually shielded low-field resonance observed for the formyl H(17) at δ 8.60 (54).

A new vincorine alkaloid, vincophylline (**37**), has been recently isolated from *K. singapurensis*, representing the first occurrence of a vincorine-type alkaloid in the genus (55). The NMR data of **37**, in particular the characteristic downfield resonance observed at δ 93.5 due to the quaternary C(2) which was linked to two nitrogen atoms, indicated that it possessed a vincorine-type skeleton. This was further supported by the presence of two isolated methylenes (NCH$_2$ and OCH$_2$), an NCH$_2$CH$_2$ unit, a CH$_2$CH$_2$CH unit, and a C=CHCH$_3$ fragment, as deduced from the 2D NMR data. The relative configuration at C(16) was established from the NOESY spectrum, which showed NOEs between H(17) and H(6). This interaction was only possible if the CH$_2$OH group was directed away from the dihydroindole moiety, while the ester group was placed above the dihydroindole moiety. The geometry of the ethylidene side chain was deduced to be *E* from the observed reciprocal NOEs between H(18) and H(15).

Aspidodasycarpine (**38**) is frequently encountered in *Kopsia*. Recently, two new alkaloids of the aspidodasycarpine-type, aspidophyllines A (**39**) and B (**40**) have been isolated from *K. singapurensis* (55,58). The ^{1}H NMR spectrum of **39** was generally similar to that of **38**, which was also isolated, except for several changes. First, the CH$_2$OH group at C(16) was replaced by H. This was indicated by the absence of the signals normally attributable to H(17) in both the ^{1}H and ^{13}C NMR spectra, while H(16) was observed as a doublet at δ 2.83. Another significant change was the presence of a formyl function at N(4). The ^{1}H NMR spectrum showed the formamide-H as a singlet at δ 8.15 (δ_C 164.4). The presence of the formyl group resulted in pronounced changes in the chemical shifts of H(3) and H(21) when compared with aspidodasycarpine (**38**), indicating that the site of substitution of the formyl group was at N(4). In addition, irradiation of N(1)–H resulted in NOE enhancement of H(12) and vice versa, furnishing additional proof that the formyl group was on N(4). The stereochemistry of the ester function at C(16), and that of the ethylidene side chain, were determined from NOE experiments. Irradiation of H(16) resulted in NOE enhancement of H(14), which places it above the dihydroindole moiety while the ester group is directed away from the dihydroindole moiety.

The NOEs seen for H(18)/H(15) and H(19)/H(21) indicated that the geometry of the C(19)–C(20) double bond is E (55). Aspidophylline A (39) showed moderate activity in reversing multidrug resistance in drug-resistant KB cells (55).

The ^{1}H NMR spectrum of 40 was also generally similar to that of aspidodasycarpine (38), except for the replacement of CH_2OH with CH_2OCOCH_3 at C(16), as indicated by the carbon resonances of the acetyl group at δ 20.7 and 169.8 in the ^{13}C NMR spectrum. The methyl hydrogens of the acetyl group were observed upfield (δ 1.97) indicating anisotropic interaction from the benzene ring, and confirming the relative configuration at C(16), which places the CH_2OCOCH_3 group above the aromatic ring (58).

Pleiocarpamine (41) is another alkaloid frequently encountered in *Kopsia*. A new derivative, 16-hydroxymethylpleiocarpamine (42), was recently found in a Malayan *Kopsia* (56). The NMR data were similar to those of pleiocarpamine, except for the replacement of H(16) by a hydroxymethyl group. Establishment of the relative configuration at C(16) in 42 was by a chemical correlation (NaH induced deformylation) with pleiocarpamine.

V. CONDYLOCARPINE, STEMMADENINE, AND AKUAMMICINE ALKALOIDS

Strychnos alkaloids are generally rare in *Kopsia*, but several have been recently isolated. *K. arborea* provided two alkaloids of the condylocarpine-type, which were the two epimers of 19-methoxytubotaiwine (43 and 44) (54).

The UV spectrum of both alkaloids showed the presence of β-anilinoacrylate chromophores, while the IR spectrum showed absorption bands due to NH and conjugated ester functions. The presence of the β-anilinoacrylate moiety was also indicated by the characteristic signals of C(16) and C(2) at δ 96.3 and 169, respectively, in the ^{13}C NMR spectrum. The NMR data of 43 showed a remarkable resemblance to those of lagunamine (19(S)-hydroxytubotaiwine, 45) (59), except for the replacement of the C(20)-hydroxyethyl side chain by a methoxyethyl group. Alkaloid 43 is therefore 19(S)-methoxytubotaiwine. The large coupling constant ($J = 10$ Hz) observed between H(19) and H(20) suggested that the conformation adopted about the C(19)–C(20) bond is one that places the two hydrogens at C(19) and C(20) *anti* to one another. The preferred *anti* conformation was due to the presence of both the OMe and the Me groups at C(19), which apparently causes steric hindrance to free rotation about the C(19)–C(20) bond. This was further supported by the observed NOEs between H(18)/H(21) and C(19)–OMe/H(15), and the absence of NOEs between H(18)/H(15) and C(19)–OMe/H(21). Another noteworthy observation is that the chemical shift of H(15) in 43 was relatively deshielded (δ 3.50) compared to the chemical shift of H(15) in the 19(R)-epimer 44 (δ 3.13) (*vide infra*). This is attributed to paramagnetic deshielding caused by the proximity of the OMe oxygen atom to H(15), which is only possible if rotation about the C(19)–C(20) bond is indeed restricted.

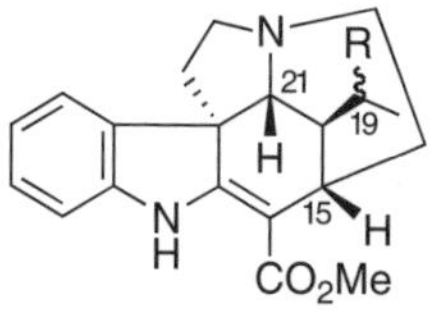

43 R = OMe, 19(*S*)
44 R = OMe, 19(*R*)
45 R = OH, 19(*S*)

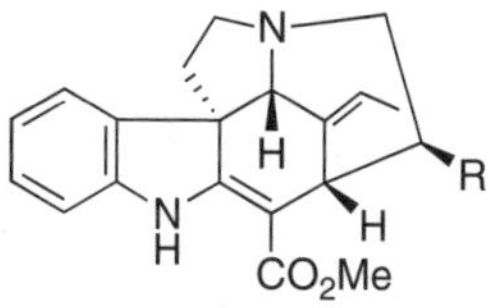

46 R = OH
47 R = H

The ^{1}H and ^{13}C NMR data of **44** were similar in all respects to those of **43**, except for the chemical shifts of H(15), H(21), and C(19)–OMe in the ^{1}H NMR spectrum. This suggested that **44** is the C(19)-epimer of **43**. The 19(*R*) configuration of **44** can be readily verified by applying the same analysis carried out for **43**. In alkaloid **44**, NOEs were observed between H(18)/H(15) and C(19)–OMe/H(21), while NOEs between H(18)/H(21) and C(19)–OMe/H(15) were not seen. In addition, the paramagnetic deshielding experienced by H(15) in **43** was now observed for H(21) in **44**. Alkaloids **43** and **44** showed moderate activity in reversing multidrug resistance in vincristine-resistant KB cells (54).

Another new condylocarpine derivative from *Kopsia* was 14-hydroxy-condylocarpine (**46**). The ^{13}C NMR spectrum was similar to that of condylo-carpine (**47**), except for the resonance due to C(14), which was observed as a deshielded methine signal at δ 69.6. The oxymethine H(14) was seen as a triplet of doublets ($J = 5, 2.5\,\text{Hz}$) at δ 4.12. The axial disposition of the OH substituent was deduced from analysis of the J_{3-14} and J_{14-15} coupling constants (56).

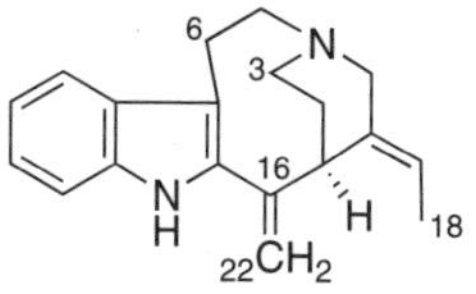

48 pericine
49 N(4)→O

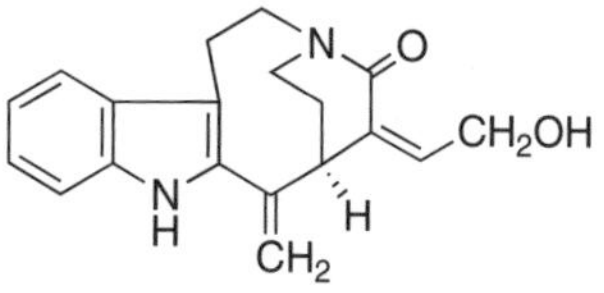

50 pericidine

The stemmadenine-type alkaloid pericine (**48**) was first isolated in 1982 from *Picralima nitida* cell suspension cultures (59), and subsequently in 2002 from *Aspidosperma subincanum* under the name, subincanadine E (60). Pericine and its *N*-oxide (**49**) have also been isolated from *K. arborea*, together with a new oxidized derivative, pericidine (**50**) (61). Pericidine showed typical indole absorptions in the UV spectrum, while the IR spectrum showed bands due to NH, OH, and lactam functions. The observation of a quaternary resonance at δ 169.8 and an oxymethylene at δ 59.4 confirmed the presence of lactam and primary alcohol functions. The ^{1}H NMR spectrum showed a characteristic pair of one-H triplets at δ 5.40 and 5.58 ($J = 1.5\,\text{Hz}$), due to the geminal hydrogens of an exocyclic double bond, reminiscent of that in apparicine (**51**) and pericine (**48**). The ^{1}H NMR spectrum was in fact similar to that of **48**, except for some differences, the most prominent of which was the replacement of the ethylidene side chain by a hydroxyethylidene. Thus, in **50**, the 18-methyl triplet of **48** has been replaced by signals due to the geminal hydrogens of an oxymethylene at δ 4.10 and 4.27 (δ_C 59.4), while the vinylic H(19) was a doublet of doublets at δ 6.09, compared to a quartet at δ 5.62 in **48**. Another departure was the absence of signals due to H(21) in **50**, suggesting that position 21 was the site of oxygenation. This was supported by the downfield shift of one of the H(5) to δ 4.38 due to aniso-tropy by the C(21) lactam carbonyl. These observations suggested oxygenation at C(21) and C(18) of pericine. Placement of the lactam carbonyl at C(21) was supported by the observed three-bond correlations from H(3), H(15), and H(19) to the lactam carbonyl in the HMBC spectrum. The geometry of the C(19)–C(20) double bond was deduced to be Z from the observed NOE between the vinylic H(19) and H(15). Another oxidized derivative of pericine, 15-hydroxypericine (subincanadine D), was also previously obtained from *A. subincanum*. Pericidine (**50**) represents the third member belonging to this small group of tetracyclic indoles characterized by the presence of the C(16)–C(22) exocyclic double bond.

K. arborea also furnished a new cytotoxic pericine-type alkaloid, valparicine (**52**) (62). The UV spectrum (228 and 297 nm) indicated the presence of an unsubstituted indolenine chromophore, which was also supported by the presence of the characteristic imine resonance at δ 186 for C(2) in the ^{13}C NMR spectrum. The HREIMS of **52** showed that it differed from pericine (**48**) by the loss of two hydrogens. The ^{13}C NMR spectrum showed, in addition to the imine resonance at δ 186.4, two other downfield quaternary resonances at δ 139.2 and 144.6. The former was associated with an ethylidene side chain, as shown by the characteristic H signals at δ 5.52 (qd) and 1.78 (d). The other was associated with an exocyclic double bond, from the two broad singlets observed at δ 5.39 and 6.02, due to the geminal hydrogens of C(22) (δ_C 116.4). The remaining quaternary carbon resonance at δ 65.1 was attributed to the indole C(7), which was supported by the observed three-bond correlations from H(9) to C(7) and from H(6) to C(2), C(8) in the HMBC spectrum. The NMR data in fact showed a similarity with those of **48**, except for the formation of a bond between C(3) and the indole C(7), and the change from an indole to an indolenine chromophore. Valparicine (**52**) therefore represents the first member of the pericine-type alkaloids, characterized by a C(16)–C(22) exocyclic double bond, in which bond formation has occurred between C(3) and C(7). Valparicine (**52**) was found to display pronounced cytotoxicity toward drug-sensitive (IC$_{50}$ 3.6 µg/mL), as well as drug-resistant, KB cells (IC$_{50}$ 0.75 µg/mL against KB/VJ300) and Jurkat cells (IC$_{50}$ 0.25 µg/mL) (54).

51 apparicine   **52** valparicine   **53** stemmadenine

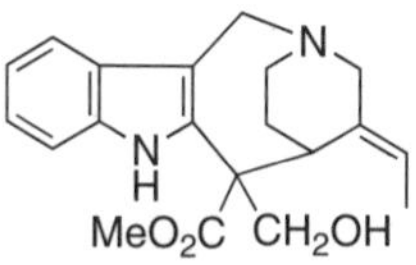

54 vallesamine 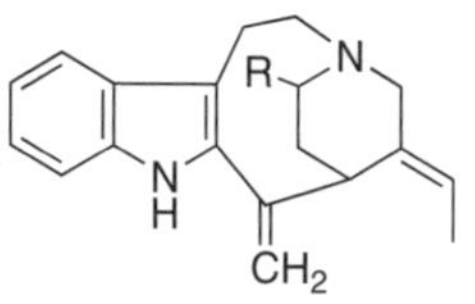   **56** R = CH$_2$COCH$_3$

A partial synthesis of valparicine (**52**) was achieved from pericine-*N*-oxide (**49**) in the context of a general plan to implement a biomimetic transformation of pericine to apparicine (**51**) (62).

The biogenetic relationship between stemmadenine (**53**) and the 5-*nor*-indole derivatives, vallesamine (**54**) and apparicine (**51**), was first suggested by Kutney, who showed that the one-carbon bridge in apparicine was C(6), following excision of C(5) from the original two-carbon tryptamine bridge (63,64). An attractive pathway from stemmadenine (**53**) to apparicine (**51**) was put forward by Potier and coworkers based on a route featuring the Potier–Polonovski fragmentation of the indole alkaloid *N*-oxide precursor (65). A subsequent demonstration of the stemmadenine (**53**) to vallesamine (**54**) transformation by Scott and coworkers provided strong support for the Potier proposal, requiring, however, a modification that the decarboxylation or deformylation step need not be synchronous with fragmentation to the iminium ion (66).

The availability of pericine (**48**) (and its *N*-oxide, **49**), as one of the major alkaloids from *K. arborea*, presented the opportunity to test whether **48** might be a viable precursor to apparicine (**51**), based on the Potier model for C-ring contraction (Scheme 6). In the event it was found, after optimization of the

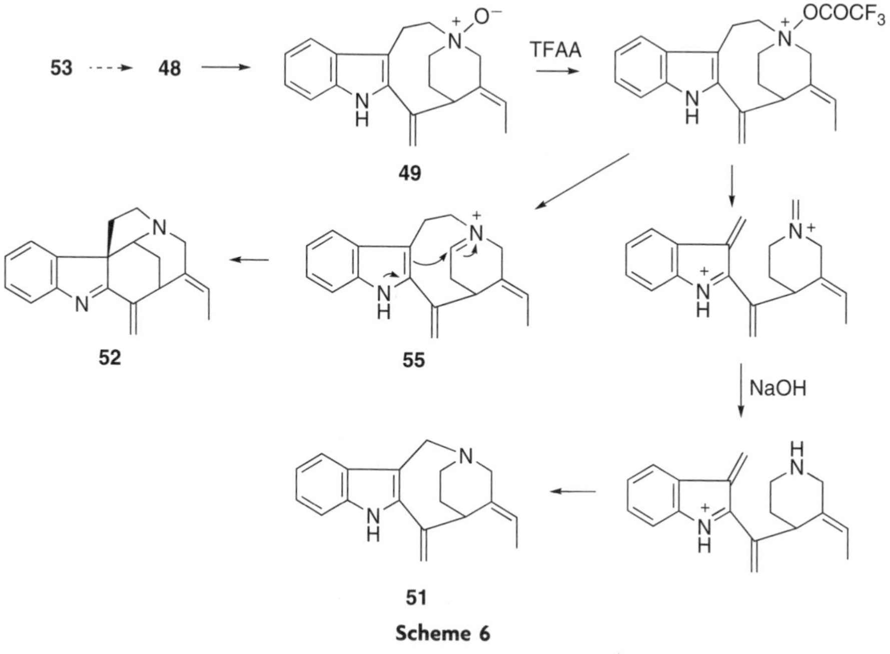

Scheme 6

reaction parameters, that treatment of **49** at 10°C with a 4 equiv excess of trifluoroacetic anhydride (TFAA) added dropwise and at high dilution (100 mL CH$_2$Cl$_2$) for 10 min, followed by hydrolysis (NaOH), gave the two major products apparicine (**51**) and valparicine (**52**), in an overall yield of about 36% (**51**, 26%; **52**, 10%). The formation of valparicine (**52**) is via the alternative cleavage of the N-oxide to the iminium ion **55**. This iminium ion is in equilibrium with valparicine (**52**) in protic media and can be trapped by NaBH$_4$. Indeed when **52** was dissolved in MeOH and NaBH$_4$ was added, **48** was the sole product isolated. The iminium ion **55** could also be trapped as the 3-acetonyl derivative **56**, on exposure of **52** to SiO$_2$ and acetone, in the presence of a trace of concentrated ammonia.

The above partial synthesis of apparicine (**51**) via the Potier–Polonovski reaction has shown that pericine (**48**) can be considered as a viable intermediate in the biogenetic pathway to apparicine (**51**), deriving from stemmadenine (**53**) following deformylation or decarboxylation, and preceding one-carbon extrusion (Scheme 6). Such an alternative would be consistent with both the Kutney (one-carbon extrusion preceding decarboxylation unlikely) (64) and Scott (one-carbon extrusion and deformylation or decarboxylation steps not necessarily synchronous) (66) results. In addition, it has also been shown that the new indole valparicine (**52**) is in all probability biogenetically related to pericine (**48**). It was, however, somewhat puzzling that apparicine (**51**) was not detected among the many alkaloids obtained from *K. arborea* (54), although both **48** and **51** have been previously found in *A. subincanum*.

VI. EBURNANE ALKALOIDS

Alkaloids of the eburnane group are frequently encountered in plants of the genus *Kopsia* (67). *K. larutensis* gave predominantly alkaloids of the eburnane-type (68,69), including (+)-eburnamonine (**57**), (+)-eburnamonine N-oxide, (−)-eburnamine (**58**), (−)-O-ethyleburnamine (**59**), (+)-isoeburnamine (**61**), (+)-eburnamenine (**64**), (−)-kopsinine, and two new alkaloids, larutensine (68,69) and eburnaminol (68). Larutensine (**73**) is isomeric with (+)-eburnamonine (**57**), the predominant alkaloid found in the leaves. The UV spectrum indicated an unsubstituted indole, and the IR spectrum indicated the absence of NH/OH functions, and the presence of an ether function. The ^{1}H and ^{13}C NMR data indicated an eburnane derivative oxygenated at C(16) (δ_C 77.5; δ_H 5.83), but differing from the other eburnane alkaloids occurring in the plant in that the C(20) ethyl substituent was missing. Instead the presence of carbon resonances at δ 58.5 (–CH$_2$O–) and 40.6 (–CH$_2$CH$_2$O–) suggested that ring formation had occurred in which an ether oxygen now links C(18) to C(16). This was supported by the C(18) hydrogens which were shifted to δ 3.80 and 3.95. The proposed structure was in accord with the 2D NMR data. The configurations at C(20) and C(21) were assumed to be similar to those in the other eburnane alkaloids isolated on biogenetic grounds. This being the case, the stereochemistry of the C(16) ether oxygen has to be β to permit formation of the six-membered ring. The likely precursor of larutensine, eburnaminol (**74**), was also isolated from *K. larutensis*, but

Table I Absolute configuration of the eburnane alkaloids

(+)-Eburnamonine (**57**)	(−)-Eburnamonine (**65**)
(−)-Eburnamine (**58**)	(+)-Eburnamine (**66**)
(−)-*O*-Ethyleburnamine (**59**)	(+)-*O*-Ethyleburnamine (**67**)
(−)-*O*-Methyleburnamine (**60**)	(+)-*O*-Methyleburnamine (**68**)
(+)-Isoeburnamine (**61**)	(−)-Isoeburnamine (**69**)
(+)-*O*-Ethylisoeburnamine (**62**)	(−)-*O*-Ethylisoeburnamine (**70**)
(+)-*O*-Methylisoeburnamine (**63**)	(−)-*O*-Methylisoeburnamine (**71**)
(+)-Eburnamenine (**64**)	(−)-Eburnamenine (**72**)

as pointed out by Lounasmaa (70,71) and Kam (72), the original structure proposed (C(16)-α-OH) required amendment. The absolute configuration of C(16) of the eburnane group of alkaloids has also been established based on X-ray analysis of (−)-*O*-ethyleburnamine (**59**) and (+)-isoeburnamine (**61**), representing the eburnamine and isoeburnamine (*epi*-eburnamine) series, respectively (Table I) (72). It has also been pointed out that the coupling constants for the H(16) doublet of doublets can be of diagnostic value, since the pentacyclic alkaloids of the eburnamine series invariably have $J = 9$ and $5\,Hz$, while the corresponding coupling constants in the diastereomeric isoeburnamine (*epi*-eburnamine) series are invariably 4 and $2\,Hz$ due to H(16) being axial and equatorial, respectively, when ring E is in the preferred chair conformation (72).

The structures of eburnaminol (**74**) and larutensine (**73**) have been confirmed by a synthesis reported by Lounasmaa from the previously available indoloquinolizidine ester **75** (Scheme 7). Successive reduction, acetylation, and Fujii oxidation of **75** yielded the enamine **76**, which was alkylated with iodoacetic ester, followed by NaBH$_4$ reduction, to give a mixture of four products. Treatment of two of these, the epimeric esters **77**, with ethanolic sodium ethoxide, resulted in cyclization to 18-hydroxyeburnamonine (**78**) accompanied by its C(20) epimer. Reduction of 18-hydroxyeburnamonine furnished (±)-eburnaminol and 16-*epi*-eburnaminol (**79**), which on overnight treatment with acid gave (±)-larutensine (**71**).

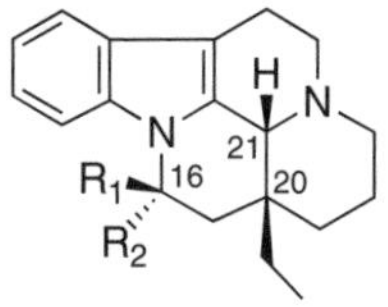

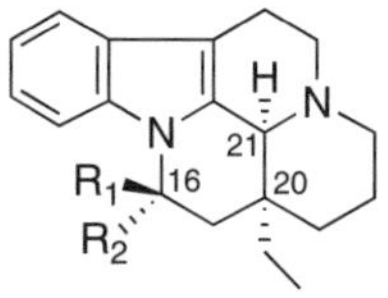

57 R^1, R^2 = O	**65** R^1, R^2 = O
58 R^1 = OH, R^2 = H	**66** R^1 = H, R^2 = OH
59 R^1 = OEt, R^2 = H	**67** R^1 = H, R^2 = OEt
60 R^1 = OMe, R^2 = H	**68** R^1 = H, R^2 = OMe
61 R^1 = H, R^2 = OH	**69** R^1 = OH, R^2 = H
62 R^1 = H, R^2 = OEt	**70** R^1 = OEt, R^2 = H
63 R^1 = H, R^2 = OMe	**71** R^1 = OMe, R^2 = H
64 R^1 = H, R^2 = nil, $\Delta^{16,17}$	**72** R^1 = H, R^2 = nil, $\Delta^{16,17}$

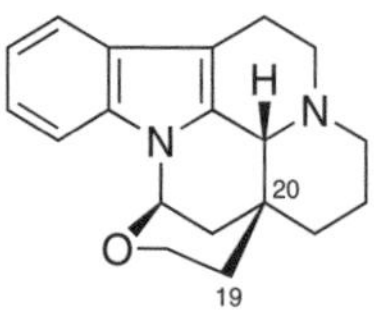

73 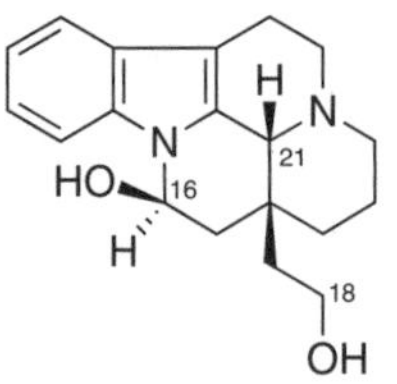   **74** 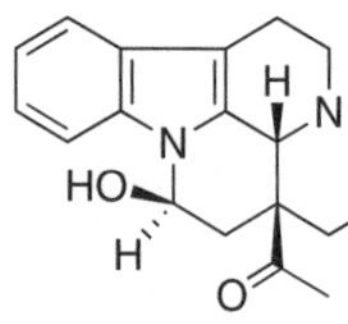   **80**

Another new eburnane alkaloid recently obtained is (+)-19-oxoeburnamine (**80**) from *K. pauciflora* (**73**). The presence of the C(20)-acetyl side chain was indicated by the observed base peak due to loss of water and the acetyl side chain in the mass spectrum, and by the replacement of signals due to the C(20)-ethyl group by signals due to an acetyl group in the ^{1}H and ^{13}C NMR spectra when compared with eburnamine. The configurations at C(21) and C(20) were assumed to be similar to those of (−)-eburnamine (**58**), (+)-isoeburnamine (**61**), and (+)-eburnamonine (**57**), which were also isolated, while the observed coupling constants for the H(16) doublet of doublets ($J = 9, 5$ Hz) allowed the configuration at C(16) (C(16)-β-OH) to be established (**73**).

The Malaysian Borneo species *K. dasyrachis* also furnished several eburnane alkaloids, including (+)-eburnamonine (**57**), (+)-isoeburnamine (**61**), (−)-19(*R*)-hydroxyisoeburnamine (**81a**), and (+)-19(*R*)-hydroxyeburnamine (**81b**). The latter was a new alkaloid, and X-ray diffraction was undertaken to establish the configuration at C(19) (**74**). The occurrence of (+)-eburnamonine and (+)-isoeburnamine in the same plant indicated that **81b** belongs to the same enantiomeric series possessing the 20β, 21β configuration. The observed coupling constants of the H(16) doublet of doublets of 10 and 5 Hz indicated that the orientation of the C(16)–OH is β. This alkaloid also occurs in *K. pauciflora* (**4**) and constitutes the eburnane half of the bisindole alkaloid kopsoffinol, which also occurs in *K. dasyrachis* (**74**).

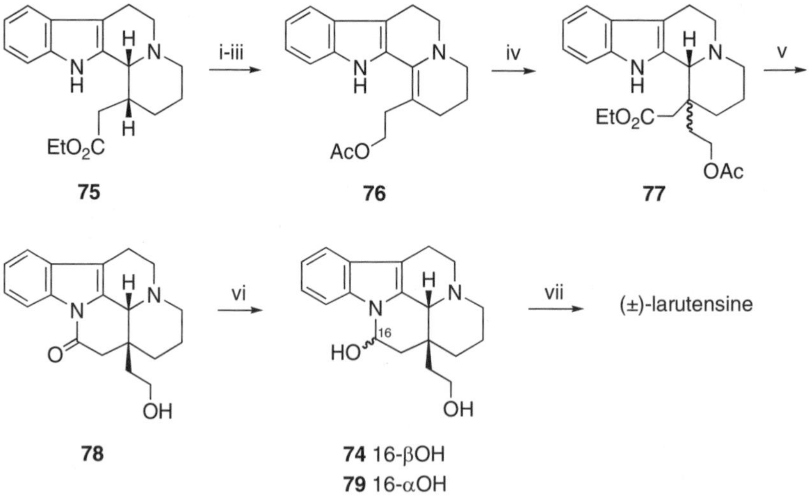

75   **76**   **77**

78   **74** 16-βOH
               79 16-αOH

Scheme 7   Reagents: (i) LiAlH$_4$, THF; (ii) Ac$_2$O, py; (iii) EDTANa$_2$, Hg(OAc)$_2$, EtOH–H$_2$O, Δ; (iv) ICH$_2$CO$_2$Et; NaBH$_4$; (v) NaOEt/EtOH; (vi) LiAlH$_4$, THF; (vii) 5% HCl, rt, overnight.

Recently, two new dihydroeburnane alkaloids, terengganensines A (**82**) and B (**83**), in addition to quebrachamine, isoeburnamine, eburnaminol, and larutensine were obtained from *K. profunda* (*K. terengganensis*) (75). The UV spectrum indicated the presence of dihydroindole chromophores, and, in common with eburnaminol and larutensine, the NMR spectra indicated the absence of NH and ethyl groups, suggesting the presence of an oxidized ethyl side chain. The aromatic C(7) (ca. δ 78) and C(2) (ca. δ 92.5) in these two alkaloids were quaternary centers, and the downfield shifts suggested that the former was α to an oxygen, while the latter was linked to both oxygen and nitrogen. Analysis of the 2D COSY and HMQC spectral data revealed fragments which were in accord with the proposed structures, while long-range C–H correlations (H(18)/C(2), C(16)) in the HMBC spectrum of **82** supported the presence of the two ether bridges linking C(18)/C(2) and C(16)/C(2). The observation of Bohlmann bands in the IR spectrum and the observed NOE between H(19β) and H(21) indicated *trans*-fused C/D and *cis*-fused D/E rings, as in larutensine. The configuration at the centers C(21) and C(20) must be similar to the other eburnane alkaloids present, and furthermore, the formation of the ether bridges required the C(16)–O and C(2)–O bonds to be on the same side of ring E. Finally, a *cis* B/C junction was required to allow a chair conformation for ring C, which fixed the stereochemistry of the C(7)–OH. Terengganensine B (**83**) showed NMR spectral data similar to those of terengganensine A, except for the absence of the C(16) and C(18) oxymethines. Instead, an oxymethylene signal assigned to C(18) and signals due to two olefinic hydrogens typical of 16,17-dehydroeburnamine alkaloids were observed. These data, and the UV spectrum (228, 283, and 307 nm), which was consistent with the presence of an *N*-arylenamine chromophore, supported the proposed structure of terengganensine B (**83**).

81a R^1 = H, R^2 = OH
81b R^1 = OH, R^2 = H

82

83

It is of interest to compare the occurrence of the eburnane alkaloids in Malaysian *Kopsia* with that in the Chinese species (4,67). From such a comparison, it would appear that the Malaysian *Kopsia* (and *Leuconotis*) species elaborate exclusively eburnane alkaloids of one enantiomeric group (20*R*, 21*R* or 20β, 21β configuration) (4,40,68,69,72–75), while the Chinese species appear to elaborate eburnane alkaloids of the opposite enantiomeric group (20*S*, 21*S* or 20α, 21α configuration) (4,76–78). Notable exceptions are the bisindole alkaloids kopsoffinol and kopsoffine isolated from the Malaysian Borneo species *K. pauciflora* (79,80), which were reported to be constituted from union of kopsinine and dihydroeburnamenine units with the 20α, 21α configuration (*vide infra*).

VII. ASPIDOSPERMINE–ASPIDOFRACTININE AND RELATED ALKALOIDS

A. Aspidospermine Alkaloids

Alkaloids of the rhazinilam–leuconolam group are frequently found in *Kopsia*, as well as other genera of the Apocynaceae, such as *Leuconotis* (4,67). Rhazinilam (**84**) (in fact derived from its natural precursor 5,21-dihydrorhazinilam (**85**) as an artifact of the isolation procedure) was first isolated in 1965 from *Melodinus australis* (81) and subsequently from the Indian plant, *R. stricta* (82,83). Since then, rhazinilam, 5,21-dihydrorhazinilam, and the related leuconolam (**86**) and leuconoxine (**87**) have been frequently encountered in *Kopsia* (as well as *Leuconotis*).

A new oxo-derivative of rhazinilam, rhazinicine (**88**), has been isolated from the North Borneo species *K. dasyrachis* (84,85). It was also subsequently found in *K. arborea* (54). The UV spectrum and mass spectral fragmentation of **88** were characteristic of rhazinilam alkaloids, as were the ^{1}H and ^{13}C NMR spectra. The ^{13}C NMR spectrum, however, showed the presence of an additional lactam carbonyl which must be at position 3 since the two adjacent aromatic H of the pyrrole ring of a rhazinilam-type compound were still intact, and, furthermore, the characteristic H(3) resonances of rhazinilam were absent in **88**, while the H(14) and C(14) signals were shifted downfield. The location of the lactam function at position 3 was also consistent with the observed downfield shift of H(5) compared with **84** due to anisotropy exerted by the proximate carbonyl function. Rhazinicine represents a new example of oxygenation of the rhazinilam skeleton. A previous example, 3-oxo-14,15-dehydrorhazinilam (**89**), was obtained from cell suspension cultures of *Aspidosperma quebracho blanco* (86).

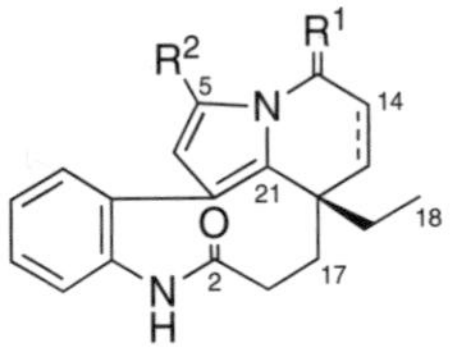

84 R^1 = H$_2$, R^2 = H
88 R^1 = O, R^2 = H
89 R^1 = O, R^2 = H, $\Delta^{14,15}$
90 R^1 = H$_2$, R^2 = CHO

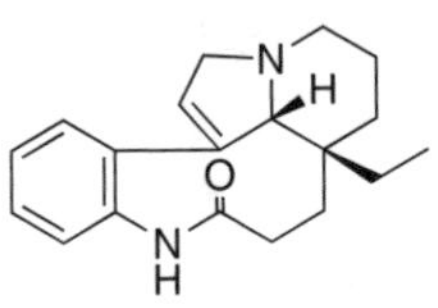

85 5,21-dihydrorhazinilam

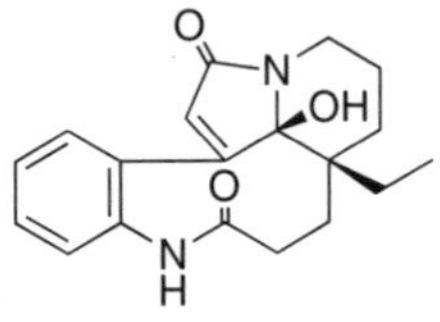

86 leuconolam

Another new rhazinilam derivative found in *Kopsia* is rhazinal (**90**) (55,87). The molecular formula of **90** indicated that it differed from rhazinilam by 28 mass units, indicating substitution of H by CHO, while the NMR data showed a general similarity with those of rhazinilam, except for the presence of a formyl group (δ_H 9.39, δ_C 178.7). Except for the carbon resonances of the pyrrole ring, which have been shifted downfield, the other carbon resonances of **90** were essentially unchanged when compared to those of rhazinilam, indicating that the site of substitution of the formyl group is at C(5) or C(6). The ^{1}H NMR spectrum of **90** showed a 1-H singlet at δ 6.54, which was assigned to H(6) from the observed three-bond correlations to C(8) and C(21) in the HMBC spectrum. Further confirmation of C(5) substitution was provided by the observed H(6)/C=O long-range correlation and the strong H(6)/CHO NOE (84). Rhazinal (**90**) therefore represents the first instance of a naturally occurring rhazinilam derivative which has incorporated an additional carbon in the form of a formyl group, although semisynthetic formyl- and diformylrhazinilam derivatives have been previously reported in connection with a study of the antitubulin activity of rhazinilam analogues (88).

Several new alkaloids related to the pentacyclic diazaspirocyclic alkaloid leuconoxine (**87**), first found in *Leuconotis eugenefolia* (89), have also been isolated recently from *Kopsia* species. These alkaloids, though in a strict sense not part of this group, are considered here for convenience as the parent alkaloid leuconoxine may be considered as having been derived from leuconolam (**86**) via an intramolecular transannular closure. The first alkaloid, 6-oxoleuconoxine (**91**), was isolated from *K. griffithii* (90). It showed UV absorption maxima at 202, 234, 251, and 349 nm, which were somewhat similar to those of leuconoxine (89). The EIMS of **91** showed a molecular ion at m/z 324, which analyzed for $C_{19}H_{20}N_2O_3$. The observed quaternary carbon resonances at δ 192.5, 157.5, and 172.2 were consistent with the presence of ketone and lactam functionalities. The ^{1}H NMR spectrum displayed some similarities to that of leuconoxine, with the characteristically deshielded H(12) due to anisotropy by the proximate lactam carbonyl. Similarity with leuconoxine was further reinforced by the presence of the characteristic quaternary carbon resonance at δ 88.0, corresponding to the doubly spirocyclic C(21), as well as the characteristic resonance due to the benzylic H(7) in the ^{1}H NMR spectrum of **91** (89). In leuconoxine, H(7) is a doublet at δ 3.81, while in **91**, the H(7) was observed as a singlet at δ 4.23. Analysis of the COSY and HMQC data of **91** revealed the presence of some fragments also present in leuconoxine, such as $NCH_2CH_2CH_2$ and CH_2CH_2, corresponding to the C(3)–C(14)–C(15) and C(16)–C(17) units, respectively. Conspicuously absent was the $CHCH_2$ fragment corresponding to the C(7)–C(6) unit of leuconoxine. This, coupled with the observation of H(7) as a singlet, suggested C(6) as the site of oxygenation and accounted for the ketone resonance observed at δ 192.5. Further verification was provided by the HMBC data which showed the key two-bond correlation from H(7) to C(6). The HMBC data also permitted differentiation of the two lactam carbonyl resonances, the resonance at δ 157.5 was assigned to C(5) from the observed three-bond correlations from H(3) and H(7) to C(5), while that at δ 172.2 was due to C(2) from the observed three-bond correlation from H(17) to C(2).

87 R = H, H
91 R = O

92 arboloscine

93

The second alkaloid, arboloscine (**92**) was obtained from the stem-bark extract of *K. arborea* (61). The UV spectrum showed absorption maxima at 210, 247, 265, and 321 nm (log ε 3.83, 4.14, 3.99, and 3.38, respectively), which resembled those of an *N*-acyl dihydroindole (such as leuconoxine **87**), but with additional bands due possibly to extended conjugation from an α,β-unsaturated ester function. The IR spectrum showed bands at 3312, 1726, and 1662 cm^{-1}, due to NH, ester, and lactam functions, respectively. The carbon resonances observed at δ 168.9 and 169.6 in the ^{13}C NMR spectrum confirmed the presence of the ester and lactam functionalities, respectively. In addition, two olefinic signals due to a trisubstituted double bond were seen at δ 140.8 and 110.4, the downfield shift of the former signal being characteristic of the β-carbon of an α,β-unsaturated carbonyl function. The ^{1}H NMR spectrum of **92** showed the presence of an ethyl side chain, a methyl ester group (δ 3.83), and an isolated vinylic hydrogen (δ 6.51).

The spectrum was somewhat similar to that of leuconoxine (**87**), with the characteristically deshielded H(12) due to anisotropy by the proximate lactam carbonyl. The affinity with leuconoxine was further reinforced by the presence of the characteristic quaternary carbon resonance at δ 88.1 corresponding to the spirocyclic C(21). There were, however, several notable differences in the NMR data of the two alkaloids. First, a methoxy signal associated with a methyl ester function was present at δ 3.83, which was absent in the spectrum of **87**. Likewise, the vinylic signal at δ 6.51 present in the spectrum of **92** was absent in that of **87**. On the other hand, the signals due to H(6), as well as the characteristic doublet due to the adjacent H(7) in leuconoxine (**87**), were not seen in the spectrum of **92**. Analysis of the COSY and HMQC data revealed the presence of some fragments also present in **87**, such as NCH$_2$CH$_2$CH$_2$ and CH$_2$CH$_2$, corresponding to the C(3)–C(14)–C(15) and C(16)–C(17) units, respectively. Conspicuously absent was the CHCH$_2$ fragment corresponding to the C(7)–C(6) unit of **87**. Since the lactam associated with the indolic nitrogen remained intact from the HMBC data, as were the fragments associated with rings B, D, and E of leuconoxine, the ester group in **92** must be in some way associated with an altered ring C. Furthermore, since the degree of unsaturation for both compounds are the same (DBE, 10), but an additional double bond was present in **92** compared to **87**, the loss of one ring was indicated.

Further clues to the structure of arboloscine were provided by the heteronuclear correlations from the HMBC spectrum, which indicated cleavage of the N(4)–C(5) bond, resulting in a *seco*-leuconoxine as shown in **92**. The observed correlation from H(9) to the quaternary olefinic carbon at δ 140.8 indicated that this carbon corresponds to C(7). The two-bond correlation from the vinylic H to C(7) and the three-bond correlations to C(8) and the spirocyclic C(21) were consistent with the branching of the exocyclic double bond of the acrylic ester unit from C(7). Finally, the NOE observed between the aromatic H(9) and the vinylic H(6), provided additional confirmation for the structure assignment, and also revealed the geometry of the double bond as Z. Arboloscine (**92**) represents the first example of a *seco*-leuconoxine. A possible origin is from a leuconolam (or *epi*-leuconolam) precursor (**86**), which on transannular cyclization leads to a dehydroleuconoxine derivative **93**; hydrolytic cleavage followed by methylation furnishes **92**. Arboloscine (**92**) showed moderate cytotoxicity toward KB cells (54).

There have been a number of syntheses of rhazinilam (**84**), prompted in large measure by its intriguing architecture as well as biological activity (*vide infra*). The first was by Smith (91), which commenced with the alkylation of the pyrrole **94** (as the sodium salt) with the γ-lactone **95** (prepared from diethyl 4-ketopimelate **96**) to give the alkylated product **97**, which on cyclization induced by AlCl₃ in MeNO₂ gave the acid **98**. Subsequent reduction of the nitro group, followed in succession by DCC-induced lactamization, saponification of the ester function, and decarboxylation, gave (±)-rhazinilam (**84**).

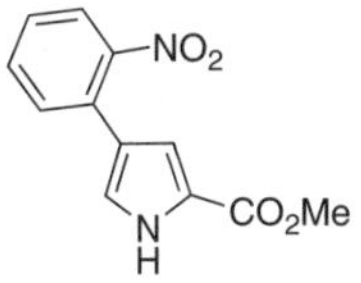

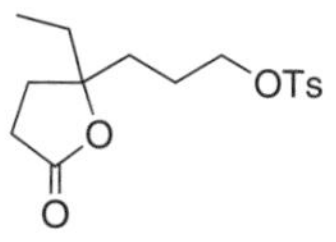

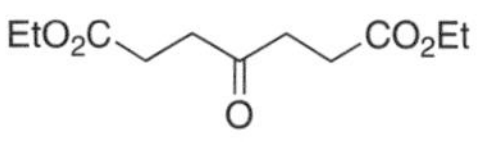

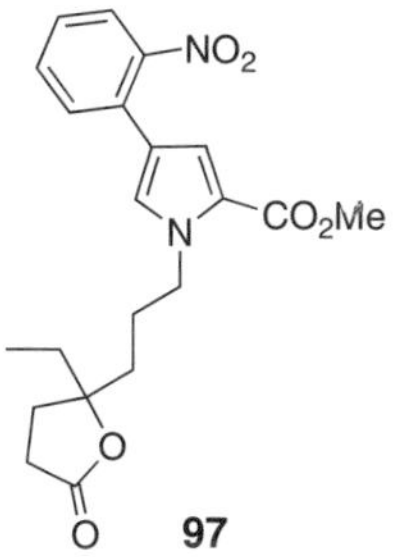

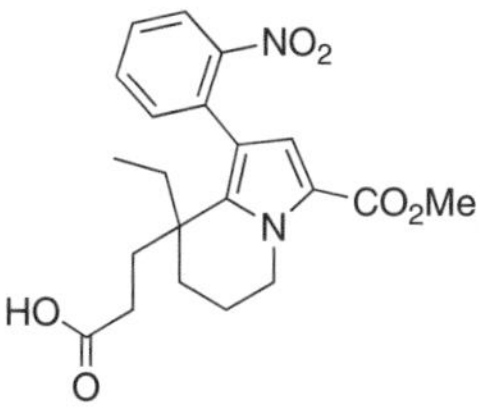

The synthesis by Sames (92,93) was based on directed C–H bond activation (94) of a hydrocarbon portion of a suitable substrate (such as the diethyl pyrrole **103**) as the key step. The idea was to exploit the proximity of the amino group to the ethyl groups in **103** by utilizing the amino function to direct an activated metal complex to the target hydrocarbon segment of the substrate (one of the enantiotopic ethyl groups in the case of **103**), in order to effect selective functionalization (dehydrogenation) of the ethyl group in question. The synthesis of the required diethyl intermediate **103** was from the readily available imine **99**, which on treatment with 2-nitrocinnamyl bromide (**100**) gave the iminium salt **101**. Heating of **101** in the presence of Ag₂CO₃ resulted in 1,5-electrocyclization to give the pyrrole **102**, which was then processed to the amine **103**, following installation of the methyl carboxylate protecting group and reduction of the nitro group. The key platinum complex **104** was obtained from **103** by sequential treatment with 2-benzoylpyridine and the dimethylplatinum reagent, [Me₂Pt(μ-SMe₂)]₂. Addition of triflic acid (1 equiv) resulted in formation of the cationic complex **105** accompanied by liberation of methane. Thermolysis of **105** in CF₃CH₂OH or CH₂Cl₂ gave the alkene-hydride **106** in high yield. Treatment of **106** with Bu₄NCl in CH₂Cl₂ led to **107** in which the platinum is σ-bonded to the methylene carbon of the ethyl group, prompting the suggestion that the initial C–H activation occurred at the methylene carbon. Decomplexation of the platinum (aqueous KCN) followed by Schiff base removal (NH$_2$OH) provided racemic alkene **108**, which was then processed to ($\pm$)-rhazinilam (**84**) on sequential one-carbon homologation/macrolactamization, followed by removal of the methyl ester protecting group (Scheme 8) (94).

Modification of the same basic approach by the use of oxazolinyl ketones (e.g., **109**) as chiral auxiliaries led to asymmetric C–H bond activation, which, in turn, facilitated the first asymmetric total synthesis of (−)-rhazinilam. Thus, oxazoline Schiff base complexes (e.g., **110**, prepared as in the case of **104**) on treatment with TfOH yielded a mixture of diastereomers **111** and **112** (diastereomeric ratio 4.4:1 when R = cyclohexyl; bulkier R, e.g., t-Bu, gave high selectivity, but very low conversion).

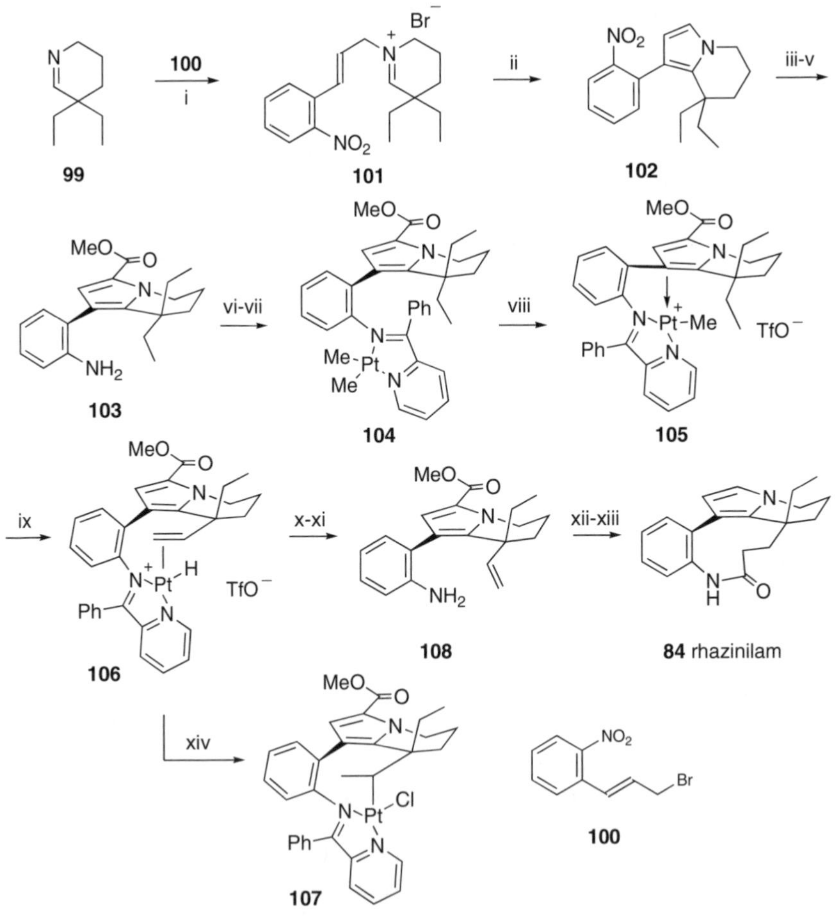

Scheme 8 Reagents: (i) DMF, 100°C; (ii) Ag$_2$CO$_3$ (2 equiv), toluene, reflux; (iii) CCl$_3$COCl; (iv) NaOMe, MeOH; (v) H$_2$, Pd/C; (vi) 2-benzoylpyridine, TsOH, PhMe; (vii) [Me$_2$Pt(μ-SMe$_2$)]$_2$; (viii) TfOH, CH$_2$Cl$_2$; (ix) CF$_3$CH$_2$OH, 70°C; (x) KCN; (xi) NH$_2$OH; (xii) 10% Pd–C (5 mol%), dppb, HCOOH, DME, CO (10 atm), 150°C; (xiii) NaOH, MeOH; HCl, 50°C; (xiv) Bu$_4$NCl, CH$_2$Cl$_2$.

Application of the C–H functionalization sequence (comprising C–H activation, decomplexation, and transamination) yielded a mixture of enantiomers of **108**. Optically pure (−)-**108** (obtained after removal of the chiral auxiliary from the major Schiff base diastereomer, following decomplexation and separation by preparative HPLC) was then processed to (−)-rhazinilam (**84**) following the same protocol as that detailed above for the racemic synthesis (93).

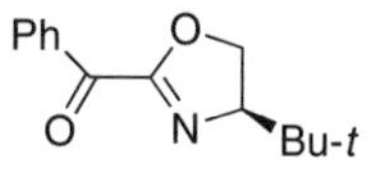

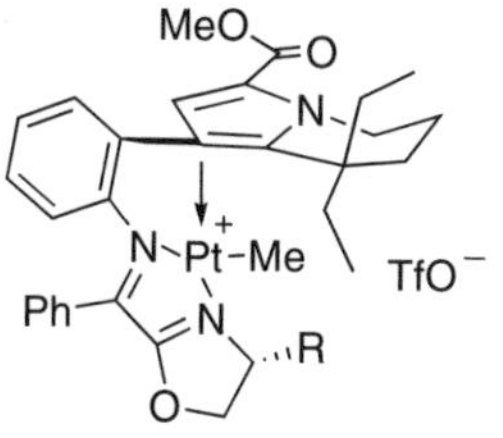

Magnus's concise racemic synthesis was based on a strategy in which the gem-alkyl groups on the piperidine ring of the key building block **114** (prepared by stepwise alkylation of 2-piperidone **113**) were differentiated at an early stage of the synthesis (95). Conversion of **114** to the thiophenyl imino ether **115**, followed in succession by treatment with 2-nitrocinnamyl bromide (**100**) and DBU-induced electrocyclization with concomitant thiophenol elimination, gave the pyrrole **116**. Hydroboration–oxidation of **116** to give **117** was followed by oxidation to the aldehyde **118**, which was converted to the acid **119** with alkaline AgNO$_3$. Ensuing reduction of the nitro group, followed by cyclization under Mukaiyama conditions, provided (±)-rhazinilam (**84**) (Scheme 9).

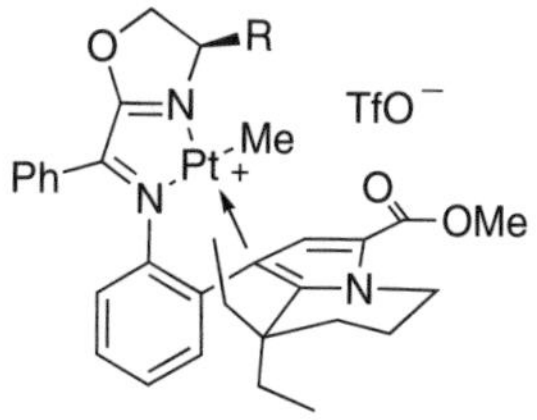

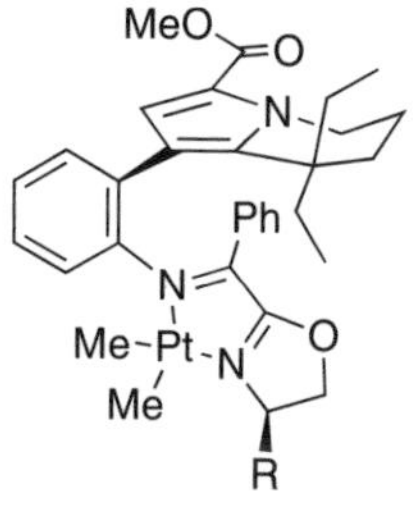

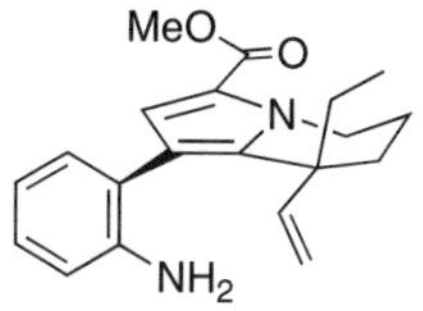

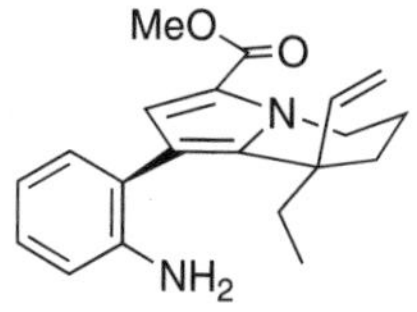

The γ-lactone **95** employed in the earlier Smith synthesis was the starting compound in Trauner's racemic synthesis of rhazinilam, which was based on direct palladium-catalyzed intramolecular coupling as the key step (96). Treatment of **95** with the sodium salt of 2-carbomethoxy pyrrole **120** gave the N-alkylated pyrrole **121**, which underwent intramolecular Friedel–Crafts alkylation on exposure to AlCl$_3$ to afford the acid **122**. Mukaiyama coupling of **122** with 2-iodoaniline gave the amide **123**, which was then protected as the MOM derivative **124**, thus setting the stage for the implementation of the crucial direct coupling step. The coupling was achieved by heating **124** with 10 mol% of

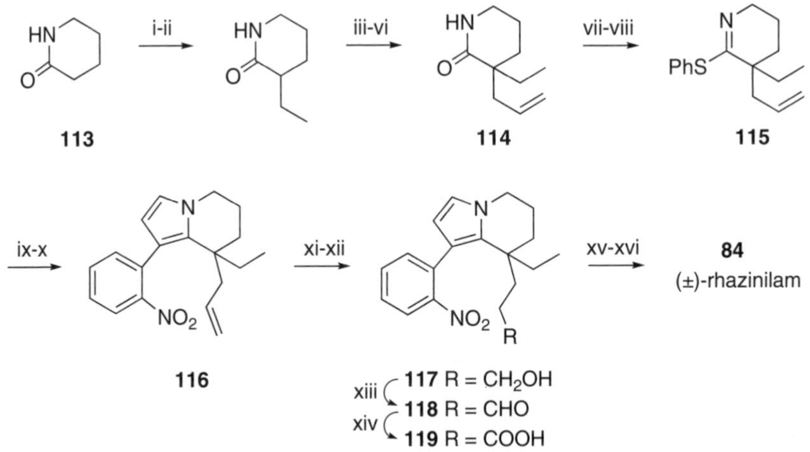

Scheme 9 Reagents: (i) n-BuLi; (ii) EtI; (iii) n-BuLi; (iv) Me₃SiCl; (v) LiN(Pr-i)₂; (vi) allyl bromide; (vii) PCl₅; (viii) PhSH/Et₃N; (ix) **100**, 100°C; (x) DBU, 0°C; (xi) BH₃ · SMe₂; (xii) H₂O₂, NaOH; (xiii) py · SO₃, DMSO, Et₃N; (xiv) AgNO₃/KOH; (xv) Raney-Ni/H₂/MeOH; (xvi) 2-chloro-1-methylpyridinium iodide/Et₃N/PhMe.

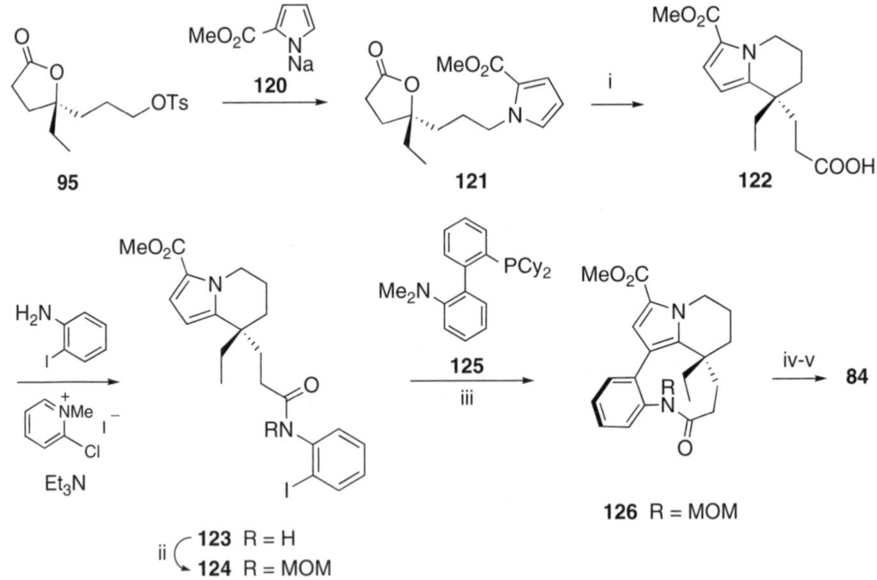

Scheme 10 Reagents: (i) AlCl₃; (ii) NaH, MOMCl; (iii) 10 mol% Pd(OAc)₂, K₂CO₃; (iv) BCl₃; (v) NaOH, H₂O; HCl.

Buchwald's "DavePhos" ligand **125** and Pd(OAc)₂ in the presence of a base (Fagnou's conditions), which yielded the strained nine-membered lactam **126**. Deprotection, followed by saponification, and then decarboxylation, gave (±)-rhazinilam (**84**) (Scheme 10).

The most recent asymmetric synthesis of (−)-rhazinilam was that of Nelson, who employed a strategy based on Au(1)-catalyzed annulation of enantio-enriched allenes to gain entry to the key optically active heterocyclic inter-mediate **130**, which was then transformed into (−)-rhazinilam (**84**) (97). The required allene substrate **129** was prepared from the 3,4-*cis*-disubstituted β-lactone **127**, in turn obtained in 99% ee from the *O*-trimethylsilylquinine (TMSQ*n*)-catalyzed cyclocondensation of propionyl chloride and 2-pentynal.

Lactone ring opening with **128**, followed by methylation, provided allene **129**, which was then subjected to Au(1)-catalyzed annulation employing Ph$_3$P·AuOTf as annulation catalyst to provide the required tetrahydroindolizine **130** with virtually complete transfer of allene chirality (94% de, 92% yield). Pyrrole protection, followed in succession by oxidative olefin cleavage, Horner–Wittig homologation, and hydrogenation, gave ester **131**, which on regioselective pyrrole iodination, followed by Suzuki–Miyaura cross-coupling of iodide **132** with the boronic ester **133** using Buchwald's SPhos ligand, forged the biaryl bond of the aryl-pyrrole **134**. Finally, ester saponification, followed in turn by *N*-Boc removal, HATU-mediated lactamization, and pyrrole deprotection, gave (−)-rhazinilam (**84**) (Scheme 11).

There has been to date only one synthesis of rhazinal (**90**), namely, that by Banwell (98), who employed a route based on intramolecular Michael addition of pyrrole through C(2) to an *N*(1)-tethered acrylate moiety (such as in the substrate **138**) to construct the CD ring system of rhazinal. The synthesis of **138** commenced with the reaction of the potassium salt of pyrrole with γ-butyrolactone to give the previously known acid **135**, which was then converted to the Weinreb amide **136** using a modified Mukaiyama amide coupling procedure (Scheme 12). The amide **136** was then converted to the ketone **137** by treatment with ethyl magnesium bromide, followed by acidic workup. Wadsworth–Emmons reaction of **137** with methyl diethylphosphonoacetate gave a 1:1 mixture of *E*- and *Z*-acrylates, **138**. Treatment of **138** with five molar equivalents of AlCl$_3$ in Et$_2$O gave the tetrahydroindolizine **139**. One-carbon homologation of **139** to the homologous ester **140**, followed successively by Vilsmeier–Haack formylation and iodination, gave **141**. Suzuki–Miyuara cross-coupling of iodide **141** with 2-(4,4,5,5-tetramethyl-1,3,2-dioxaborolan-2-yl)benzenamine gave the aryl-pyrrole **142**, which on lactamization via ester hydrolysis, followed by EDCI–DMAP-promoted coupling of the resulting amino acid, afforded racemic rhazinal (±)-**90**. With a view to compare its biological activity with that of rhazinal (**90**), the conformationally more constrained B-norrhazinal analogue **143** was also obtained from the tetrahydroindolizine **139** in essentially the same manner as that described for **90**, but without one-carbon homologation (98,99).

X-ray diffraction analysis of synthetic rhazinal (98) has confirmed the structure previously proposed based on spectroscopic analysis of the natural product (87). The results revealed that as with rhazinilam (**84**), the amide unit in rhazinal (**90**) adopts an *s*-cissoid conformation. The dihedral angle between the planes of the phenyl and pyrrole rings in rhazinal, however, is about 89°, compared to 56° in the case of **143** (99), and 96° in the case of rhazinilam (**84**)

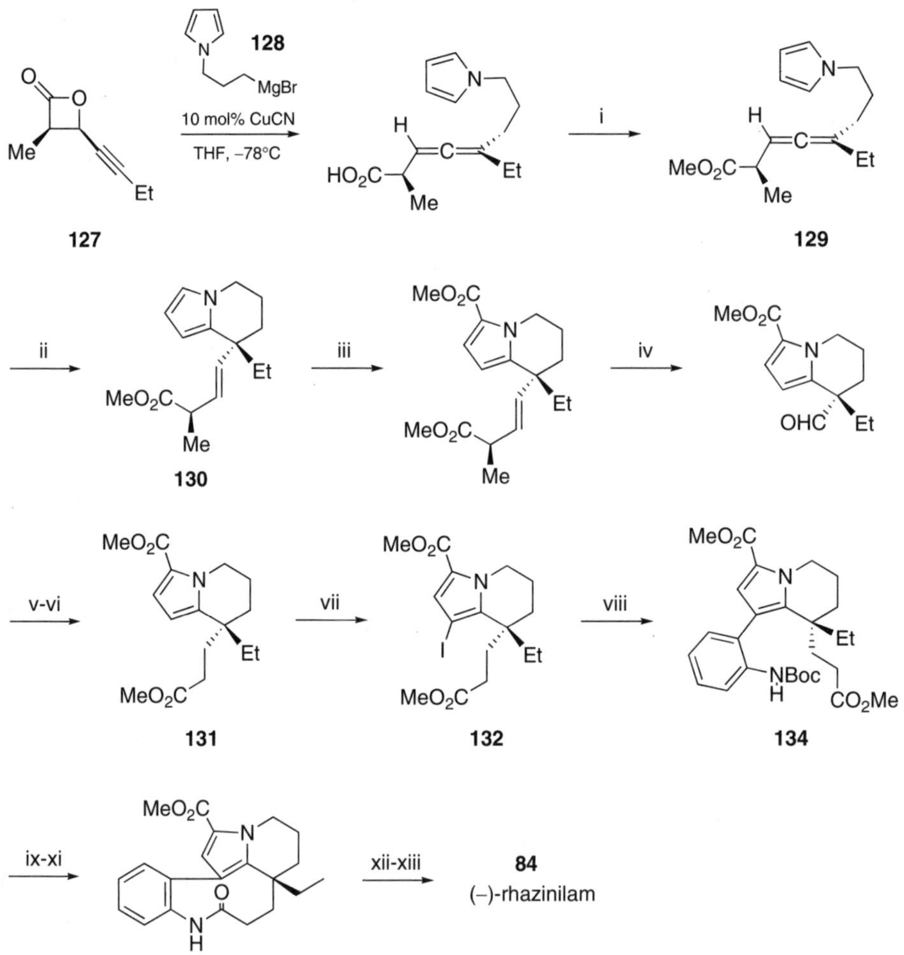

Scheme 11 Reagents: (i) TMSCHN$_2$; (ii) Ph$_3$P · AuOTf, 23°C; (iii) Cl$_3$COCl; NaOMe; (iv) 10 mol% OsO$_4$, NMO; NaIO$_4$; (v) Ph$_3$P=CHCO$_2$Me; (vi) 10% Pd/C, H$_2$, MeOH; (vii) I$_2$, AgCO$_2$CF$_3$; (viii) 2-C$_6$H$_4$(Bpin)NHBoc (**133**), 2.5 mol% Pd$_2$(dba)$_3$, 10 mol% SPhos, K$_3$PO$_4$, THF; (ix) Ba(OH)$_2$; (x) TFA; (xi) HATU, (Pr-i)$_2$NEt; (xii) 50% NaOH; (xiii) HCl.

(100). All three compounds, however, showed comparable *in vitro* cytotoxicities toward cancer cells (55,88,98,99,101), implying that a certain degree of flexibility in the conformation adopted by the molecule is tolerated without compromising its biological activity.

Rhazinilam was initially found to inhibit the polymerization of tubulin to microtubules by inducing spiralization or nonreversible assembly of tubulin, thus mimicking the action of vinblastine. Rhazinilam also resembles Taxol by protecting microtubules from cold-induced disassembly. In addition, like Taxol, it also possesses the ability to induce formation of asters in mitotic cells and microtubule bundles in interphase cells (102–105). In view of these tubulin-binding properties, natural (−)-rhazinilam (**84**) (55,88,101), (−)-rhazinicine (**88**) (55), and (−)-rhazinal (**90**) (55) (all

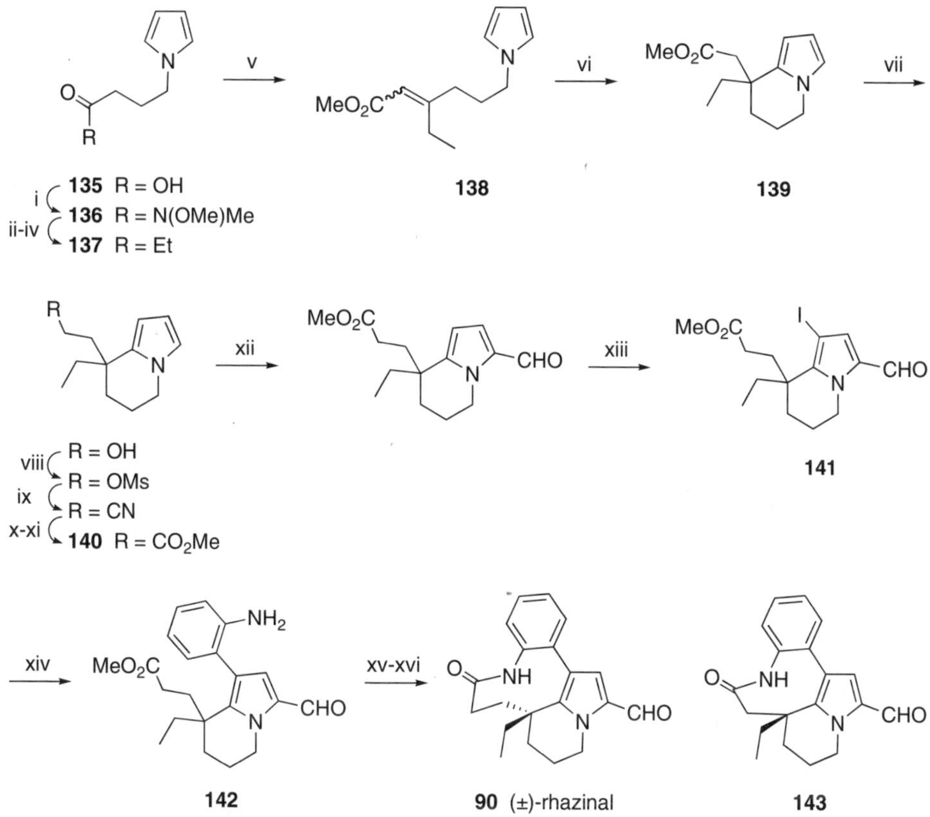

Scheme 12 Reagents: (i) MeNHOMe · HCl, Et$_3$N, (2-pyridine *N*-oxide)disulfide, Bu$_3$P; (ii) EtMgBr, Et$_2$O; (iii) KHSO$_4$, $-40°$C; (iv) NaHCO$_3$, -40 to 18°C; (v) NaH, (EtO)$_2$POCH$_2$CO$_2$Me; (vi) AlCl$_3$, Et$_2$O; (vii) DIBALH, hexane, $-78°$C; (viii) MeSO$_2$Cl, Et$_3$N, CH$_2$Cl$_2$, 0°C; (ix) NaCN, DMPU; (x) KOH, H$_2$O–MeOH, reflux; HCl; (xi) DCC, DMAP, CH$_2$Cl$_2$–MeOH; (xii) DMF, POCl$_3$, Et$_2$O, 0°C; (xiii) I$_2$, AgOCOCF$_3$, CHCl$_3$; (xiv) 2-(4,4,5,5-tetramethyl-1,3,2-dioxaborolan-2-yl)benzenamine, Pd(PPh$_3$)$_4$, PhMe, Na$_2$CO$_3$, MeOH; (xv) KOH, EtOH; HCl; (xvi) EDCI, DMAP, CH$_2$Cl$_2$.

obtained from *Kopsia*), as well as synthetic ($\pm$)-B-norrhazinal (**143**) (98,99), have been reported to show moderately pronounced *in vitro* cytotoxicity (in the low micromolar range) toward various cancer cell lines. Rhazinilam, however, despite its novel mode of interaction with tubulin, has been found to be inactive *in vivo* (103–106). As a result of its diminished potential for pharmacological applications, many studies on the syntheses and biological evaluation of analogues of rhazinilam have been carried out in a quest for analogues with improved pharmacological properties (88,98,99,101,104–119). These studies have converged on the biaryl analogue **144** as the most active analogue generated to date, possessing tubulin-binding activity which is twice that of rhazinilam, as well as *in vitro* cytotoxicities toward human cancer cell lines which are comparable to those of rhazinilam (101,116,119). Analogue **144** also encapsulates the required structural features that are necessary for biological activity, namely, a rigid biaryl/nine-membered lactam structure similar to rhazinilam, with the B-ring existing in a boat-chair conformation and containing a quaternary carbon atom. In addition, the biologically active atropisomeric enantiomer is one that possesses the R_a absolute configuration of the biaryl axis, while replacement of the amide function by carbamate also appears to enhance the biological activity. An asymmetric synthesis of (−)-**144**, based on an intermolecular asymmetric Suzuki coupling as the key step, has also been reported (119).

(−)-**84**   (−)-**144**   **87**   **145**

A recent study has moreover shown that the lack of *in vivo* activity of rhazinilam could be due to its rapid bioconversion to markedly less active metabolites, which were shown to be the oxidized derivatives of rhazinilam, namely, 3-(*S*)-hydroxyrhazinilam (**145**) and leuconolam (**86**), as well as the diazaspiroleuconolam derivative, leuconoxine (**87**) (106). These results prompted the suggestion that analogues with the 3 and 5 positions blocked to prevent oxidation may present viable candidates for evaluation of *in vivo* efficacy (106).

B. Aspidofractinine Alkaloids

The aspidofractinine alkaloids constitute by far the overwhelmingly predominant alkaloids occurring in *Kopsia*. Kopsinine (**146**), first isolated from the Australian *K. arborea* (*K. longiflora*) (120), has since been frequently encountered in *Kopsia*, as well as in other genera of the Apocynaceae. The acid derivative kopsininic acid (**147**) was isolated as a minor alkaloid from the same plant (*K. officinalis*) occurring in China, although the possibility that it was an artifact of the isolation process cannot be discounted (120–122). Kopsamidine B (**148**), readily deduced to be the 15α-methoxy derivative of **146**, was isolated from the Malayan *K. arborea* (54). The NMR spectral data were similar to those of **146**, except for the presence of an additional OMe group in the piperidine ring. The correlations observed between H(15) and H(17α), as well as H(19β), in the NOESY spectrum established the orientation of the C(15)–OMe group as α. This was further confirmed by the significant downfield shift of H(21) (from ca. δ 3.00 in **146** to δ 3.44 in **148**) observed in the ¹H NMR spectrum, due to its proximity to the α-oriented OMe group at C(15). The related 15α-hydroxykopsinine (**149**) which was also present, was also found in *K. singapurensis* (*K. fruticosa*) (123), as well as in *Catharanthus ovalis* (124) and *Melodinus guillauminii* (125). Several new pleiocarpine derivatives which have been reported from *Kopsia* include 12-methoxypleiocarpine (**150**) from *K. griffithii* (40) and the two 17β-hydroxy derivatives (**151** and **152**) from *K. deverrei* (126). *K. singapurensis* (55,123), as well as *K. teoi* (90), also provided 16-*epi*-kopsinine (**153**). The ¹H NMR spectrum of **153** largely resembled that of **146**, except for changes involving the signal due to H(16). In **153**, the H(16) signal was observed as a doublet of triplets at δ 3.20, while in kopsinine (**146**), the H(16) resonance was observed as a triplet of doublets at δ 2.89. The ^{13}C NMR spectrum of **153** was also almost identical with that of **146**, except for the C(18) signal which was shifted upfield by ca. 7.5 ppm. Further confirmation for the change in the configuration at C(16) was provided by the observed NOE interaction between H(16) and H(6β), which was only possible if H(16) is β. Pleiocarpine (**154**) from *K. griffithii* was found to display moderate leishmanicidal activity against *L. donovani* (41).

146 R = H
148 R = OMe
149 R = OH

150 R^1 = OMe, R^2 = CO$_2$Me, R^3 = H
151 R^1 = H, R^2 = CO$_2$Me, R^3 = OH
152 R^1 = H, R^2 = CO$_2$Me, R^3 = OH, Δ^{14,15}
154 R^1 = R^3 = H, R^2 = CO$_2$Me

147 R^1 = CO$_2$H, R^2 = H
153 R^1 = H, R^2 = CO$_2$Me

K. deverrei is also notable for furnishing several 17-oxoaspidofractinine alkaloids. Four such kopsinone derivatives were obtained, namely, kopsinone (**155**), 10-methoxykopsinone (**156**), 12-methoxykopsinone (**157**), and 14, 15-dihydro-10-methoxykopsinone (**158**) (126,127). A distinguishing feature of these alkaloids was the presence of the C(17) ketone resonance at δ 213. Reduction (LiAlH$_4$/THF) of kopsinone (**155**) gave the alcohol **159**, which displayed the retro-Diels-Alder fragment at *m/z* 280 in its mass spectrum, thus confirming the presence of the OH group on C(17) in the alcohol **159** (126).

Two additional alkaloids with a ketone function at C(17) are kopsonoline (**160**) and kopsofinone (**161**), from *K. teoi* (90) and *K. singapurensis* (128), respectively. The NMR data of kopsonoline (**160**) indicated that it possesses the basic aspidofractinine carbon framework, except for oxygenation at C(17) in the form of a ketone function (δ$_C$ 213.6). The indolic NH was seen as a broad singlet at δ 3.56 in the ^{1}H NMR spectrum, which also revealed an isolated methylene at δ 3.03 and 2.59. The chemical shifts of the methylene suggested that it was adjacent to a carbonyl function, and, together with the COSY data, indicated either position 17 or 19 as the site of oxygenation. The absence of W-coupling to H(21) indicated that oxygenation was likely at C(17), which was further confirmed on comparison of the ^{13}C NMR spectral data with synthetic racemic 17-oxoaspidofractinine (129) and 19-oxoaspidofractinine (130,131), which revealed a close correspondence with the former. Kopsonoline (**160**) is therefore 17-oxoaspidofractinine, and while the racemic form constitutes one of the intermediate compounds in the synthesis of various aspidofractinine alkaloids (129–131), it was in this instance encountered as an optically active natural product for the first time. Another kopsinone congener is kopsofinone (161) isolated from *K. singapurensis* (128). The NMR spectral data revealed, in addition to a ketone function at C(17), as indicated by the carbonyl resonance at δ_C 209.6, the presence of an 11,12-methylenedioxy function, as well as a quaternary C(16) substituted by carbomethoxy and hydroxy groups ($\delta_{C(16)}$ 79.2).

155 $R^1 = R^2 = R^3 = R^5 = R^6 = H$, $R^4 = CO_2Me$, $R^7 = \Delta^{14,15}$

156 $R^1 = OMe$, $R^2 = R^3 = R^5 = R^6 = H$, $R^4 = CO_2Me$, $R^7 = \Delta^{14,15}$

157 $R^1 = R^2 = R^5 = R^6 = H$, $R^3 = OMe$, $R^4 = CO_2Me$, $R^7 = \Delta^{14,15}$

158 $R^1 = OMe$, $R^2 = R^3 = R^5 = R^6 = H$, $R^4 = CO_2Me$, $R^7 = nil$

160 $R^1 = R^2 = R^3 = R^4 = R^5 = R^6 = H$, $R^7 = nil$

161 $R^1 = R^4 = H$, $R^2, R^3 = OCH_2O$, $R^5 = OH$, $R^6 = CO_2Me$,

     $R^7 = \alpha$-epoxide

162 $R^1 = OH$, $R^2 = H$

163 $R^1 = H$, $R^2 = OH$, $\Delta^{14,15}$

159

Two other aspidofractinine alkaloids, oxidized at C(16) or C(17), are 162 and kopsinginol (163), respectively. Alkaloid 162 was isolated from *K. arborea* (*K. officinalis*), while 163 was isolated from *K. teoi*. In the case of 162, hydroxy substitution at C(16) was deduced from the HMBC data, while the observed NOE between H(16) and H(18β) indicated that the orientation of the OH substituent is β (122). In the case of kopsinginol (163), the position, as well as the stereochemistry, of the hydroxy substitution on the C(16)/C(17) ethylene bridge was inferred from the *W*-coupling observed between H(16β) and H(18α) on the one hand, and between H(17α) and H(21) on the other (132).

The aspidofractinine alkaloids, kopsiflorine (164), kopsilongine (165), and kopsamine (166), were also first reported from the Australian plant *K. arborea* (*K. longiflora*) (27,120). These alkaloids, as well as their *N*(4)-oxides, have been encountered in several other *Kopsia*. The 11-methoxy and 11-hydroxy derivatives of kopsilongine (167 and 168, respectively) were isolated from *K. arborea* (*K. officinalis*), together with 12-hydroxy-11-methoxykopsiflorine (*N*-carbo-methoxy-12-hydroxy-11-methoxykopsinaline 169) (77,133). Alkaloids 168 and 169 were distinguished primarily by the diagnostic (M–15) peak in the EI-mass spectra, which was observed in 168, but absent in 169. Further support was provided by an X-ray diffraction analysis of 168 and *O*-methylation of 169 to yield 167 (133,134). Other new kopsilongine derivatives include 12-methoxykopsinaline (170) from the Chinese species *K. arborea* (*K. officinalis*) (133) and kopsinginine (171) from the Malaysian *K. teoi* (135). Two new aspidofractinine alkaloids which are related to kopsamine (166) are 11,12-methylenedioxykopsinaline (172) (133) and kopsamidine A (173) (54).

Two aspidofractinine-type alkaloids, lahadinines A (**174**) and B (**175**), isolated from *K. pauciflora* occurring in Malaysian Borneo are remarkable for having a cyano-substituent at C(21) (136). The mass spectra of these alkaloids are characterized by a strong molecular ion with the odd mass indicating the presence of a third nitrogen. Fragments attributable to loss of CN and HCN were also detected in the mass spectrum and the IR spectrum showed a weak band at ca. $2250\,cm^{-1}$. The characteristic H(21) signal was absent in the ^1H NMR spectrum and the ^{13}C NMR spectrum showed an additional quaternary resonance at δ 118 due to the cyano group. The location of the cyano group at C(21) was supported by the observed three-bond correlations from H(5), H(17), and H(19) to C(21) in the HMBC spectrum. Alkaloids containing the unusual cyano group have previously been obtained from the Indonesian *Alstonia angustiloba* (137). Lahadinines A and B are C(21)-cyano derivatives of kopsamine (**166**) and 11-methoxykopsilongine (**167**), respectively, both of which also occur in the plant. The C(21)-oxygenated derivative of **166**, paucifinine (**176**) and its *N*-oxide **177**, were also isolated (136). Kopsiflorine (**164**), kopsamine (**166**), 11-methoxykopsilongine (**167**), and lahadinine A (**174**) displayed comparatively strong potency in reversing multidrug resistance in drug-resistant KB cells (IC$_{50}$ 1–5 µg/mL in the presence of 0.25 µg/mL of vincristine) (138).

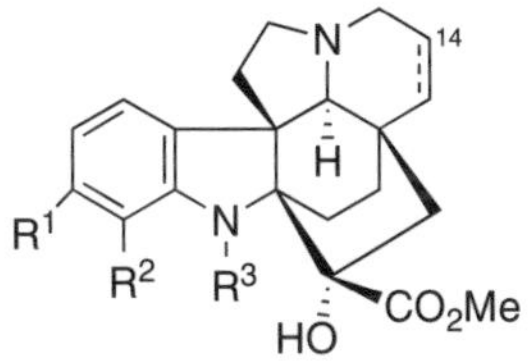

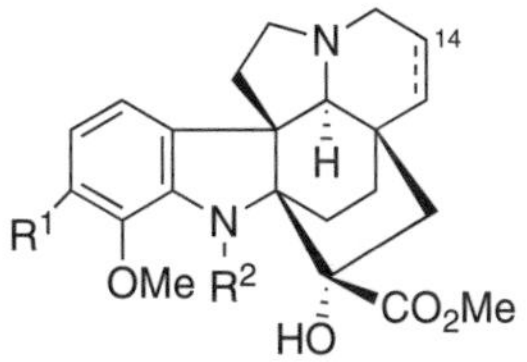

164 $R^1 = R^2 = H$, $R^3 = CO_2Me$
166 R^1, $R^2 = OCH_2O$, $R^3 = CO_2Me$
169 $R^1 = OMe$, $R^2 = OH$, $R^3 = CO_2Me$
172 R^1, $R^2 = OCH_2O$, $R^3 = H$
173 R^1, $R^2 = OCH_2O$, $R^3 = H$, $\Delta^{14,15}$

165 $R^1 = H$, $R^2 = CO_2Me$
167 $R^1 = OMe$, $R^2 = CO_2Me$
168 $R^1 = OH$, $R^2 = CO_2Me$
170 $R^1 = R^2 = H$
171 $R^1 = H$, $R^2 = CO_2Me$, $\Delta^{14,15}$

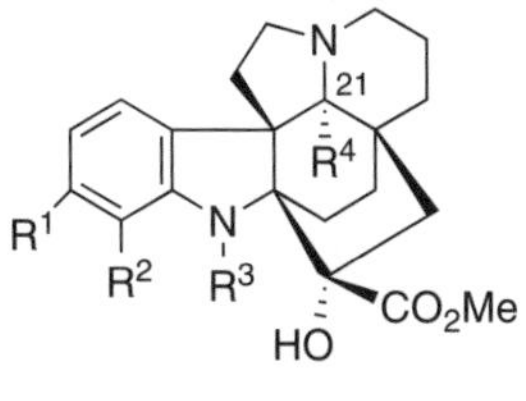

174 R^1, $R^2 = OCH_2O$, $R^3 = CO_2Me$, $R^4 = CN$
175 $R^1 = R^2 = OMe$, $R^3 = CO_2Me$, $R^4 = CN$
176 R^1, $R^2 = OCH_2O$, $R^3 = CO_2Me$, $R^4 = OH$
177 R^1, $R^2 = OCH_2O$, $R^3 = CO_2Me$, $R^4 = OH$, N→O

Alkaloid **178** ($C_{21}H_{24}N_2O_3$) isolated from *K. teoi* had spectral data which indicated it was 17α-hydroxy-$\Delta^{14,15}$-kopsinine (135). The NMR data indicated an aspidofractinine-type alkaloid with an unsubstituted aromatic ring, the presence of a C(16) methyl ester and a C(17)-hydroxyl function, unsaturation at C(14) and C(15), and the absence of carbamate and C(16)–OH groups. The orientation of the C(17)–OH was assigned as α from the observed *W*-coupling (2 Hz) between

H(17β) and H(19α), while the absence of similar coupling between H(16) and one of the H(18) indicated that the C(16) methyl ester group was β, which was also in accord with the observed H(16)/H(17) coupling of 7.5 Hz (135). In another study of the same species, an alkaloid **179** (16-*epi*-17α-hydroxy-$\Delta^{14,15}$-kopsinine) with similar spectral data was reported, but in which *W*-coupling between H(16) and H(18) was apparently detected, requiring the methyl ester function to be α (139). A reinvestigation of this alkaloid has been carried out by detailed analysis (COSY, homonuclear decoupling, NOE) of the ^{1}H NMR spectrum obtained at 400 MHz, which resolved the H(18) and H(16) signals as follows: δ 1.30, H(18β), ddd ($J_{18-18} = 13$ Hz, $J_{18\beta-19\beta} = 11$ Hz, and $J_{18\beta-19\alpha} = 2.1$ Hz); δ 1.88, H(18α), dddd ($J_{18-18} = 13$ Hz, $J_{18\alpha-19\alpha} = 11$ Hz, $J_{18\alpha-19\beta} = 7.5$ Hz, and ${}^4J_{18\alpha-16} = 1.1$ Hz); δ 2.77, H(16α), dd ($J_{16-17} = 7.7$ Hz and ${}^4J_{16-18\alpha} = 1.1$ Hz). In addition, the observed NOE interaction between H(18β) and H(16) provided incontrovertible evidence for the α orientation of H(16) (140). These results have vindicated the original assignment of this alkaloid as 17α-hydroxy-$\Delta^{14,15}$-kopsinine (**178**) (135).

An entire family of such alkaloids, kopsiloscines A–J (**180–189**), characterized by the presence of an α-oriented OH at C(17) has been subsequently isolated from *K. singapurensis* (53,58,128). The α-OH substitution at C(17) in the kopsiloscines was deduced, either from the observed $J_{17\beta-19\alpha}$ *W*-coupling of 2 Hz, or from the observed H(17)/H(5β), H(6β) NOEs, or both (55,58,128). In kopsiloscines C, E, and F (**182, 184**, and **185**, respectively), the α-OH substitution at C(15) in the piperidine ring D was confirmed by the observed reciprocal NOEs observed for H(15)/C(17)–αOH, as well as the paramagnetic deshielding of the proximate H(21) by the α-oriented C(15)–OH (55). Kopsiloscines A (**180**), B (**181**), D (**183**), and J (**189**) were found to reverse drug resistance in drug-resistant KB cells, while kopsiloscines C (**182**), E (**184**), and F (**185**), which are characterized by the presence of a C(15)–OH substituent, were ineffective (IC$_{50}$ > 25 μg/mL) (55,128).

178 R^1 = CO$_2$Me, R^2 = H
179 R^1 = H, R^2 = CO$_2$Me

189

180 R^1 = R^2 = H, R^3 = CO$_2$Me, R^4 = OH, R^5 = $\Delta^{14,15}$
181 R^1 = R^2 = H, R^3 = CO$_2$Me, R^4 = OH, R^5 = nil
182 R^1 = R^2 = H, R^3 = CO$_2$Me, R^4 = OH, R^5 = 15α-OH
183 R^1 = OMe, R^2 = H, R^3 = CO$_2$Me, R^4 = OH, R^5 = nil
184 R^1 = OMe, R^2 = H, R^3 = CO$_2$Me, R^4 = OH, R^5 = 15α-OH
185 R^1 = H, R^2 = OMe, R^3 = CO$_2$Me, R^4 = OH, R^5 = 15α-OH
186 R^1 = R^2 = R^3 = R^4 = H, R^5 = nil
187 R^1 = R^2 = R^3 = H, R^4 = OH, R^5 = $\Delta^{14,15}$
188 R^1 = R^2 = R^3 = H, R^4 = OH, R^5 = nil

In 1959, Kiang and Amarasingham reported the isolation of kopsingine (**4**) and kopsaporine (12-demethoxykopsingine) (**190**) from *K. singapurensis* (141,142). The structures of kopsingine and kopsaporine were then established based on degradative experiments carried out on kopsingine, but the configuration at C(17) could not be assigned with certainty (26). Kopsingine and kopsaporine, as well as many new aspidofractinines, have been recently obtained from the Malayan *Kopsia* species, *K. teoi*. The relative configuration at C(17) of kopsingine (and kopsaporine) can be established from the ^{1}H NMR spectrum which showed *W*-coupling between H(17) and H(21) (2 Hz), requiring H(17) to be α (135). The structure has also been confirmed by an X-ray analysis (135). Several new derivatives of kopsingine or kopsaporine with different aromatic substitution which were also isolated from the same plant include 11-hydroxykopsingine (**191**), 11-methoxykopsingine (**192**) (143), and 11,12-methylenedioxykopsaporine (**193**) (139). Another new aspidofractinine alkaloid, kopsinol (**194**), was readily shown to be the *N*(1)-decarbomethoxy derivative of kopsaporine (132), while alkaloid **195** was assigned as 12-hydroxy-11-methoxykopsinol from the spectral data (143). The position of the aromatic methoxy group in **195** was established to be at C(11) from the observed NOE between H(10) and the aromatic OMe.

4　R^1 = H, R^2 = OMe, R^3 = CO$_2$Me
190　R^1 = R^2 = H, R^3 = CO$_2$Me
191　R^1 = OH, R^2 = OMe, R^3 = CO$_2$Me
192　R^1 = R^2 = OMe, R^3 = CO$_2$Me
193　R^1, R^2 = OCH$_2$O, R^3 = CO$_2$Me
194　R^1 = R^2 = R^3 = H kopsinol
195　R^1 = OMe, R^2 = OH, R^3 = H
196　R^1, R^2 = OCH$_2$O, R^3 = H

197　R^1 = CO$_2$Me, R^2 = H$_2$, R^3 = 15-αOH
198　R^1 = CO$_2$Me, R^2 = H$_2$, R^3 = 15-oxo
199　R^1 = CO$_2$Me, R^2 = O, R^3 = Δ^{14,15}
200　R^1 = CO$_2$Me, R^2 = H$_2$, R^3 = nil

Another kopsinol derivative, kopsinicine (11,12-methylenedioxykopsinol, **196**), was obtained from *K. singapurensis* (128), while *K. teoi* gave kopsinganol (**197**), which possesses additional hydroxy substitution at C(15). The position of hydroxy substitution in **197** was deduced to be at C(15) from the major product **198** ($\delta_{C(15)}$ 209.0) obtained from oxidation (Collins reagent) of **197** (132). The orientation of the C(15)–OH was assigned as α from the absence of NOE enhancement on irradiation of H(21) (irradiation of H(17) was not feasible due to overlap). Additional confirmation of this assignment was provided by the subsequent observation that reduction (NaBH$_4$) of the oxo-bridged alkaloid kopsidine C gave kopsinganol (**197**), indicating that the C(15)–OH in **197** is α (*vide infra*). *K. singapurensis* also provided kopsidarine (**199**), which was shown to be the 3-oxo derivative of kopsingine. The NMR data showed the presence of a conjugated lactam carbonyl absorption at δ_C 162.9, while signals normally due to H(3) were absent. The location of the lactam carbonyl at C(3) was further confirmed from the substantial downfield shift of the H(15) olefinic resonance to δ_H 6.11, which is characteristic of the β-hydrogen of an α,β-unsaturated carbonyl moiety. The observation of W-coupling between H(17α) and H(21) in **199** confirmed that the orientation of the C(17)–OH is β (58).

A study of the antihypertensive activity of the alkaloids of *K. teoi* was carried out, which was prompted by positive results from preliminary screening of alkaloidal extracts. The major aspidofractinine alkaloid kopsingine (**4**) was found to produce dose-related decreases in mean arterial blood pressure and heart rate in anesthetized, spontaneously hypertensive rats (SHR) which were similar to those elicited in normotensive controls. The same depressor response was shown by the 12-demethoxy derivative (kopsaporine, **190**) and the semisynthetic 14,15-dihydro derivative of kopsingine **200**, indicating that minor modifications to the basic structure of kopsingine do not significantly alter the hypotensive responses. A more drastic change in the structure, as in the heptacyclic kopsidine A and the semisynthetic 3-to-17 oxo-bridged compound (*vide infra*), resulted in an increase in blood pressure. Based on experiments involving pretreatment with various blockers such as hexamethonium, atropine, and phentolamine, it would appear that the depressor, as well as the pressor, effects produced by these compounds could be ascribed to both central and peripheral actions (144).

A group of related aspidofractinine alkaloids, possessing in common an epoxide function at C(14), C(15) are the kopsimalines (**201–206**), which were obtained from *K. singapurensis* (58,128). The presence of an epoxide function was indicated by the characteristic signals due to H(14) and H(15) at δ_H 3.5 and 3.2, respectively, and the corresponding C(14) and C(15) carbon signals at δ_C 54 and 57, respectively. The stereochemistry of the epoxide function was deduced to be α from the observed reciprocal NOEs between H(17) and H(15).

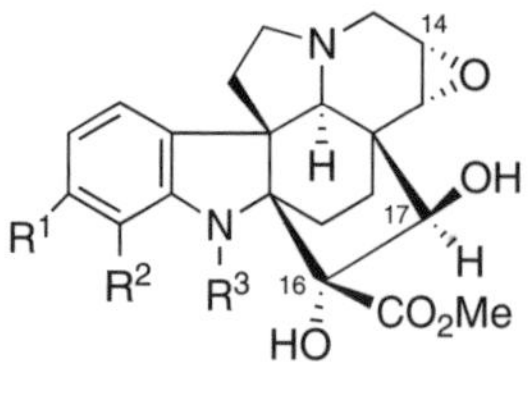

201 $R^1, R^2 = OCH_2O, R^3 = CO_2Me$
202 $R^1 = H, R^2 = OH, R^3 = CO_2Me$
203 $R^1 = OMe, R^2 = OH, R^3 = CO_2Me$
205 $R^1 = H, R^2 = OMe, R^3 = CO_2Me$

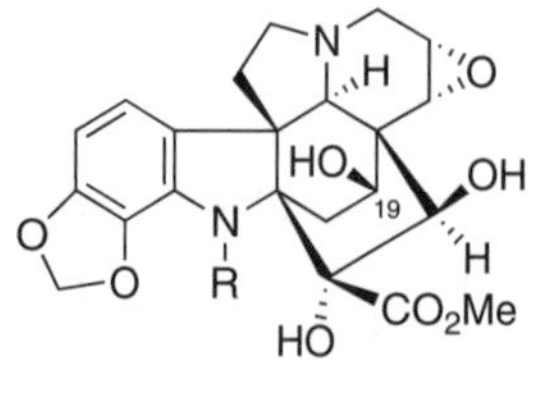

204 R = CO_2Me

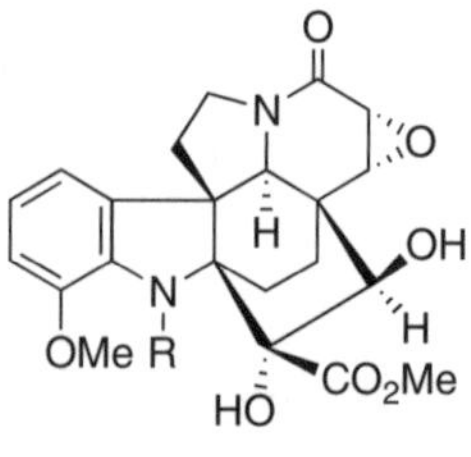

206 R = CO_2Me

The NMR data of kopsimaline B (**202**) indicated hydroxy substitution in the aromatic ring, from the presence of three contiguous aromatic hydrogens and an OH signal at δ 9.28. Since H(9) was characterized by its NOE interaction with H(21), the position of OH substitution was readily deduced to be at C(12). In kopsimaline C (**203**), the NMR data indicated similarity to **202**, except for the presence of additional aromatic substitution in the form of a methoxy group. The observed NOE between H(10) and the aromatic OMe provided confirmation for methoxy substitution at C(11).

The NMR and MS data of kopsimaline D (**204**) indicated substitution of H by an OH when compared with kopsimaline A (**201**). Establishment of the position of OH substitution was by HMBC which showed ^3J correlations from H(15) and H(21) to C(19), while the orientation of the C(19)–OH function was determined to be β from the observed reciprocal NOEs between H(19) and H(21) (128).

Kopsimaline E (**205**) was first reported from the leaves of *K. teoi*, and was then identified as 14,15-β-epoxykopsingine (143) based on the apparent NOE observed between H(21) and H(14) (^{1}H NMR at 270 MHz). In the subsequent study (128), extensive NOE/NOESY experiments were repeated at 400 MHz and the results indicated that a revision of the epoxide stereochemistry was necessary. In this instance, irradiation of H(21) resulted only in enhancement of H(9) and *vice versa*, while irradiation of H(14) resulted in enhancement of H(3) and H(15). Irradiation of H(17), on the other hand, caused enhancement of H(15) and *vice versa*. Based on these results the orientation of the 14,15-epoxide function in **205** was clearly shown to be α (128).

Kopsimaline F (**206**) was readily deduced to be the 3-oxo derivative of **205** from the NMR data (58). In addition to the signals due to the 14,15-epoxy function, a lactam resonance due to C(3) was observed at δ 167. Kopsimalines A (**201**), B (**202**), C (**203**), D (**204**), and E (**205**), were found to reverse multidrug resistance in vincristine-resistant KB cells, with **201** showing the highest potency (IC$_{50}$ 3.9 µg/mL in the presence of 0.1 µg/mL of vincristine) (128).

Several *Kopsia* species are sources of new aspidofractinine-type alkaloids which are characterized by the presence of a double bond across the C(16)/C(17) bridge. The simplest member of this group is kopsijasminine (**207**) from *K. teoi* (90), which is the *N*(1)-decarbomethoxy derivative of kopsijasmine (**208**), previously obtained from *K. jasminiflora* (145). Two other related dehydrokopsinine derivatives are venacarpines A (**209**) and B (**210**) from *K. singapurensis* (*K. fruticosa*) (123). Other new dehydropleiocarpine alkaloids include kopsidasine (**211**) and its *N*-oxide (**212**) from *K. dasyrachis* (146), and *N*(1)-methoxycarbonyl-11,12-methylenedioxy-$\Delta^{16,17}$-kopsinine (**213**), *N*(1)-methoxycarbonyl-12-methoxy-$\Delta^{16,17}$-kopsinine (**214**), *N*(1)-methoxycarbonyl-12-hydroxy-$\Delta^{16,17}$-kopsinine (**215**), and the *N*-oxides of **213** and **214** (**216** and **217**, respectively) from *K. profunda* (147–149). These alkaloids showed the typical olefinic carbon resonances, as well as the vinylic-H resonance, associated with the acrylic ester moiety. Pleiocarpine (**154**) and the new dehydropleiocarpine derivative *N*-methoxycarbonyl-11,12-methylenedioxy-$\Delta^{16,17}$-kopsinine (**213**) displayed strong potency in reversing multidrug resistance in drug-resistant KB cells (IC$_{50}$ 3.1 and 4.4 µg/mL in the presence of 0.25 µg/mL of vincristine), while kopsijasminine (**207**) showed only moderate activity (IC$_{50}$ 13 µg/mL in the presence of 0.1 µg/mL of vincristine); N(1)-methoxycarbonyl-12-methoxy-$\Delta^{16,17}$-kopsinine (**214**), on the other hand, was found to be appreciably cytotoxic to KB cells (90,138).

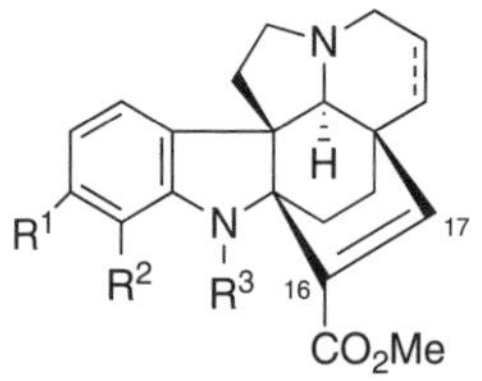

207 R^1 = R^2 = R^3 = H
208 R^1 = R^2 = H, R^3 = CO$_2$Me
209 R^1, R^2 = OCH$_2$O, R^3 = H, $\Delta^{14,15}$
210 R^1 = R^3 = H, R^2 = OMe, $\Delta^{14,15}$

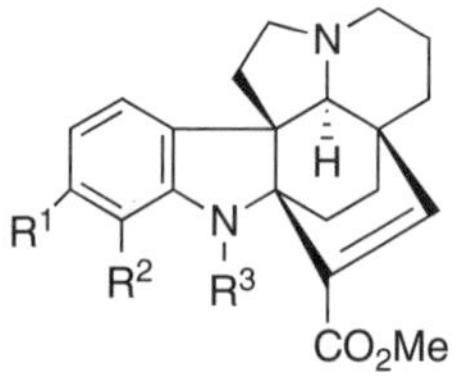

213 R^1, R^2 = OCH$_2$O, R^3 = CO$_2$Me
214 R^1 = H, R^2 = OMe, R^3 = CO$_2$Me
215 R^1 = H, R^2 = OH, R^3 = CO$_2$Me
216 R^1, R^2 = OCH$_2$O, R^3 = CO$_2$Me, N→O
217 R^1 = H, R^2 = OMe, R^3 = CO$_2$Me, N→O

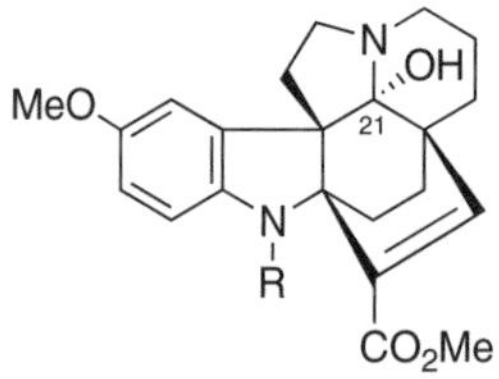

211 R = CO$_2$Me
212 R = CO$_2$Me, N→O

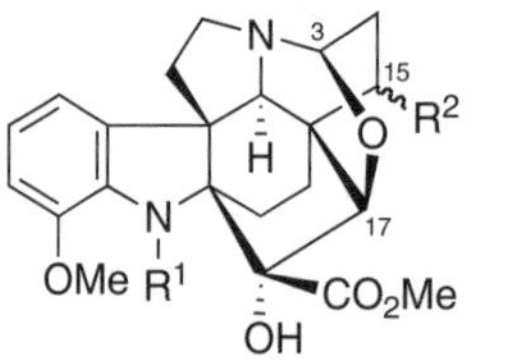

218 R^1 = CO$_2$Me, R^2 = α-OMe
219 R^1 = CO$_2$Me, R^2 = α-OEt
220 R^1 = CO$_2$Me, R^2 = α-OH
221 R^1 = CO$_2$Me, R^2 = β-OH

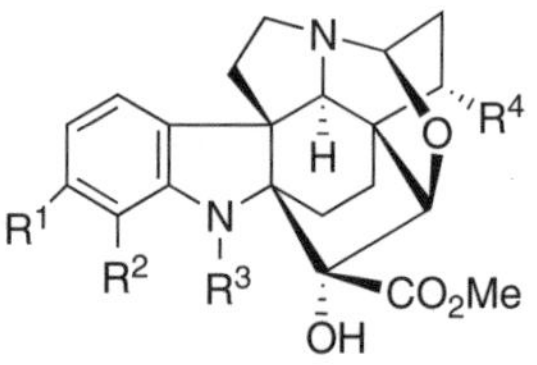

222 R^1 = R^2 = H, R^3 = CO$_2$Me, R^4 = OH
223 R^1 = R^2 = H, R^3 = CO$_2$Me, R^4 = OMe
224 R^1, R^2 = OCH$_2$O, R^3 = CO$_2$Me, R^4 = OH
225 R^1, R^2 = OCH$_2$O, R^3 = CO$_2$Me, R^4 = OMe

Several new heptacyclic alkaloids, kopsidines A, B, C, and D (**218–221**), were obtained from *K. teoi* (90,150,151) and *K. singapurensis* (55,58), which are characterized by formation of an oxygen bridge linking C(17) and C(3) of the aspidofractinine framework. The NMR spectra of these alkaloids indicated similarity with those of kopsingine, except that the 14,15 double bond was absent and two of the piperidine ring carbons were oxygenated, of which one was substituted by a methoxy group (in the case of kopsidine A, **218**). The H(3) signal in these alkaloids appeared as a low-field triplet at ca. δ 4.5 and LRCOSY showed long-range coupling between H(3) and H(17). Likewise, the H–C COLOC spectrum showed three-bond correlation between H(3) and C(17), further confirming the C(3) to C(17) linkage. The methoxy substituent was thus at C(15) and the configuration at this center was deduced from the observed NOE interaction between H(15) and H(17), indicating that the orientation of the C(15) alkoxy group was α. Formation of the additional ring has forced the piperidine ring into a boat conformation which was reflected in the observed coupling pattern of the H(3) (t, $J_{3-14} = 2.5\,\text{Hz}$) and H(15) (d, $J_{14-15} = 7.5\,\text{Hz}$) resonances (150,151). Several other alkaloids of the same type, singapurensines A–D (**222–225**), have also been subsequently obtained from *K. singapurensis* (152).

226 R = CO$_2$Me   **227** R = CO$_2$Me   **228** R = CO$_2$Me

Two new alkaloids, paucidactines A (**226**) and B (**227**), were obtained from *K. pauciflora* from Sabah, Malaysian Borneo, which have as the novel feature, a heptacyclic ring system containing a lactone moiety (153). The ^{13}C NMR spectrum showed, in addition to a lactam carbonyl (δ 166), another carbonyl signal due to a lactone function (δ 169). A conspicuous feature of the ^{1}H NMR spectrum was the presence of a low-field one-H singlet at δ 4.74 which was attributed to H(6). The corresponding C(6) signal was also observed at δ 84, its downfield shift being consistent with its being α to both an oxygen and a carbonyl group of a lactam function. The 2D COSY and HMQC data revealed the presence of an isolated methylene, an isolated oxymethine, a CH$_2$CH$_2$ unit, and a CH$_2$CH$_2$CH$_2$ fragment, which were consistent with the proposed structure, especially the location of the lactam carbonyl at C(5), since the alternative location of the lactam carbonyl at C(3) would require the presence of two CH$_2$CH$_2$ fragments. The structure was supported by the HMBC data, which showed three-bond correlation from H(6) to C(22), and was also confirmed by an X-ray diffraction analysis. Paucidactine B (**227**), as well as a new paucidactine congener, paucidactine C (**228**), differing from **227** in aromatic substitution have been isolated recently from *K. arborea* (54).

C. Kopsine, Fruticosine, and Related Alkaloids

A recent NMR study of the heptacyclic bridged alkaloids, kopsine (**1**), fruticosine (**2**), and fruticosamine (**3**), isolated in 1960s by Battersby (14,25) and Schmid (21,22) from *K. fruticosa* has allowed complete ^{1}H and ^{13}C NMR assignments of these alkaloids (154). Recently, several new derivatives of kopsine, namely, kopsifine

(229) (35,54,84,85), decarbomethoxykopsifine (230) (85), kopsinarine (231) (85), kopsorinine (232) (123), 11,12-methylenedioxykopsine (233) (54,85), kopsinidines A (234), and B (235) (54), were obtained from several Malaysian *Kopsia*. In the oxokopsine derivatives 229–232, the ^{13}C NMR data indicated the presence of a lactam carbonyl, in addition to the ketonic carbonyl group bridging C(6) and C(16). The presence of the characteristic H(3) and H(21) resonances of kopsine-type alkaloids in these alkaloids ruled out oxygenation at these positions. Oxygenation at position 5 was supported by the observation of H(6) as a singlet which had been shifted downfield as a result of anisotropy exerted by the proximate carbonyl function, as well as from the observed correlations from H(21) to C(5) and C(6) in the HMBC spectrum in the case of kopsifine (229) (84,85). In kopsorinine (232), the carbamate and 16-OH groups have been replaced by H, while the D ring showed evidence of unsaturation at C(14) and C(15) as indicated by the appearance of the characteristic olefinic signals in the NMR spectra (123). A derivative of fruticosine, jasminiflorine (236), has been reported from the Thai species, *K. jasminiflora* (145).

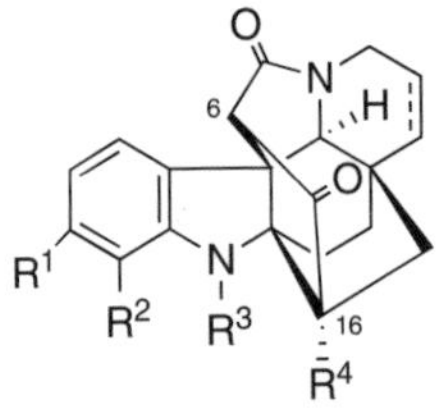

229 R^1, R^2 = OCH$_2$O, R^3 = CO$_2$Me, R^4 = OH
230 R^1, R^2 = OCH$_2$O, R^3 = H, R^4 = OH
231 R^1 = R^2 = OMe, R^3 = CO$_2$Me, R^4 = OH
232 R^1 = R^2 = R^3 = R^4 = H, $\Delta^{14,15}$

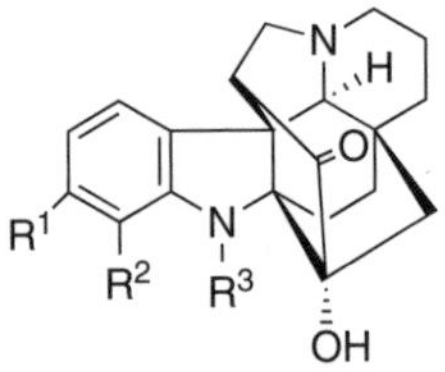

233 R^1, R^2 = OCH$_2$O, R^3 = CO$_2$Me
234 R^1, R^2 = OCH$_2$O, R^3 = H
235 R^1 = R^2 = OMe, R^3 = CO$_2$Me

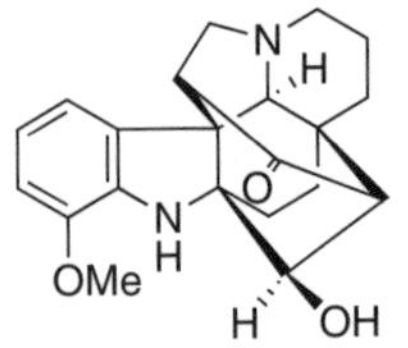

236

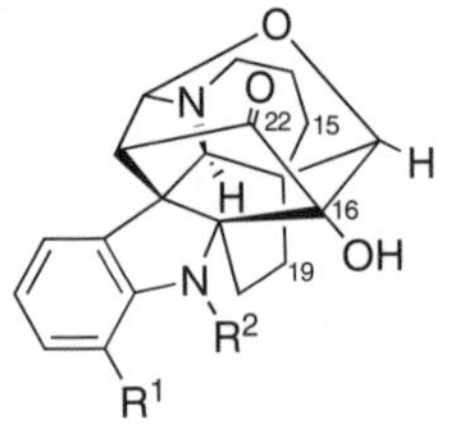

237 R^1 = OMe, R^2 = CO$_2$Me, $\Delta^{14,15}$
238 R^1 = OMe, R^2 = H, $\Delta^{14,15}$
239 R^1 = OMe, R^2 = H, 15-αOH
240 R^1 = OMe, R^2 = CO$_2$Me, 15-αOH
241 R^1 = H, R^2 = CO$_2$Me

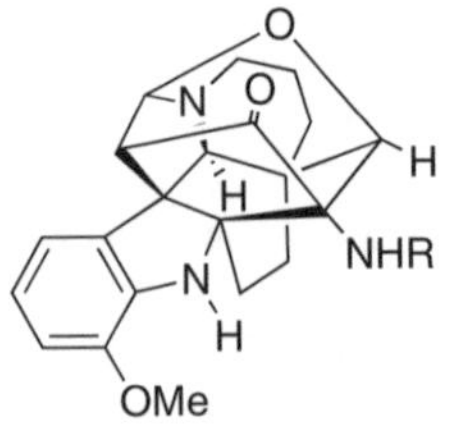

242 R = H, $\Delta^{14,15}$
243 R = H, 15-αOH
244 R = COMe, $\Delta^{14,15}$

Another new group of alkaloids, kopsinitarines A–E (**237–241**), were obtained from *K. teoi*, which are characterized by formation of a cage-like structure as a consequence of formation of oxo and carbonyl bridges, across C(17) and C(5) and across C(16) and C(6), respectively (90,155,156). These alkaloids were also found in *K. singapurensis* (58).

The distinguishing feature of the ^{1}H NMR spectra of these alkaloids was the significant downfield shift of the H(5) and H(6) resonances, which were seen as a pair of AX doublets at δ 5.2 and 2.7 with a coupling constant of 4.9 Hz. The corresponding ^{13}C NMR resonances have also undergone a similar downfield shift to δ 95 (C(5)) and 57 (C(6)). In addition, the ^{13}C NMR spectrum showed the presence of a ketonic function (δ 207) corresponding to the C(22) carbonyl bridge. The proposed structure was supported by LRCOSY, which showed long-range coupling between H(5) and H(17), and by HMBC, which revealed three-bond correlations between H(17)/C(5) and H(5)/C(17) and a two-bond correlation from H(6) to C(22). The subsequent isolation of kopsinitarine D (**240**) in sufficient amounts has allowed confirmation of the structure by X-ray analysis (156).

In addition to the cage kopsinitarines, two other alkaloids, mersingines A (**242**) and B (**243**), were also obtained in trace amounts (156,157). These alkaloids had NMR spectral data which indicated their structural affinity to the kopsinitarines but differed in possessing an additional nitrogen atom, as shown by HRMS. Furthermore, these alkaloids gave positive tests with ninhydrin, which indicated the presence of a primary amino group, and **242** was also readily acetylated to the amide derivative **244**. The orientation of the C(15)–OH function in **243** was assigned as α based on the observed NOE interaction between H(15) and H(17). These compounds are probably artifacts which originated from the kopsinitarines in the basic conditions under which extraction of the alkaloids were carried out. These alkaloids possibly originate from initial formation of an isokopsine-like precursor **245** under basic conditions via a reversible acyloin rearrangement of the kopsinitarine precursor (**238** or **239**). This could then be followed by a reversible condensation with ammonia resulting in the unstable imine **246**, which could then subsequently rearrange back to the original kopsine-like alkaloid (**242** or **243**), which now incorporates a C(16)-amino function as shown in Scheme 13. Paucity of material in the case of kopsinitarines A–C prevented verification of the proposed pathway, and in the case of kopsinitarine D, attempted reaction did not furnish the corresponding C(16)-amino compound,

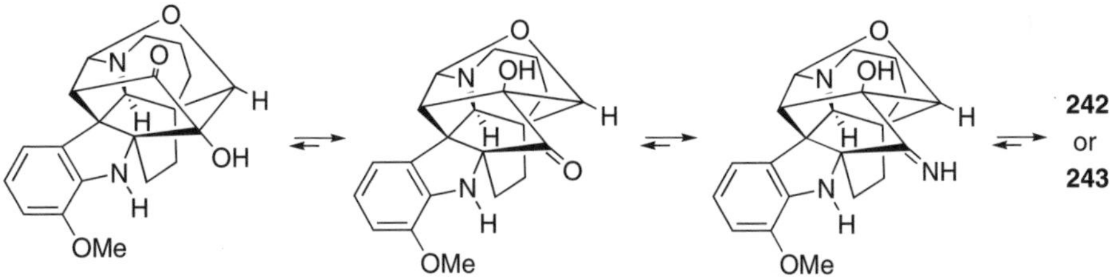

238 or **239**   **245**   **246**   **242** or **243**

Scheme 13

probably as a result of intramolecular H-bonding involving the C(16)–OH group and the proximate carbamate function, which was detected in the solid state (156).

K. *arborea* and K. *dasyrachis* furnished a number of alkaloids belonging to the methyl chanofruticosinate series (35,158–162). The parent alkaloid methyl chanofruticosinate (247) was first obtained from fruticosine by Guggisberg *et al.* (22). It was isolated as a natural product relatively recently from the Chinese species, K. *arborea* (K. *officinalis*), together with two other derivatives, methyl 11,12-methylenedioxychanofruticosinate and methyl N-decarbomethoxy-chanofruticosinate (248 and 249, respectively). The structure was firmly established by an X-ray diffraction analysis performed on 249 (158). A further 14 alkaloids (250–263), possessing the same carbon skeleton, have been isolated subsequently from K. *arborea* (including K. *officinalis* and K. *flavida*). The structures of these alkaloids were readily established based on the spectroscopic data, and by comparison with the previously known alkaloids (35,159–162).

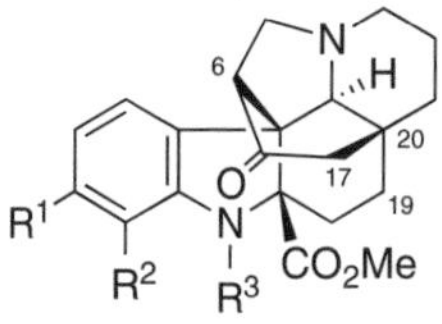

247 $R^1 = R^2 = H$, $R^3 = CO_2Me$

248 R^1, $R^2 = OCH_2O$, $R^3 = CO_2Me$

249 $R^1 = R^2 = R^3 = H$

250 $R^1 = H$, $R^2 = OMe$, $R^3 = CO_2Me$

251 $R^1 = R^2 = OMe$, $R^3 = CO_2Me$

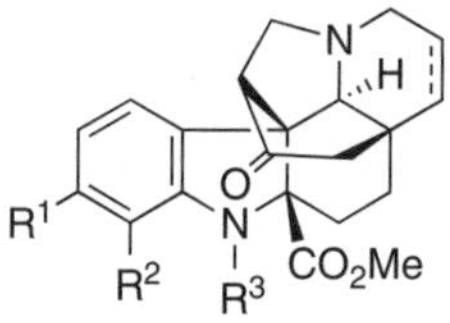

252 R^1, $R^2 = OCH_2O$, $R^3 = H$

253 R^1, $R^2 = OCH_2O$, $R^3 = H$, $\Delta^{14,15}$

254 $R^1 = R^3 = H$, $R^2 = OMe$

255 $R^1 = R^3 = H$, $R^2 = OMe$, $\Delta^{14,15}$

256 $R^1 = R^2 = OMe$, $R^3 = CO_2Me$, $\Delta^{14,15}$

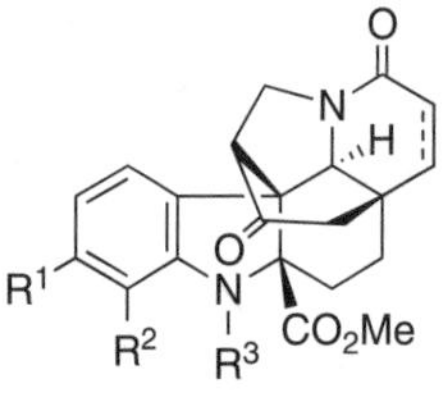

257 $R^1 = R^2 = R^3 = H$, $\Delta^{14,15}$

258 R^1, $R^2 = OCH_2O$, $R^3 = H$, $\Delta^{14,15}$

259 $R^1 = R^3 = H$, $R^2 = OMe$, $\Delta^{14,15}$

260 $R^1 = H$, $R^2 = OMe$, $R^3 = CO_2Me$

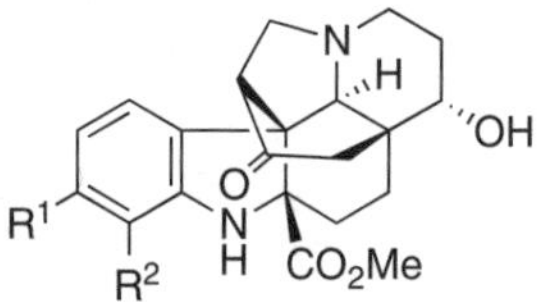

261 $R^1 = R^2 = H$

262 R^1, $R^2 = OCH_2O$

263 $R^1 = H$, $R^2 = OMe$

Only one isokopsine-type alkaloid, namely, dasyrachine (264) has been reported, in the first instance from K. *dasyrachis* (85), and subsequently from K. *arborea* (54). Dasyrachine (264) showed an M$^+$ at m/z 366 in EIMS, which analyzed for $C_{21}H_{22}N_2O_4$. The ^{13}C NMR spectrum showed a total of 21 carbon resonances, while the UV spectrum was characteristic of a dihydroindole chromophore. The NMR spectral data indicated the presence of a methylene-dioxy substituent, the absence of carbamate or lactam functions, the presence of the characteristic H(21) of aspidofractinine-type alkaloids (δ 2.82), an oxygenated quaternary carbon (δ_C 82.4), and a ketonic function (δ_C 217.3). Application of COSY and HMQC methods revealed the presence of the same partial structures as those in kopsine (**1**) and methylenedioxykopsine (**233**).

Comparison of the ^{1}H and ^{13}C NMR spectral data of dasyrachine (**264**) (85) with those of 11,12-methylenedioxykopsine (**233**) showed that although there were many similarities, some differences were noted in the chemical shifts of dasyrachine compared to those of **233**. Of even greater significance was the observation that the HMBC data for **264** and **233** were different, suggesting a change in the structure. In methylenedioxykopsine (**233**), H(5) and H(17) showed heteronuclear correlations to the ketonic C(22), whereas in the case of **264**, only correlations from H(17) to the equivalent carbon, C(16), were detected. In **233**, H(6), H(18) and H(17), showed correlations to the oxygenated quaternary carbon C(16), whereas in **264**, the equivalent carbon, C(22), showed, in addition to the correlations from H(6) and H(17), a correlation from H(5), but not from H(18). These changes are consistent with an isokopsine-type carbon skeleton for dasyrachine (**264**) (21,163). The carbonyl function is now at C(16), which is consistent with the observed three-bond correlations from both the H(17) in the HMBC spectrum, and the noted absence of any correlation with H(5). The oxygenated quaternary carbon is at C(22), which is consistent with the observed HMBC correlations from H(17) and H(6) (^2J), as well as from H(5) (^3J).

Isokopsine (**265**) has been obtained from kopsine (**1**) via a thermally induced acyloin rearrangement, or by the action of dilute alkali (reflux) on kopsine or decarbomethoxykopsine, through which an equilibrium mixture of decarbomethoxykopsine and decarbomethoxyisokopsine was obtained (27,163). Since kopsine, decarbomethoxykopsine, and decarbomethoxyisokopsine are known to occur in *K. fructicosa* (21,27), and since the possible precursor of **264**, 11,12-methylenedioxykopsine (**233**), was also isolated, the possibility that **264** was an artifact formed from **233** cannot be completely discounted, although this was rendered unlikely under the mild conditions of the extraction procedure employed. This was further confirmed by the fact that attempted base-induced reaction of **233** (refluxing in 0.1 M NaOH for 8 h) resulted in the formation of the decarbomethoxylated derivative as the major product, with **264** detected only as a minor product (ratio 17:1) (85). Dasyrachine (**264**) represents the second, naturally occurring isokopsine alkaloid isolated to date.

264 $R^1, R^2 = OCH_2O$

265 $R^1 = R^2 = H$

266 $R^1, R^2 = OCH_2O, R^3 = CO_2Me$

270 $R^1 = R^2 = H, R^3 = CO_2Me$

267 $R = CH_2OH$

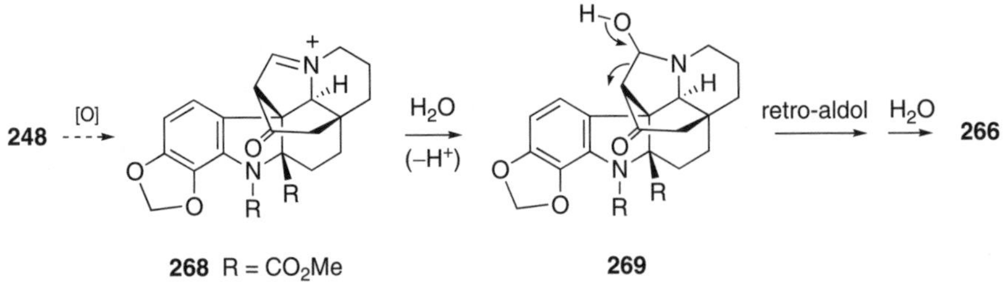

248 $\xrightarrow{\text{[O]}}$ **268** R = CO$_2$Me $\xrightarrow[\text{(–H}^+\text{)}]{\text{H}_2\text{O}}$ **269** $\xrightarrow{\text{retro-aldol}}$ $\xrightarrow{\text{H}_2\text{O}}$ **266**

Scheme 14

The leaf extract of *K. dasyrachis* provided a minor alkaloid, danuphylline (**266**), characterized by a novel pentacyclic skeleton (164,165). The structure elucidation was based on spectral analysis of danuphylline as well as of the diol **267** derived from NaBH$_4$ reduction of danuphylline. The structure of danuphylline represents a novel skeletal arrangement in which a new six-membered ring has been formed by cleavage of the C(5)–C(6) bond of the precursor alkaloid **248**, which was the predominant alkaloid present. Danuphylline (**266**) can thus be considered a *"seco-methylchanofruticosinate,"* and represents the first member of this group isolated as a natural product. An unusual feature of the ^{1}H NMR spectrum of danuphylline is the rather high-field shift of the formamide-H (δ 6.68), which is rationalized based on a preferred conformation in which the formamide carbonyl is directed away from the aromatic ring in order to avoid unfavorable repulsive interactions between the π-electron density of the formamide C=O, as well as that of the lone pairs of the oxygen, and the π-electron density of the aromatic system, thus placing the formamide-H within the shielding zone of the aromatic ring current. The structure of danuphylline has also been subsequently confirmed by an X-ray analysis (35).

A possible origin of this ring-opened alkaloid is from the methyl chanofruticosinate **248**, which on oxidation provides the iminium ion **268**. Hydrolysis of this iminium ion gives the presumably unstable carbinol amine **269**, which could then undergo a retro-aldol-type reaction to provide the *seco*-alkaloid, danuphylline (**266**) (Scheme 14). An electrochemically mediated semisynthesis based on such a biomimetic route has been successfully implemented (*vide infra*). A second danuphylline congener (**270**) has been recently isolated from the Chinese species, *K. arborea* (*K. officinalis*) (122).

D. Syntheses of Aspidofractinine and Kopsane Alkaloids

It is expedient to consider the syntheses of these two groups of closely related alkaloids together in the same section. The total syntheses of these alkaloids have been virtually dominated by Magnus who released a series of papers starting with the synthesis of the kopsane alkaloids, 5,22-dioxokopsane (**271**) and kopsanone (**272**), both not originally isolated from *Kopsia*, and encountered in the genus less frequently compared to kopsine and kopsinine. The construction of the heptacyclic carbon framework of these kopsane alkaloids provided a ready platform for the implementation of the subsequent syntheses of other *Kopsia* alkaloids by Magnus and his coworkers.

The essential approach adopted involved the construction of the key homoannular diene intermediate **279** using, in the early steps, the indole-2,3-quinodimethane strategy (166–168), and subsequent intramolecular Diels–Alder cycloaddition, following *endo* face allylation of the diene at C(6) (169,170). The racemic tetracyclic amine **277** (used later in the synthesis of kopsinine) was assembled from the imine **275** obtained by condensation of the *N*-protected 2-indole-3-carboxaldehyde **273** and the chloroamine **274** (Scheme 15). Conversion of the imine to the carbamate **276**, followed by removal of the trichloroethyl group with concomitant decarboxylation gave **277**. Acylation of **277**, followed by oxidation, provided the sulfoxides **278** as a mixture of diastereomers. Treatment of **278** with TFAA, followed in succession by addition of the resulting solution to hot chlorobenzene and methanol gave crystals of the homoannular diene **279**. Treatment of **279** with potassium hexamethyldisilazide (KN(SiMe$_3$)$_2$), followed by allyl bromide, gave exclusively the *endo*-allyl product **280**, which upon heating at 100°C furnished the Diels–Alder adduct **281**. Reduction of the double bond by *in situ* generated diimide, followed by oxidation, yielded a mixture of diastereomeric sulfoxides. Heating of the mixture resulted in conversion of the major sulfoxide **283** to **285**, which upon treatment with TFAA yielded the epimeric sulfoxide **286**. Sulfoxide **286** on warming to 130°C gave the *N*-protected 5,22-dioxokopsane derivative **287**, which was then transformed to 5,22-dioxokopsane (**271**) and kopsanone (**272**) in straightforward steps. The steps from **282** to **285** exemplify the application of an oxidation–elimination–addition sequence involving the intermediacy of the anti-Bredt compound **284** to effect the transfer of oxidation level from C(6) to C(22) (169,170).

The successful construction of the 5,22-dioxokopsane framework provided an advanced intermediate for implementation of the kopsine synthesis (Scheme 16) (171). Thus, basic cleavage of the *N*(1)-protected 5,22-dioxokopsane **287** gave the acid **288**. Adjustment of the indolic nitrogen-protecting group led to the carbamate **289**, following which NaBH$_4$ reduction of the acid functionality via the derived mixed anhydride gave the primary alcohol **290**. Application of the Grieco procedure installed the exocyclic double bond in **293** (via **291** and **292**). Osmylation of **293** followed by Swern oxidation yielded the α-hydroxy aldehyde **294**. Treatment of **294** with LDA/THF at −78°C gave the intramolecular aldol addition product **295** as a single diastereomer. Reduction of the amide carbonyl with BH$_3$·THF, followed by hydrolytic workup with HCl gave the diol **296**, which on Swern oxidation furnished kopsine (**1**) (171).

The synthesis of kopsijasmine (**208**) (172) commenced with the homoannular diene **279** (prepared earlier in the 5,22-dioxokopsane synthesis), and followed essentially the same route to the diastereomeric sulfoxides **297** and **298**, except for the use of 1-iodo-2-chloroprop-2-ene in place of allyl bromide to effect the *endo* face allylation at C(6). Only **297** can undergo a thermally induced *syn* elimination of PhSOH to give the α,β-unsaturated anti-Bredt amide **299**. Capture of the

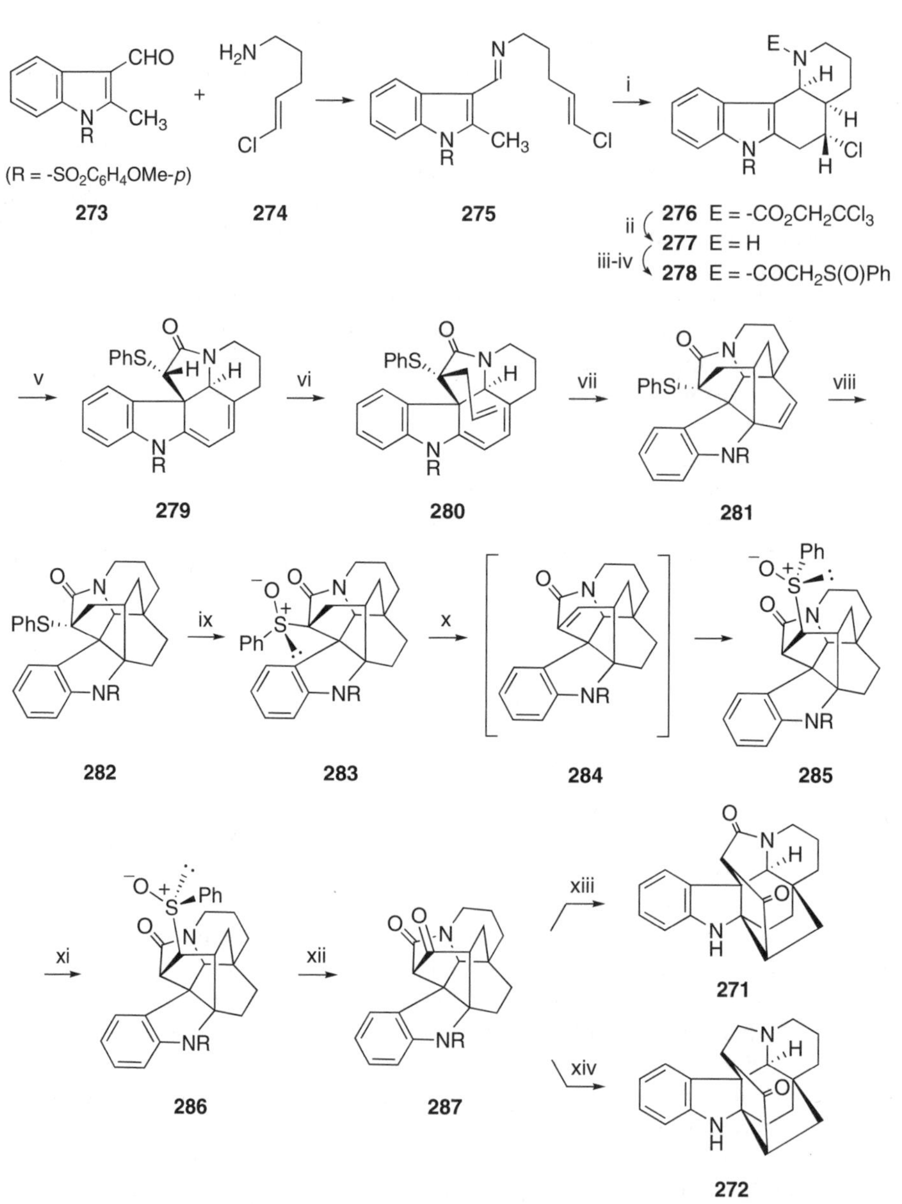

Scheme 15 Reagents: (i) Cl₃CCH₂OCOCl, *i*-Pr₂NEt; (ii) Zn/AcOH/THF/H₂O; (iii) PhSCH₂COCl; (iv) *m*-CPBA; (v) Tf₂O, CH₂Cl₂, 0°C; PhCl, 135°C; MeOH; (vi) KN(SiMe₃)₂/THF; allyl bromide; (vii) 100°C; (viii) TsNHNH₂/NaOAc/EtOH; (ix) *m*-CPBA, −70°C; (x) 215°C; (xi) TFAA/CH₂Cl₂, 0°C; (xii) PhCl/TFAA, 130°C; (xiii) Li/NH₃; Me₂SO/1-cyclohexyl-3-(2-morphinoethyl)-carbodiimide metho-*p*-toluenesulfonate; (xiv) LiAlH₄/THF; Me₂SO/1-cyclohexyl-3-(2-morphinoethyl)-carbodiimide metho-*p*-toluenesulfonate.

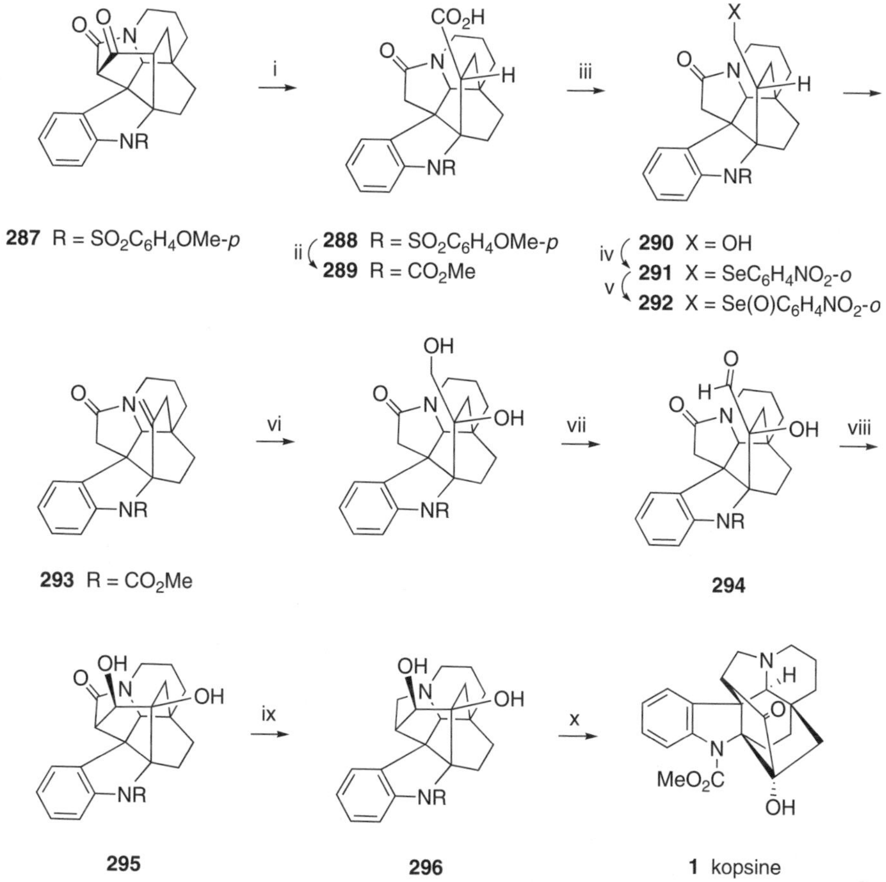

Scheme 16   Reagents: (i) NaOH/MeOH; (ii) sodium naphthalenide, MeOCH$_2$CH$_2$OMe, $-30°C$; K$_2$CO$_3$, PhCH$_2$Et$_3$NCl, ClCO$_2$Me; (iii) Et$_3$N/THF, ClCO$_2$Bu-i; NaBH$_4$; (iv) NCSeC$_6$H$_4$NO$_2$-o/PBu$_3$/THF; (v) 50% H$_2$O$_2$; (vi) OsO$_4$/N-methylmorpholine N-oxide/t-BuOH/THF/H$_2$O; (vii) DMSO/(COCl)$_2$/CH$_2$Cl$_2$, $-60°C$; (viii) LDA/THF, $-78°C$; (ix) BH$_3$ · THF; HCl; (x) DMSO/(COCl)$_2$/CH$_2$Cl$_2$, $-60°C$.

anti-Bredt alkene by nucleophilic addition by AcOH gave **300**, which upon hydrolysis and Jones oxidation provided the β-ketoamide **301**. The non-enolizable **301** was cleaved with NaOH/MeOH to give **302**, which on treatment with DBN/DME gave the α,β-unsaturated acid **303**, which was then processed to kopsijasmine (**208**) following removal of the arylsulfonyl group, installation of the carbamate, methylation, and reduction of the amide carbonyl as shown in Scheme 17 (172).

Magnus was also the first to report the synthesis of kopsinine (**146**), which was in the context of implementing a synthesis of the bisindole, pleiomutine (173). The synthesis commenced with the racemic tetracyclic amine **277** prepared earlier in the synthesis of 5,22-dioxokopsane and followed essentially the same procedure to the diketone **287**, but with some modifications. Thus,

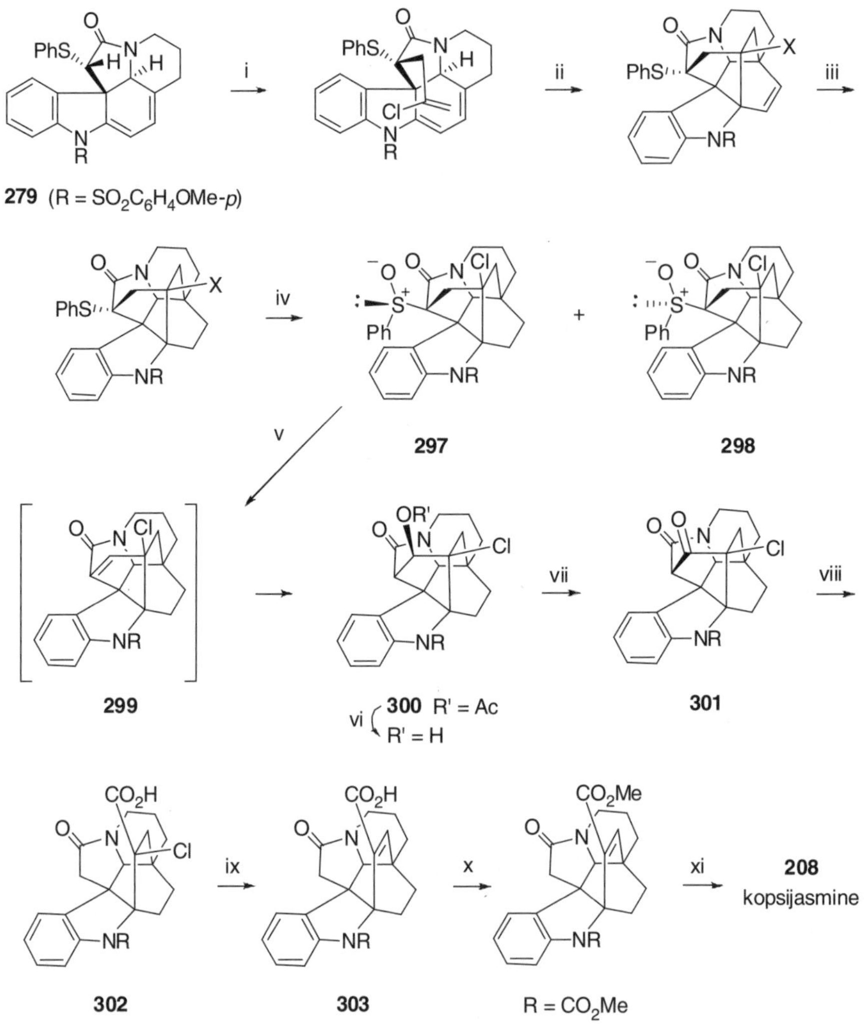

279 (R = SO$_2$C$_6$H$_4$OMe-p)

Scheme 17 Reagents: (i) KH; CH$_2$=CClCH$_2$I; (ii) 111°C; (iii) TsNHNH$_2$/NaOAc/EtOH/H$_2$O; (iv) m-CPBA, NaHCO$_3$; (v) AgOAc/AcOH, 205°C; (vi) LiOH aq; (vii) Jones reagent; (viii) NaOH/MeOH; (ix) DBN/DME; (x) Na/anthracene; ClCO$_2$Me/K$_2$CO$_3$; CH$_2$N$_2$; (xi) BH$_3$·THF.

treatment of amine **277** with (R)-(+)-p-tolylsulfinylacetic acid gave the readily separated diastereomers **304** and **305**. Subsequent treatment of **304** with TFAA at 0°C, followed by heating to 130°C, gave the homoannular diene **306**. Confirmation of its absolute configuration was by X-ray analysis of the sulfoxide derivative **309**. Allylation of **306** gave **307**, which upon thermolysis (100°C, 4 h) yielded the intramolecular Diels–Alder adduct **308**. The alkene **308** was then processed to the diketone **287** via the sulfoxide **309**. Subsequent treatment of the diketone **287** with KOH–MeOH as before resulted in cleavage of the β-ketoamide functionality to

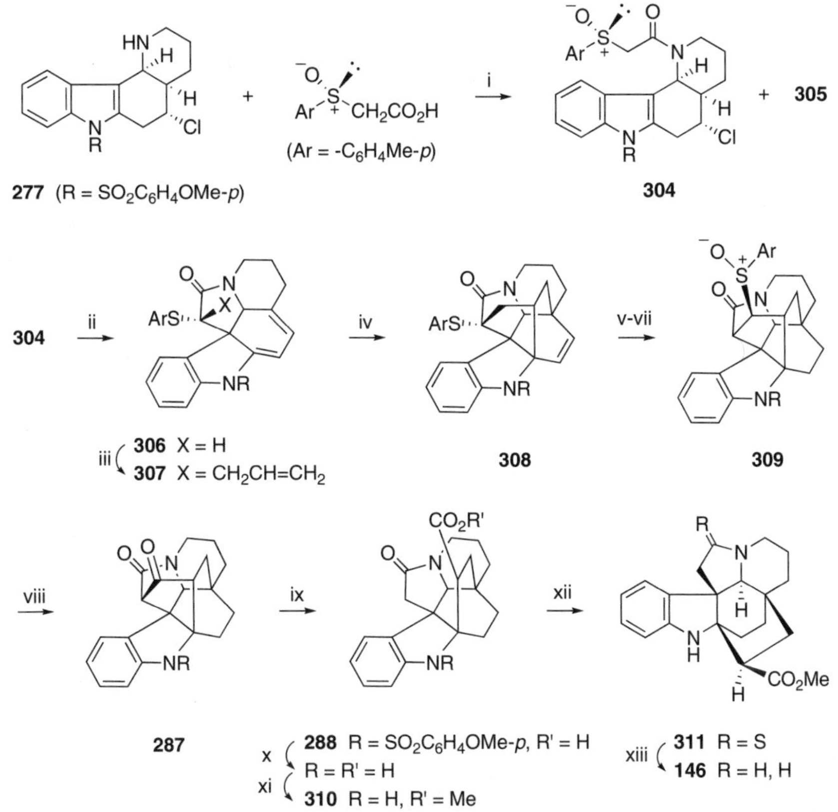

277 (R = SO$_2$C$_6$H$_4$OMe-*p*)

(Ar = -C$_6$H$_4$Me-*p*)

304

305

ii **304** →

iii (**306** X = H
     307 X = CH$_2$CH=CH$_2$

308

309

viii →

287

ix →

x (**288** R = SO$_2$C$_6$H$_4$OMe-*p*, R' = H
  (R = R' = H
xi (**310** R = H, R' = Me

xii →

xiii (**311** R = S
     (**146** R = H, H

Scheme 18   Reagents: (i) 1-cyclohexyl-3-(2-morpholinoethyl)-carbodiimide metho-*p*-toluenesulfonate; (ii) TFAA/CH$_2$Cl$_2$, 0°C; 130°C, PhCl; (iii) KN(SiMe$_3$)$_2$, allyl bromide; (iv) 100°C, 4 h; (v) HN=NH; (vi) *m*-CPBA; (vii) 240°C; (viii) TFAA; (ix) KOH/MeOH; (x) Li–NH$_3$, −78°C; (xi) MeOH–HCl; (xii) Lawesson's reagent; (xiii) Raney-Ni.

provide the acid **288**, which was then transformed to kopsinilam (**310**), and thence to kopsinine (**146**), following reduction of the amide carbonyl via conversion to the thioamide **311** by Lawesson's reagent, followed by Raney nickel desulfurization (Scheme 18).

There are three other reported syntheses of kopsinine. The one by Kuehne (174) was based on generation of the pentacyclic diene **317**, followed by Diels–Alder reaction with phenyl vinyl sulfone (Scheme 19). The synthesis commenced with preparation of the α-phenylselenyl aldehyde **312**, which on condensation with the indoloazepine **313** furnished a diastereomeric mixture of the bridged azepines, **314**. Heating of the mixture in CH$_2$Cl$_2$ resulted in thermally induced intramolecular *N*-alkylation and rearrangement of **314** to yield a single diastereomer of the pentacyclic derivative **315**. Oxidation of **315** by *m*-CPBA gave the *N*-oxide **316** as the major product, which on reduction with triphenylphosphine gave the diene **317**. Heating of **317** with phenyl vinyl sulfone led to the

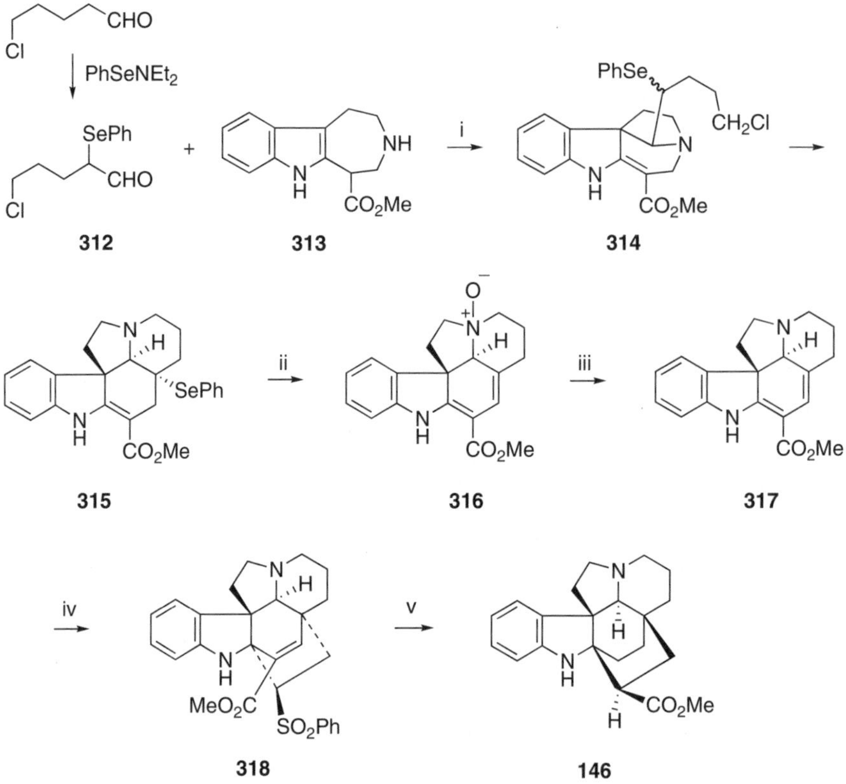

Scheme 19   Reagents: (i) boric acid, CH_2Cl_2, reflux, 14 h; (ii) m-CPBA, CH_2Cl_2; (iii) Ph_3P, PhH, 100°C, 5 h; (iv) $CH_2\!=\!CHSO_2Ph$, PhH, 120°C, 16 h; (v) Raney-Ni.

Diels–Alder adduct **318**, which on Raney nickel desulfurization gave racemic kopsinine (**146**). The method can be readily adapted to the preparation of pleiocarpinine, aspidofractinine, pleiocarpine, and kopsanone (174).

Wenkert's formal synthesis (via the Kuehne homoannular diene **317**) was based on a modification of a general route previously employed in the syntheses of various *Aspidosperma* alkaloids starting from β-acetylpyridine (**319**). The transformation of the ketoester **320** (prepared previously in a synthesis of deethylvincadifformine) to **323** via **321** and **322**, and thence to the ring C diene **317**, constitutes a formal synthesis of kopsinine (Scheme 20) (175,176).

Another formal synthesis based on preparation of the homoannular diene **317** was that by Natsume and coworkers (177). The key step was the double cyclization of the β-ketoester chloride **324** by treatment with t-BuOK in THF–HMPA. The reaction proceeded by initial base hydrolysis of the carbamate function, followed by formation of the 1-indolyl anion, which then initiates the first cyclization to forge the* pyrrolidine ring E. A second, base-induced cyclization yielded the pentacyclic ketoester product **325** (as an equilibrium

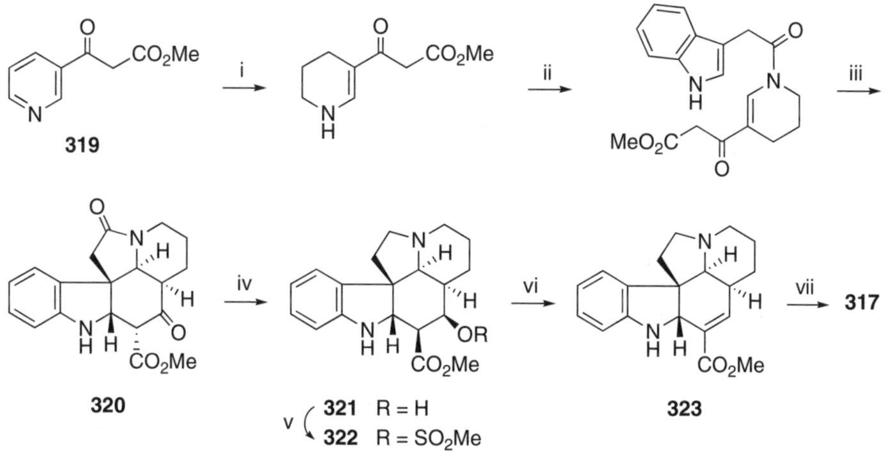

Scheme 20 Reagents: (i) H$_2$, Pd/C; (ii) NaH, indoleacetic anhydride, THF; (iii) PPA, 100°C, 20 Torr (iv) BH$_3$ · THF; NaBH$_4$; (v) MeSO$_2$Cl, *i*-Pr$_2$NEt, Cl(CH$_2$)$_2$Cl, 0°C; (vi) KH/THF; (vii) Pb(OAc)$_4$, CH$_2$Cl$_2$.

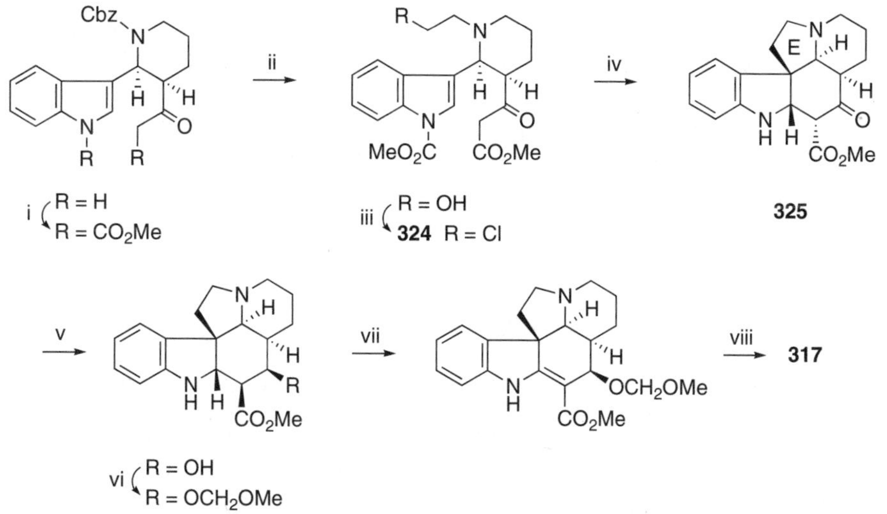

Scheme 21 Reagents: (i) (TMS)$_2$NLi/THF/HMPA, −70°C; MeCO$_2$CN, −70°C; (ii) H$_2$, 10% Pd/C, MeOH; ethylene oxide, MeOH; (iii) MeSO$_2$Cl, 3 equiv, K$_2$CO$_3$, CH$_2$Cl$_2$; (iv) *t*-BuOK/THF/HMPA, −70 to 20°C; (v) NaBH$_4$, MeOH/MeOCH$_2$CH$_2$OMe; (vi) MeOCH$_2$Cl, *i*-Pr$_2$NEt; Cl(CH$_2$)$_2$Cl; HOAc/MeOH; (vii) (PhSeO)$_2$O, PhH, Et$_3$N; (viii) 2% TsOH/MeOH.

mixture with the enol form), which was then processed to the desired diene **317** in a few straightforward steps (Scheme 21).

The Magnus syntheses of kopsidasine (**211**), lahadinine B (**175**), and 11-methoxykopsilongine (**167**) are considered together with the syntheses of the pauciflorine alkaloids (*vide infra*).

E. Kopsidasinine Alkaloids

Kopsidasinine (**326**) was first isolated from *K. dasyrachis* collected from Sabah in Malaysian Borneo. The structure was established by a combination of spectroscopy and chemical correlation, in particular, the key observation that the Hofmann degradation product of kopsidasinine, **327**, was identical to the product obtained by treatment of kopsidasine (**211**) with methyl iodide (146). Only two other kopsidasinine-type alkaloids have since been reported, 10-demethoxy-12-methoxykopsidasinine (**328**) from *K. pauciflora* (73) and 10-demethoxykopsidasinine (**329**) from the Thai species, *K. arborea* (*K. jasminiflora*) (178). A distinguishing feature of these alkaloids is the presence of a relatively low field pair of AB doublets (δ 4.0 and 3.5) due to an isolated CHCH unit (corresponding to C(16)–C(17)) and an unusually low-field ketonic carbon resonance (δ 214) due to C(21) (73,178).

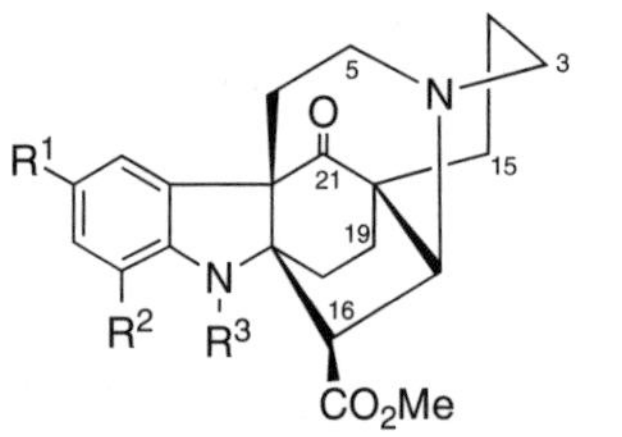

326 R^1 = OMe, R^2 = H, R^3 = CO_2Me
328 R^1 = H, R^2 = OMe, R^3 = CO_2Me
329 R^1 = R^2 = H, R^3 = CO_2Me

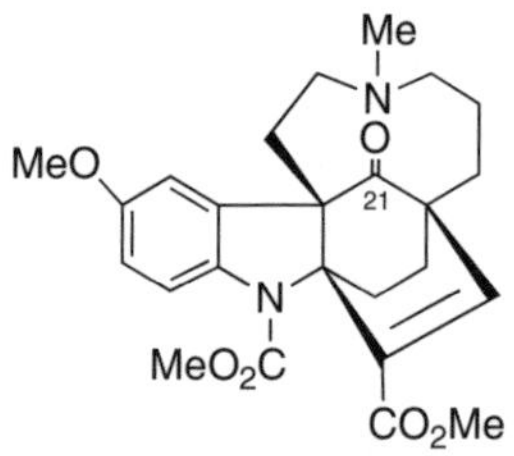

327

F. Pauciflorines and Related Alkaloids

K. pauciflora also provided two new alkaloids, pauciflorines A and B (**330** and **331**) (179), possessing a pentacyclic skeleton similar to that of kopsijasminilam (**332**), deoxykopsijasminilam (**333**), and 14,15-dehydrokopsijasminilam (**334**) from the Thai species *K. jasminiflora*. The structure of **332** was established by X-ray analysis (145). These rare alkaloids can be considered as being derived from an aspidofractinine-type precursor by cleavage of the C(20)–C(21) bond, which has been subsequently shown to be the case by the syntheses of kopsijasminilam and pauciflorine B (*vide infra*).

Pauciflorines A and B, both characterized by the presence of a bridgehead double bond, were obtained from the leaves of *K. pauciflora*. The EIMS of pauciflorine A (**330**) showed a molecular ion at *m/z* 470, which analyzed for $C_{24}H_{26}N_2O_8$, while the UV spectrum indicated the presence of a dihydroindole chromophore. The ^1H and ^{13}C NMR data showed the presence of a methylenedioxy substituent at C(11) and C(12), a CO_2Me substituent at N(1), a methyl ester and an OH at C(16), a lactam carbonyl at C(21), and a trisubstituted double bond. COSY and HMQC experiments revealed the presence of the following partial structures, namely, an isolated methylene, a CH_2CH_2, a $CH_2CH_2CH_2$, and a $CH_2CH=C$ fragment. This information suggested that the lactam carbonyl should be placed at position 21, since the alternative location of the lactam carbonyl at position 3 or 5 would result in three isolated methylenes in the former case, and two CH_2CH_2 fragments in the latter, which was clearly not the case. This would also account for the unusual deshielding observed for $H(3\alpha)$ as a result of the anisotropy due to the proximate lactam carbonyl function. Finally, the placement of the lactam carbonyl at position 21 was also supported by the observed ^3J correlation between H(6) and C(21) in the HMBC spectrum. The presence of a β-carbomethoxy function, together with an α-OH substituent on C(16), was supported by the carbon NMR shifts which are characteristic of such a substitution pattern in related aspidofractinine-type alkaloids. Furthermore, this can be confirmed from the observed long-range *W*-coupling between the intramolecularly hydrogen-bonded C(16)–OH and H(17β), as well as from the observed ^3J (C(16)–OH to C(17)) and ^2J (C(16)–OH to C(16)) correlations in the HMBC spectrum.

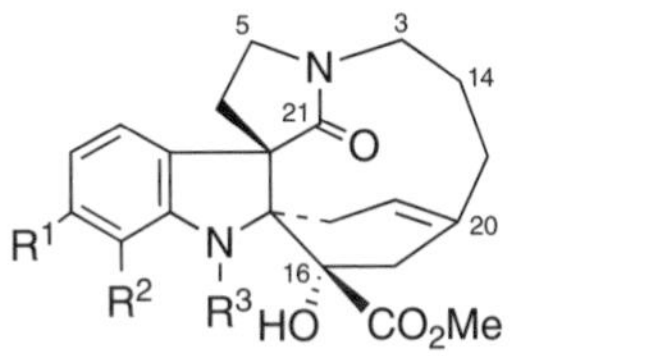

330 R^1, R^2 = OCH_2O, R^3 = CO_2Me
331 R^1 = R^2 = OMe, R^3 = CO_2Me

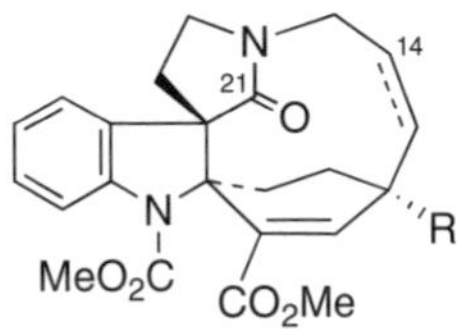

332 R = OH
333 R = H
334 R = OH, $\Delta^{14,15}$

The molecular formula of pauciflorine A yielded a DBE value of 13 and, discounting the methylenedioxy function, suggested the presence of a pentacyclic ring system incorporating a trisubstituted double bond. HMBC and NOESY data were in accord with the proposed structure of pauciflorine A as shown in **330**, and facilitated the assignment of pauciflorine B as the 11,12-dimethoxy congener of **330**. The structure of pauciflorine B has been subsequently confirmed by a total synthesis (*vide infra*). Pauciflorines A (**330**) and B (**331**) are remarkable for being rare examples of plant-derived indole alkaloids which showed potent inhibitory activity (equivalent to that of the commercial compound, arbutin) toward melanin biosynthesis in cultured B16 melanoma cells at concentrations of 13 and 25 µg/mL, respectively, without any cytotoxicity toward the cells (179).

The syntheses of kopsidasine (**211**), lahadinine B (**175**), and 11-methoxykopsilongine (**167**) were developed by Magnus in the context of the larger objective of accessing the ring systems of kopsijasminilam and the pauciflorines, and as such, the syntheses of the former three alkaloids will be discussed in the context of the syntheses of kopsijasminilam and pauciflorine B (180–182).

The strategy was centered on the construction of the key homoannular diene **346**, and from it an aspidofractinine-type precursor that might be expected to undergo a Grob fragmentation to the pauciflorine ring system.

The synthesis of kopsidasine (and kopsijasminilam) commenced with Pictet–Spengler condensation of **335** with tryptamine to furnish **336**, which was then transformed to **337** by treatment with EDCI/Et₃N/DMF (Schemes 22–24). Treatment of **337** with Belleau's reagent, followed by Raney nickel

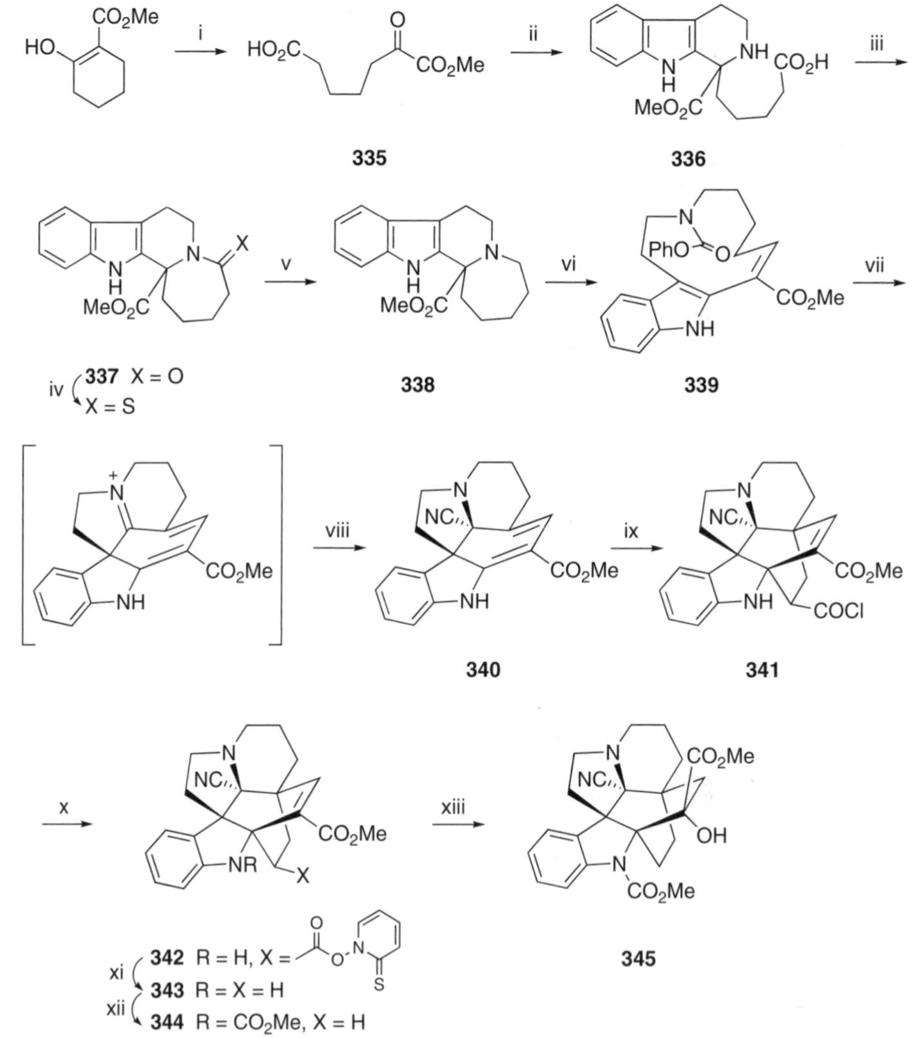

Scheme 22   Reagents: (i) O₃; Me₂S; (ii) tryptamine, Δ; (iii) EDC/HOBt; (iv) Belleau's reagent; (v) NiCl₂ · 6H₂O, NaBH₄; (vi) PhOCOCl, Cl(CH₂)₂Cl; (vii) Tf₂O/DMAP/CH₂Cl₂, reflux; (viii) Me₃SiCN; (ix) CH₂═CHCOCl, 75°C; (x) N-hydroxy-2-thiopyridone/Et₃N; (xi) hν/t-BuSH; (xii) triphosgene/pyr; MeOH; (xiii) Mn(dpm)₃, PhSiH₃/O₂/0°C, i-PrOH/Cl(CH₂)₂Cl.

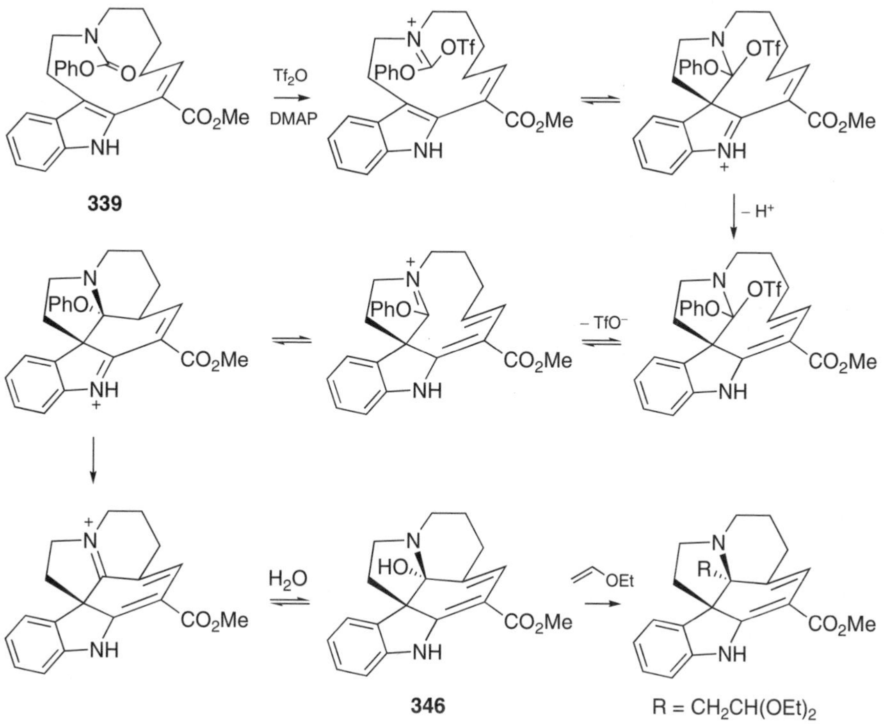

Scheme 23

desulfurization gave the tetracyclic indole derivative **338**. Subsequent treatment of **338** with phenyl chloroformate in 1,2-dichloroethane under reflux induced an isogramine-type fragmentation of the resulting cation to yield the ring-opened tricycle **339**. Addition of triflic anhydride/DMAP to **339** triggered the anticipated cascade (Scheme 23) resulting in transannular cyclization to furnish the homoannular diene **346**. It was, however, found that all attempted Diels–Alder cycloadditions with **346** were unsuccessful (Scheme 23), which led to the preparation of the more robust cyano-analogue **340**, the choice of which was prompted, in part, by the reported isolation of the cyano-substituted lahadinines from *K. pauciflora* (136). Thus, treatment of **339** with Tf$_2$O/DMAP, followed by quenching of the reaction with trimethylsilyl cyanide, gave **340**, which turned out to be well behaved in [4+2] cycloadditions with various dienophiles (Scheme 22). Diels–Alder reaction of **340** with acryloyl chloride led to the adduct **341**, which was then converted (via the thiohydroxamate ester **342**) to **343** by the Barton protocol for reductive decarboxylation. Conversion to the carbamate **344**, followed by installation of the C(16)-α-hydroxy function, gave demethoxylahadinine B (**345**) (Scheme 22). Alternatively, treatment of **343** with PhI(OAc)$_2$/MeOH, followed by reduction with Zn/AcOH, gave **347**, which following conversion to the carbamate **348**, followed by sequential treatment with AgBF$_4$/THF and aqueous NaHCO$_3$, yielded (±)-kopsidasine (**211**) (Scheme 24).

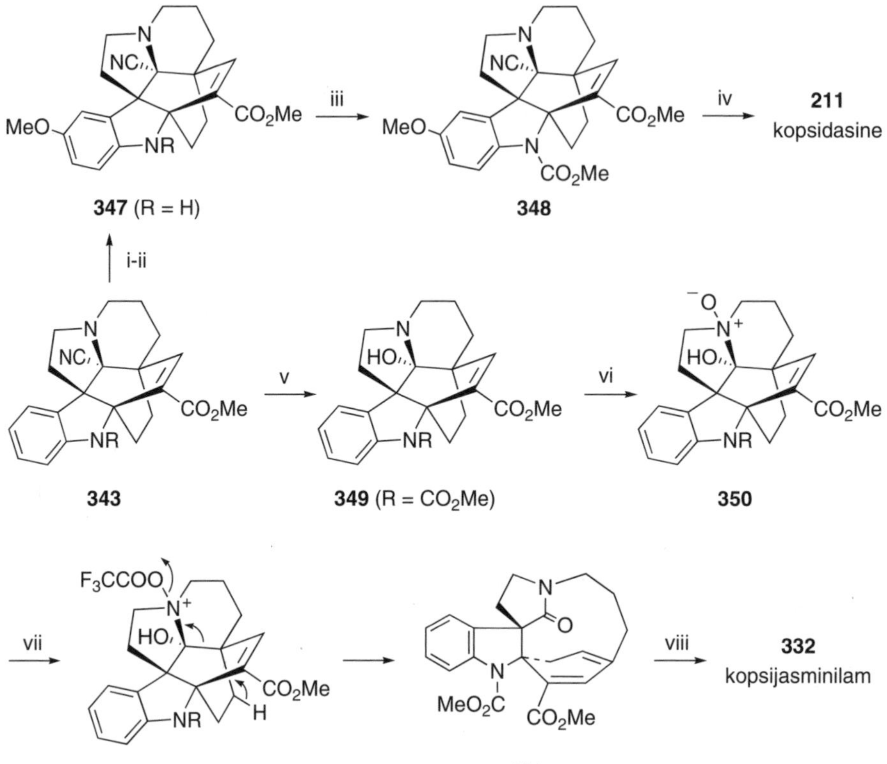

Scheme 24 Reagents: (i) PhI(OAc)$_2$, MeOH; (ii) Zn/AcOH; (iii) triphosgene/pyr; MeOH; (iv) AgBF$_4$, THF; H$_2$O; (v) AgBF$_4$, NaHCO$_3$; (vi) m-CPBA; (vii) TFAA; (viii) Mn(dpm)$_3$, PhSiH$_3$/O$_2$/ 0°C, i-PrOH, Cl(CH$_2$)$_2$Cl.

The cyano-substituted aspidofractinine derivative **343** on treatment with AgBF$_4$/THF, followed by aqueous NaHCO$_3$, gave the carbinol amine **349**, setting the stage for the implementation of the Grob (Polonovski) fragmentation. Accordingly conversion to the N-oxide **350**, followed by exposure of **350** to TFAA, resulted in the expected fragmentation to yield the ring-opened product **351**. Treatment of **351** with Mn(dpm)$_3$ (cat)/PhSiH$_3$/O$_2$ in i-PrOH at 0°C, followed by aqueous Na$_2$S$_2$O$_3$ workup, gave kopsijasminilam (**332**) (Scheme 24).

The synthesis of lahadinine B (and pauciflorine B) required preparation of the tetracycle **355** (Scheme 25), which was accordingly prepared as before, but starting with Pictet–Spengler condensation of dimethoxytryptamine **353** (in turn prepared from vanillin, **352**) and **335** to provide the acid **354**, which was then processed to **355** (182). The reaction of **355** with phenyl chloroformate in dichloroethane required prolonged reflux, and resulted in only modest yields of **356** (Scheme 25). Treatment of **356** as before with Tf$_2$O/DMAP, followed by quenching with TMSCN, gave **357**, which on reaction with acryloyl chloride, followed by Barton reductive decarboxylation, gave **358**. Installation of the

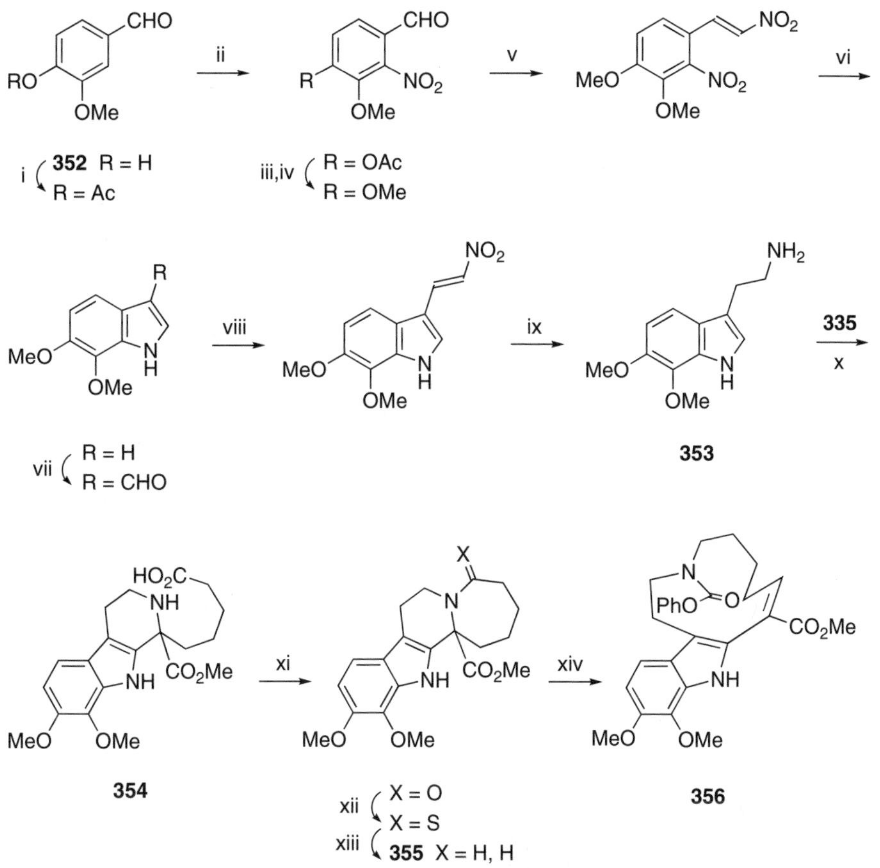

Scheme 25 Reagents: (i) AcCl/pyr, 100°C; (ii) fuming HNO₃, <6°C; (iii) K₂CO₃/MeOH;
(iv) MeI, K₂CO₃, DMF; (v) MeNO₂, KOH, DMF/EtOH; HCl, 0°C; Ac₂O/NaOAc, reflux; (vi) Fe/
AcOH/EtOH, reflux; (vii) POCl₃/DMF; (viii) MeNO₂/NH₄OAc, reflux; (ix) LiAlH₄/THF/0°C to
reflux; (x) CF₃CO₂H/CH₂Cl₂/4 Å molecular sieves; (xi) EDCI/HOBt/Et₃N/DMF; (xii) Belleau's
reagent/THF; (xiii) NiCl₂·6H₂O/NaBH₄/THF/MeOH/0°C; (xiv) PhOCOCl, Cl(CH₂)₂Cl, reflux.

carbamate group proved problematic, but was eventually achieved by treatment of the phenylselenide derivative **359** with KN(SiMe₃)₂/18-crown-6/CO₂
at −78°C. Implementation of the conjugate reduction–oxidation sequence on
360 using Mn(dpm)₃/PhSiH₃/O₂/*i*-PrOH–ClCH₂CH₂Cl gave **361**, which on
removal of the PhSe substituent gave (±)-lahadinine B (**175**). Reductive
removal of the cyano function by treatment of **175** with triethylsilane in the
presence of TFA, in turn, provided 11-methoxykopsilongine (**167**) (Scheme 26).

Treatment of lahadinine B (**175**) with AgBF₄/THF, followed by workup with
EtOAc/MeCO₃H, gave (±)-pauciflorine B (**331**) accompanied by the double
bond isomer **362** as a minor product. The structure of **331** was confirmed by X-ray
analysis.

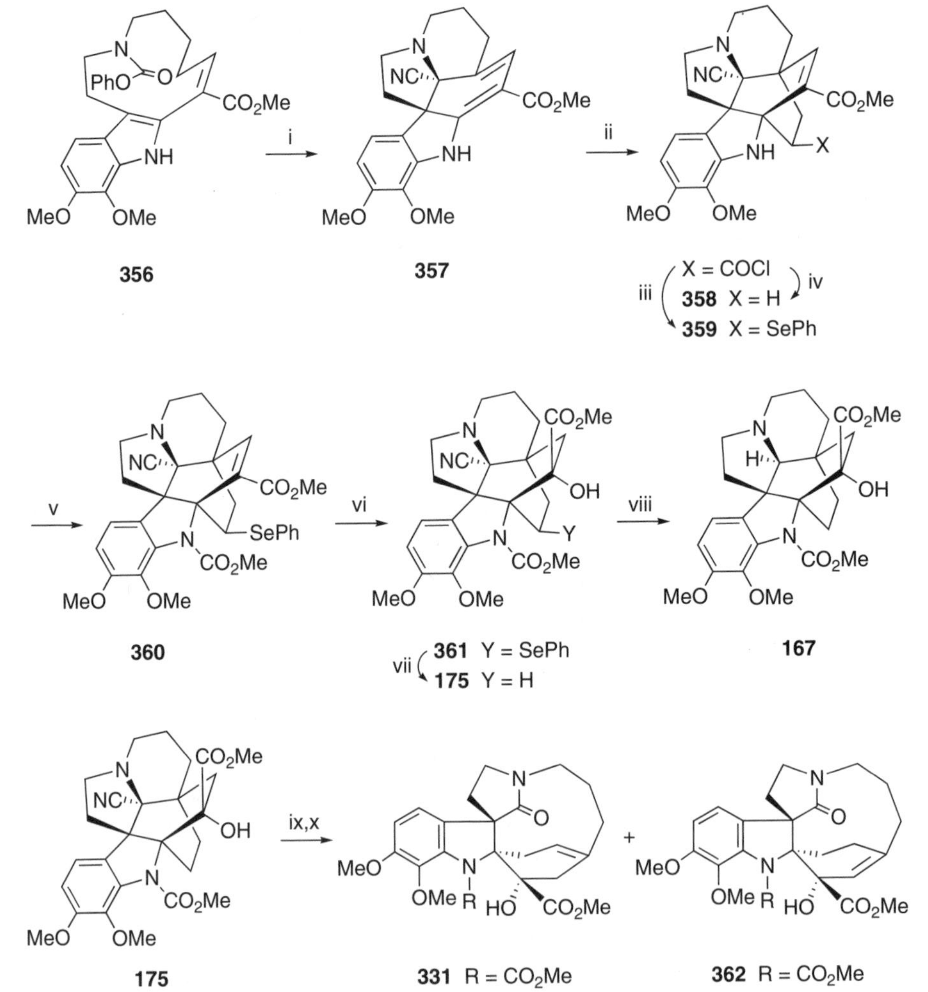

Scheme 26 Reagents: (i) Tf₂O/DMAP/CH₂Cl₂, reflux; TMSCN/DMAP/CH₂Cl₂;
(ii) CH₂=CHCOCl; (iii) 2-thiopyridone-N-oxide (Na salt)/(PhSe)₂/CH₂Cl₂/hν;
(iv) 2-thiopyridone-N-oxide (Na salt), t-BuSH, CH₂Cl₂, hν; (v) KN(TMS)₂/18-crown-6/THF,
−78°C; CO₂; Me₂SO₄; (vi) Mn(dpm)₃ 5 mol%, PhSiH₃, O₂, i-PrOH, Cl(CH₂)₂Cl; (vii) Ph₃SnH/PhMe,
reflux; (viii) CF₃CO₂H, Et₃SiH, CH₂Cl₂; (ix) AgBF₄/THF; (x) EtOAc/MeCO₃H.

The use of peracetic acid to induce the Grob fragmentation was discovered serendipitously during transformations in the demethoxypauciflorine synthesis where it was noticed that the fragmentation product **363**, in addition to the N-oxide **364**, was also obtained during purification (SiO₂, EtOAc/hexanes/NEt₃) of the carbinol amine **365**. The formation of the fragmentation product was attributed to the presence of peracetic acid in the EtOAc solvent used during chromatography, a supposition which received additional support when the deliberate use of EtOAc/MeCO₃H mixtures to quench the AgBF₄/THF-mediated conversion of **345** to **365** yielded **363** as the major product (182).

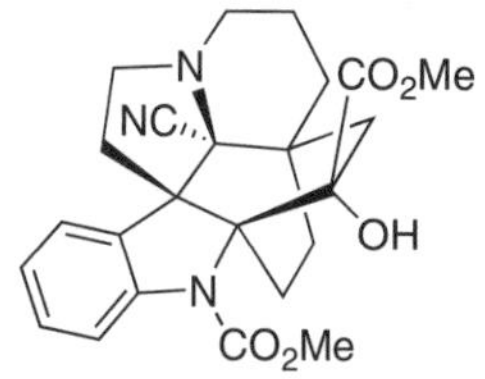

345 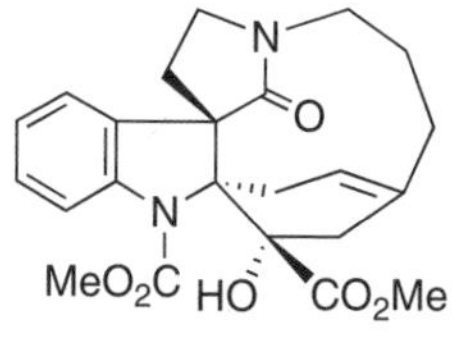   **363** 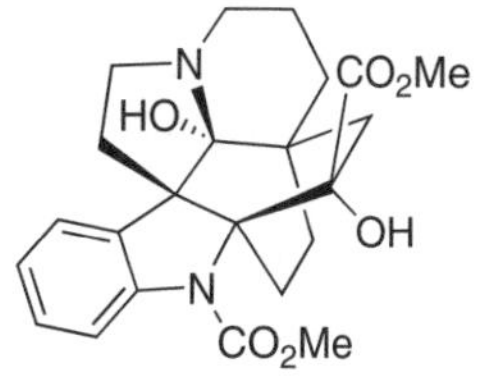   **364** N→O
                                   365

A synthesis of kopsijasminilam (**332**), based also on a fragmentation strategy involving cleavage of the C(20)–C(21) bond in an aspidofractinine precursor, was reported by Kuehne (183,184). The synthesis commenced with acid-induced cyclization of the *Aspidosperma* alkaloid (±)-minovincine (**366**) to the hexacyclic ketone **367**. Reduction of **367** with DIBALH provided the epimeric alcohols which were separable by chromatography. Reaction of the major epimer **368** with tosyl anhydride gave **369**, which on treatment with KCN provided the nitrile **370**, the result of a Grob fragmentation with cyanide trapping of the resulting iminium ion. Oxidation of **370** with potassium *t*-butoxide and oxygen gave the lactam **371**, which was then transformed to the carbamate **373** via the cyclic ketene acetal-acylal **372** (Scheme 27). The carbamate **373** was then processed to deoxykopsijasminilam (**333**) via the diene ester **374**. Alternatively, implementation of a conjugate (δ) addition–oxidation sequence on **374** provided kopsijasminilam (**332**).

G. Lapidilectines and Lundurines

K. grandifolia (*K. lapidilecta*) provided the novel pentacyclic alkaloid lapidilectine A (**375**), notable for having a carbon skeleton which incorporates a central eight-membered ring fused to an unsaturated five-membered ring (185,186). Four other alkaloids related to lapidilectine A were also obtained, namely, isolapidilectine A (**376**), lapidilectam (**377**), lapidilectinol (**378**), and *epi*-lapidilectinol (**379**). In addition to these, another new hexacyclic alkaloid, lapidilectine B (**380**), was also obtained in which a five-membered ring lactone unit has been formed by loss of a methyl ester and a methyl from lapidilectine A. These alkaloids, which are related to kopsijasminilam and the pauciflorines, were obtained in amorphous form and their structure elucidation relied mainly on analysis of NMR spectral data.

The ^{13}C NMR spectrum of lapidilectine A (**375**) showed the presence of two methyl esters and a carbamate group. One of the ester methyls (C(16)–COOMe) was relatively shielded (δ 2.93), as a result of being located above the aromatic plane in order to avoid steric hindrance. The ^{1}H NMR spectrum showed the three-carbon CH$_2$CH=CH fragment corresponding to the C(3)–C(14)–C(15) unit and the magnitude of J_{14-15} and J_{3-15} of 6 and 1 Hz, respectively, indicated the

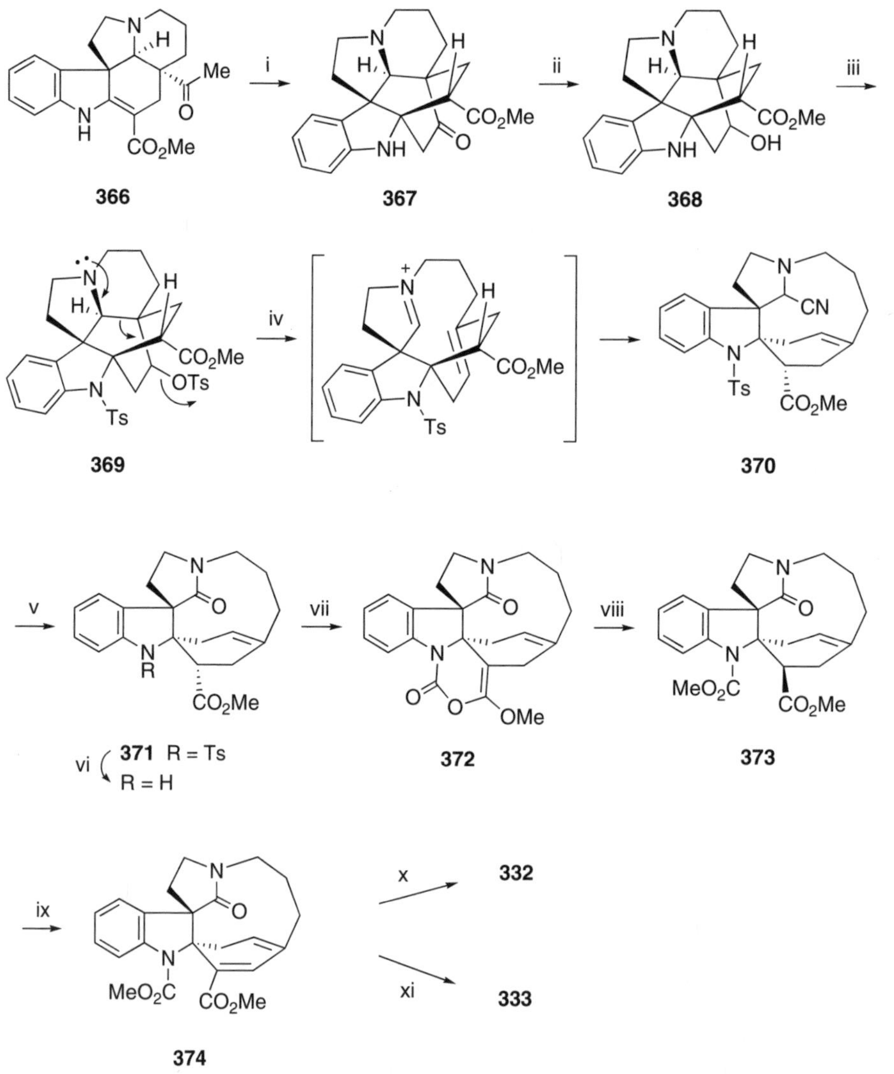

Scheme 27 Reagents: (i) HCl/MeOH, reflux; (ii) DIBALH, THF, −78°C; (iii) Ts₂O, pyr; (iv) KCN, EtOH/H₂O, reflux; (v) t-BuOK, THF, O₂; (vi) Na naphthalenide, DME, −78°C; (vii) triphosgene, pyr, CH₂Cl₂, MeOH, pyr; (viii) NaOMe, MeOH, reflux; (ix) KHMDS, 2-(phenylsulfonyl)-3-phenyloxaziridine, −78°C; (x) O₂, Mn(dpm)₂, PhSiH₃, −10°C; (xi) H₂/Pd.

presence of a five-membered unsaturated ring, while the carbon resonance of C(20) at δ 67 supported its direct attachment to N(4). The proposed structure incorporating the central eight-membered ring bearing two methyl ester groups was supported by the 2D NMR data, while the relative configurations of C(2), C(7), and C(20) were presumed to be similar to those of venalstonine, which was also isolated from the plant. Isolapidilectine A (**376**) differed from lapidilectine A in the configuration at C(16), which was reflected in the downfield shift of the C(16) ester methyl signal to the normal value of ca. δ 3.5. The mass spectral data of lapidilectam (**377**) indicated incorporation of oxygen and loss of two hydrogens compared with lapidilectine A, while the NMR spectral data revealed the presence of a conjugated, five-membered ring lactam consistent with the proposed structure (δ_C 171, δ_H 6.0, 6.8).

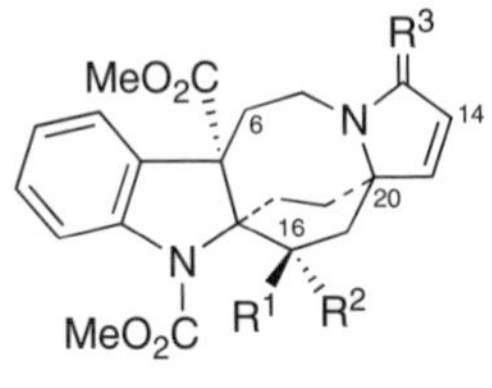

375 $R^1 = H$, $R^2 = CO_2Me$, $R^3 = H_2$

376 $R^1 = CO_2Me$, $R^2 = H$, $R^3 = H_2$

377 $R^1 = H$, $R^2 = CO_2Me$, $R^3 = O$

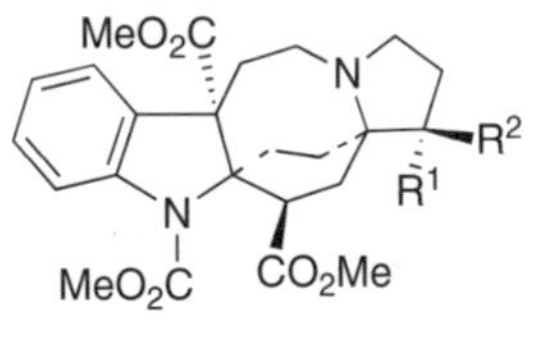

378 $R^1 = OH$, $R^2 = H$

379 $R^1 = H$, $R^2 = OH$

Lapidilectinol (**378**) and its C(15) epimer **379** were distinguished from lapidilectine A (**375**) by loss of the C(14)–C(15) unsaturation and the presence of an alcohol function at C(15). These features were consistent with the NMR spectral data which showed absence of olefinic signals and the appearance of an oxymethine (δ 81). The orientation of the C(15)–OH in **378** and **379** were determined from the respective NOESY spectrum. The hexacyclic lapidilectine B (**380**) had a HRMS which indicated it differed from lapidilectine A by loss of a methyl ester and a methyl group. The NMR spectral data were essentially similar to those of **375**, except for the loss of the two methyl ester signals, the appearance of a lactone carbonyl resonance (δ 177), and the downfield shift of the C(7) resonance to δ 91, indicating its attachment to the lactone oxygen. The configurations at C(2), C(16), and C(7) follow from those of its probable progenitor, lapidilectine A. Three other lapidilectine B congeners, tenuisines A–C (**381–383**), were isolated from *K. tenuis*. The structures of these alkaloids were revised from the dimeric structure with an axis of symmetry (187,188) to the monomeric version as represented in **381–383**, based on new HRLSIMS data, as well as LSIMS and NMR analyses of the methyl iodide salt of tenuisine A (**381**) (189).

K. tenuis also provided four new alkaloids which were obtained in minute amounts, lundurines A (**384**), B (**385**), C (**386**), and D (**387**), which possess a novel carbon skeleton incorporating a cyclopropyl unit (189,190). The NMR spectral data of lundurine A (**384**) showed the presence of an aromatic methoxy group at C(10), a carbamate function, a lactam carbonyl at C(3), and a C(14)–C(15) double bond which was deduced to be part of a five-membered ring from the observed J_{14-15} vicinal coupling constant of 6 Hz. The location of the lactam carbonyl at C(3) was indicated by the absence of signals normally attributed to H(3), the presence of an amide resonance at δ 170, as well as the substantial downfield shift of the H(15) olefinic resonance to δ 6.87, which is characteristic of a β-hydrogen of an α,β-unsaturated carbonyl moiety. COSY and HMQC data revealed the remaining partial structures to be made up of two CH_2CH_2 and one CH_2CH unit. A particularly conspicuous feature of the ^1H NMR spectrum was the presence of a high-field doublet at ca. δ 1.0, which was suggestive of cyclopropyl or cyclobutyl rings. This doublet was shown by COSY to be due to the methine of the CH_2CH fragment. HMBC data allowed the eventual structure **384** to be assembled, which has the distinctive feature of having a cyclopropyl ring system fused to a dihydroindole unit at C(2) and C(7).

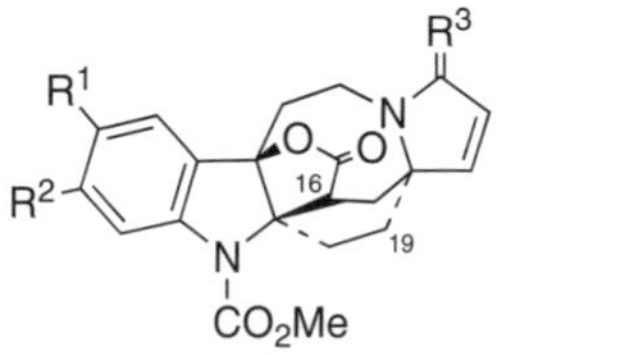

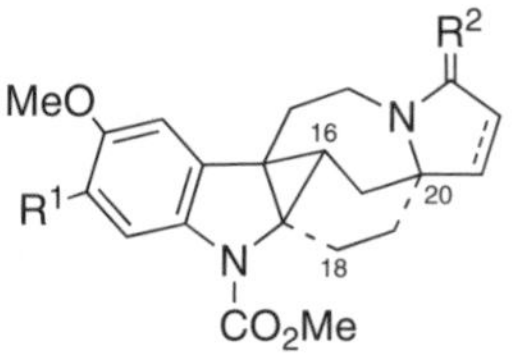

380 $R^1 = H, R^2 = H, R^3 = H_2$
381 $R^1 = OMe, R^2 = H, R^3 = H_2$
382 $R^1 = OMe, R^2 = OMe, R^3 = H_2$
383 $R^1 = OMe, R^2 = H, R^3 = O$

384 $R^1 = H, R^2 = O, \Delta^{14,15}$
385 $R^1 = H, R^2 = H_2, \Delta^{14,15}$
386 $R^1 = H, R^2 = H_2$
387 $R^1 = OMe, R^2 = H_2, \Delta^{14,15}$

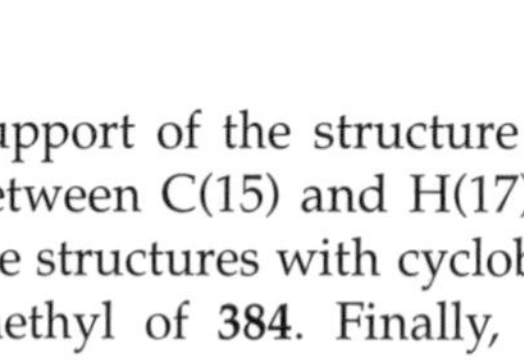

388

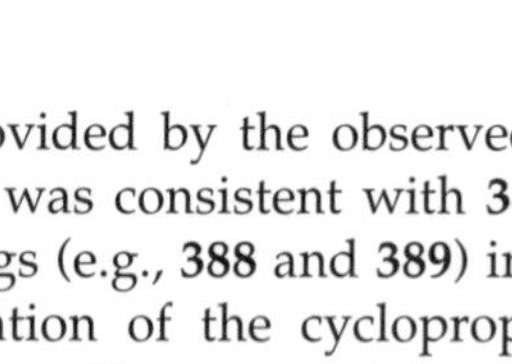

389

Further support of the structure was provided by the observed three-bond correlation between C(15) and H(17), which was consistent with **384**, but ruled out alternative structures with cyclobutyl rings (e.g., **388** and **389**) in place of the cyclopropylmethyl of **384**. Finally, confirmation of the cyclopropyl unit was obtained from the measurement of the $^1J_{C-H}$ coupling constant for C(16) from the gated decoupled spectrum which yielded a large $^1J_{C-H}$ value of 164.1 Hz. With the structure of lundurine A thus established, the structures of lundurines B, C, and D follow from the spectral data. The co-occurrence of these alkaloids in the leaves reflects the progressive stages in the oxidation level of the five-membered ring moiety in the lundurines. The novel lundurines can be envisaged to have arisen from the structurally related lapidilectine B (**380**) via decarboxylation. Scheme 28 shows a possible biogenetic relationship between the aspidofractinines, pauciflorines, lapidilectines, and lundurines.

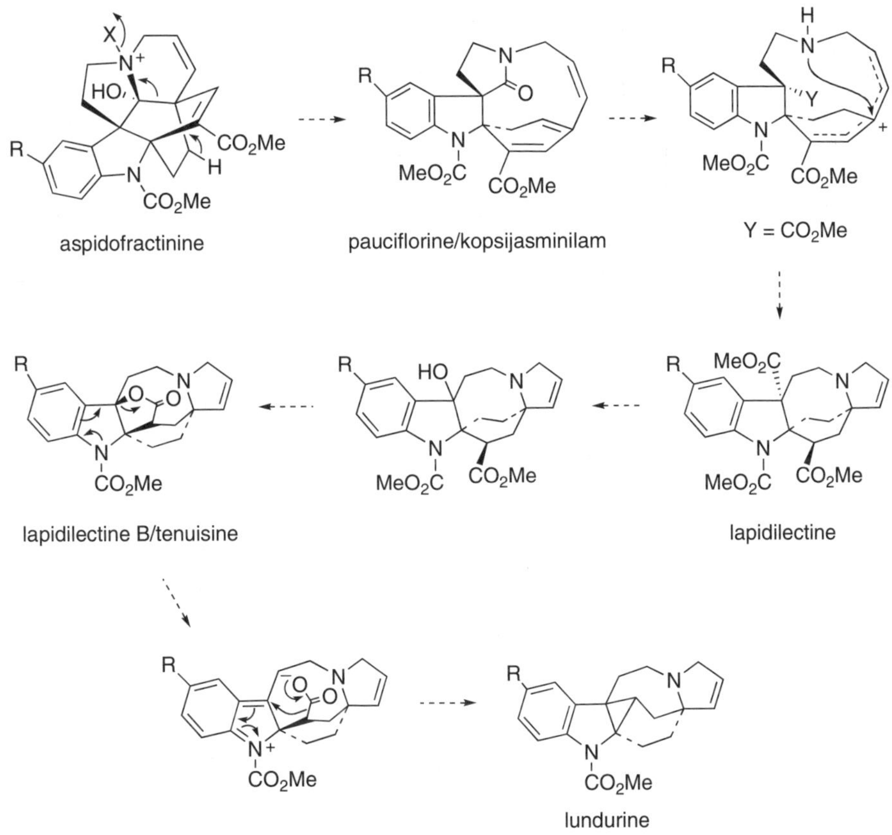
X
N+
HO
R
CO2Me
H
N
CO2Me
aspidofractinine
R
N
O
N
MeO2C
CO2Me
pauciflorine/kopsijasminilam
H
N
R
Y
N
MeO2C
CO2Me
Y = CO2Me
R
MeO2C
N
N
MeO2C
CO2Me
lapidilectine
R
HO
N
N
MeO2C
CO2Me
R
N
O
O
N
CO2Me
lapidilectine B/tenuisine
R
N
O
O
N+
CO2Me
R
N
N
CO2Me
lundurine

Scheme 28

Lundurines B (**385**) and D (**387**) showed appreciable *in vitro* cytotoxicity toward B16 melanoma cells, in contrast to lundurines A (**384**) and C (**386**) which were practically inactive. Of the two active lundurine derivatives, lundurine B (**385**) displayed the highest potency (IC_{50} 2.8 μg/mL). It would appear from the observed values that the presence of the pyrrolidine 14,15-double bond is necessary (cf. **385** vs. **386**), although oxygenation of C(3) has the effect of abolishing the cytotoxic effect altogether, despite the presence of the 14,15-unsaturation (cf. **385** vs. **384**). Surprisingly, lundurines B and D (**385** and **387**) did not display appreciable cytotoxicity toward KB cells, but were found instead to be effective in circumventing multidrug resistance (MDR) in vincristine-resistant KB cells (IC_{50} 4.6 μg/mL in the presence of 0.1 μg/mL of vincristine for both alkaloids) (189).

The total synthesis of racemic lapidilectine B (**380**) has been reported by Pearson and coworkers (Schemes 29 and 30). The synthesis featured the application of the Smalley azido-enolate cyclization to forge the indoxyl core of the alkaloid, as well as the installation of the dihydropyrrole ring by the application of a 2-azaallylithium [3+2] cycloaddition (191,192). Preparation of the

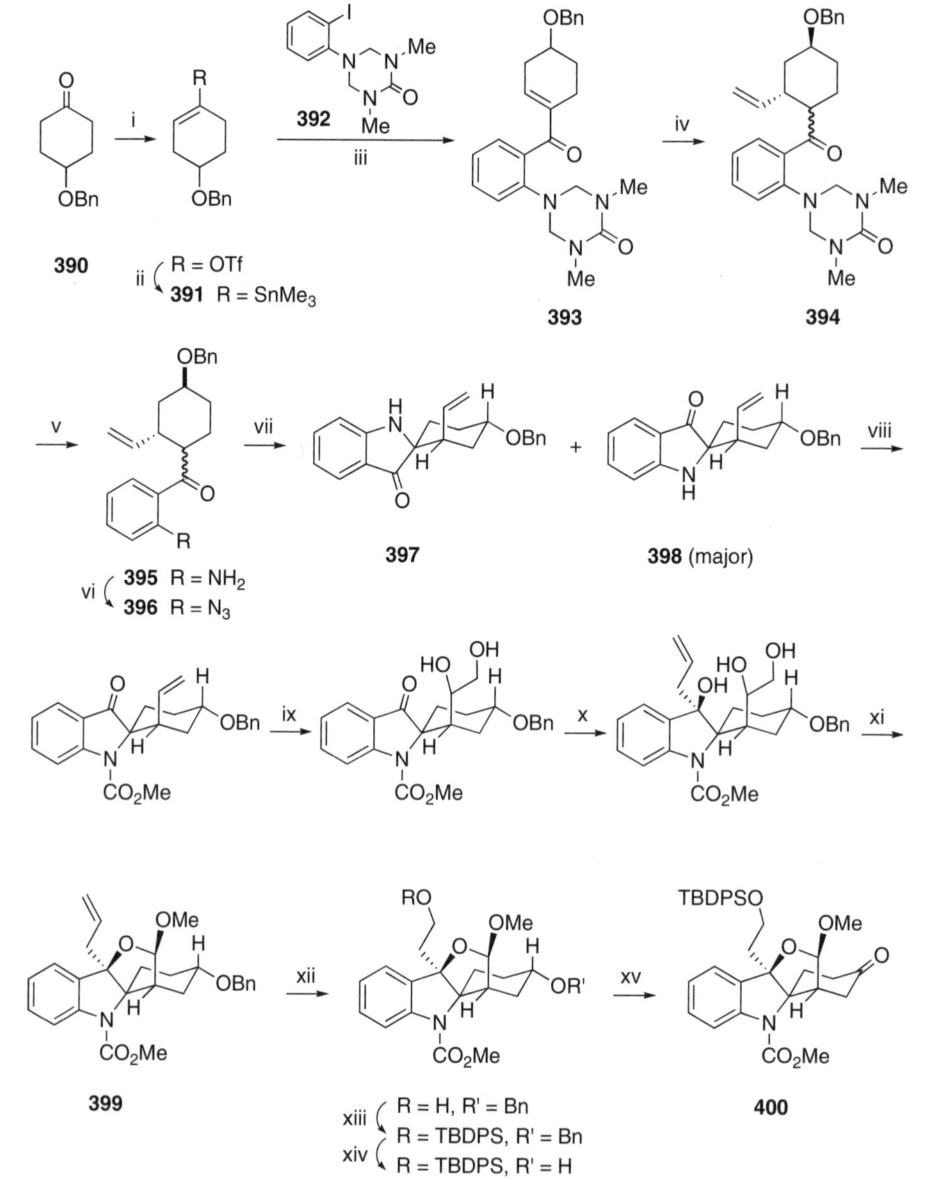

Scheme 29 Reagents: (i) LDA, −78°C; PhNTf₂; (ii) Me₃SnSnMe₃, Pd(PPh₃)₄, LiCl, THF; (iii) Pd₂dba₃ · CHCl₃, Ph₃As, LiCl, NMP, CO (70 psi); (iv) CH₂=CHMgBr, Li-2-thienylcyanocuprate, BF₃ · Et₂O, THF, −78°C; (v) HCl, MeOH; (vi) HCl, NaNO₂, H₂O; NaN₃; (vii) KOH, i-PrOH; (viii) t-BuLi, ClCO₂Me, −10°C; (ix) OsO₄, NMO, acetone; (x) allyl MgBr, THF, −40°C; (xi) NaIO₄, THF; CSA, MeOH; (xii) O₃, Sudan III, CH₂Cl₂/MeOH, −78°C; NaBH₄, MeOH; (xiii) TBDPSCl, DIEA, CH₂Cl₂; (xiv) H₂, Pd(OH)₂/C, AcOH, EtOH/THF; (xv) TPAP, NMO, 4 Å molecular sieves, CH₂Cl₂.

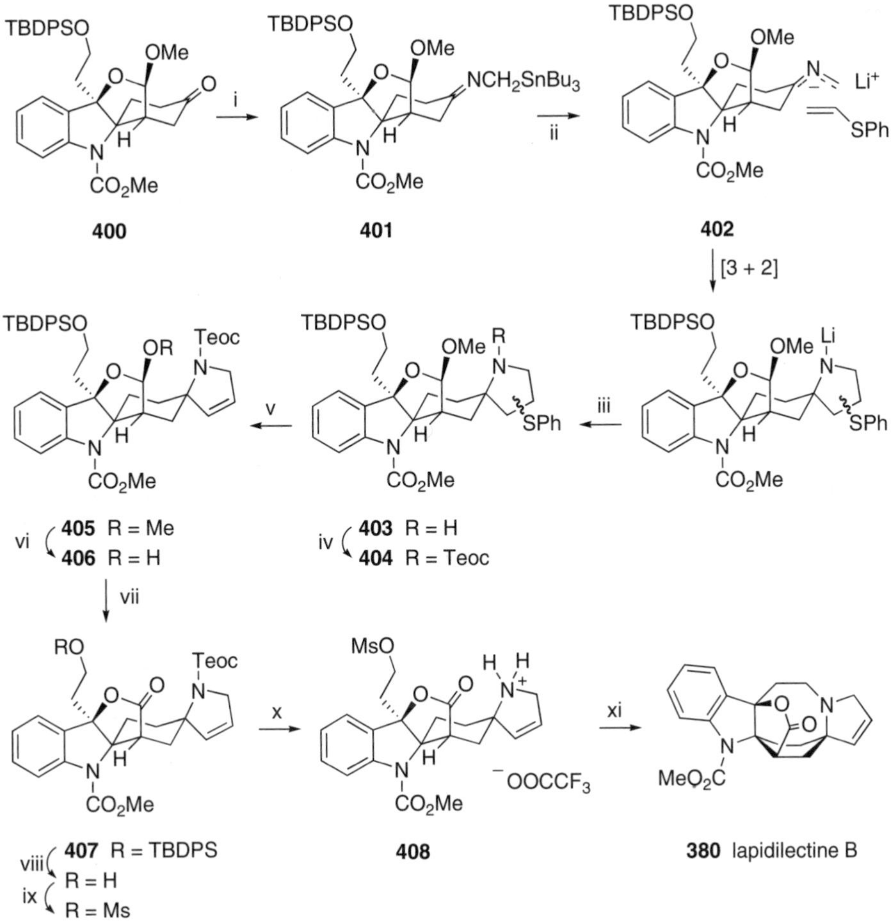

Scheme 30 Reagents: (i) (tri-*n*-butyl)-stannylmethylamine, Me₃Al, PhMe; (ii) CH₂=CHSPh, *n*-BuLi, THF, −78°C; (iii) NH₄Cl; (iv) Teoc-Cl, DIEA, CH₂Cl₂; (v) *m*-CPBA, NaHCO₃, CH₂Cl₂, −30°C to rt; Cl₂C=CCl₂, pyr, 125°C; (vi) BCl₃, CH₂Cl₂; (vii) PCC, Celite, CH₂Cl₂; (viii) HF·pyr, THF; (ix) MsCl, CH₂Cl₂, −10°C; (x) TFA/CH₂Cl₂; (xi) *i*-Pr₂NEt, MeCN, 60°C, 10 h.

Smalley cyclization precursor commenced with the known 4-benzyloxycyclohexanone (**390**), which was converted to the alkenyl stannane **391**. Carbonylative Stille coupling of **391** with the known triazinone-protected iodoaniline **392** gave enone **393**, which on conjugate addition of vinylmagnesium bromide in the presence of lithium 2-thienyl cyanocuprate (Lipshutz's conditions) provided ketone **394** as an inseparable mixture of diastereomers. Removal of the triazinone protecting group gave aniline **395**, which on diazotization, followed by treatment of the resulting diazonium salt with sodium azide, gave the Smalley cyclization precursor **396**. Treatment on **396** with KOH/2-propanol gave a mixture of cyclization products, **397** and **398**. The major cyclization product **398** was processed to the methyl acetal **399** following installation of the carbamate, dihydroxylation, addition of allyl magnesium bromide, oxidative cleavage, and treatment with CSA/MeOH. The structure of **399** was confirmed by X-ray analysis.

The methyl acetal **399** was then transformed to the 2-azaallyllithium cycloaddition precursor **400**, following ozonolysis and reductive workup with NaBH$_4$, followed in succession by primary alcohol protection, debenzylation with Pearlman's catalyst, and finally oxidation to the ketone. Treatment of **400** with Bu$_3$SnCH$_2$NH$_2$/Me$_3$Al generated stannane **401** *in situ*, which on treatment with phenyl vinyl sulfide/BuLi at $-78°C$ resulted in formation (via tin–lithium exchange) of the transient azaallyllithium **402**, which upon cycloaddition and aqueous workup provided pyrrolidine **403** as a complex mixture of diastereomers. Pyrrolidine **403** was then converted to the *N*-Teoc derivative **404**. Oxidation of **404** gave the sulfoxide which underwent thermal elimination away from the nitrogen to give dihydropyrrole **405**. Demethylation of **405** gave hemiacetal **406**, which on oxidation provided lactone **407**. Removal of the primary alcohol protecting group and installation of the mesylate, followed by *N*-Teoc removal, gave **408**, setting the stage for the final cyclization step, which was achieved by treatment of **408** with diisopropylethyl amine (DIEA, 2 h), followed by heating to $60°C$ in acetonitrile for 10 h, to give ($\pm$)-lapidilectine B (**380**) (191,192).

H. Kopsifolines

The kopsifolines (A–F, **409–414**) represent a new structural subclass of the monoterpenoid indole alkaloids and were isolated together, and for the first time from the same source (193,194). All six alkaloids, kopsifolines A–F, occur only in the leaf extract of a Malayan *Kopsia* species, which was originally identified by Middleton as *K. fruticosa*, but subsequently revised to *K. singapurensis* (1,195).

The ^{13}C NMR of kopsifoline A (**409**) (as well as **410–414**), in particular the nonaromatic portion of the molecule, showed a general similarity to that of the aspidofractinine alkaloid, venalstonine (**415**), which was also isolated, except for changes involving C(2), C(16), C(17), C(20), and C(21). The COSY spectrum revealed virtually the same partial structures as those present in an aspidofractinine skeleton, such as NCH$_2$CH=CH, NCH$_2$CH$_2$, CH$_2$CH$_2$, and an isolated methine. However, whereas a CH$_2$CH fragment was also present, in addition to the above partial structures, in **409** (and its congeners **410–414**), this fragment (which corresponds to C(16)–C(17)) was conspicuously absent in the kopsifolines, but was replaced instead by an isolated methylene and an additional quaternary center when compared with the aspidofractinines. These differences suggested a structure bearing some resemblance to the aspidofractinines, but at the same time, distinguished by a significant departure from the aspidofractinine carbon skeleton, a feature which was revealed on careful examination of the HMBC spectrum. The HMBC spectrum of **409** showed the following long-range (^3J) correlations: H(19) to C(16), H(17) to C(18), and H(18) to the ester carbonyl. These correlations represent a clear departure from those normally observed in aspidofractinine alkaloids, and clearly indicate that in kopsifoline A (**409**) (as well as in **410–414**), a novel carbon skeleton is present in which C(18) is now linked to C(16), instead of to C(2).

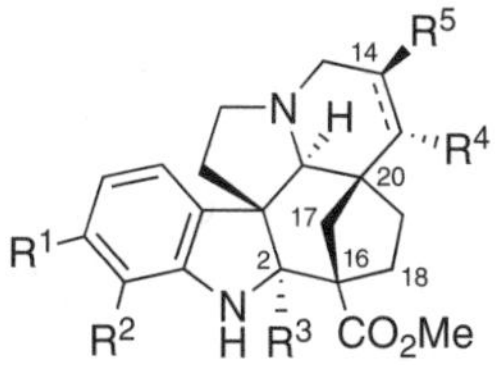

409 $R^1 = R^4 = R^5 = H$, $R^2 = OMe$, $R^3 = OH$, $\Delta^{14,15}$

410 $R^1 = R^5 = H$, $R^2 = OMe$, $R^3 = R^4 = OH$

411 $R^1 = H$, $R^2 = OMe$, $R^3 = R^4 = R^5 = OH$

414 $R^1 = OMe$, $R^2 = R^3 = R^4 = R^5 = H$

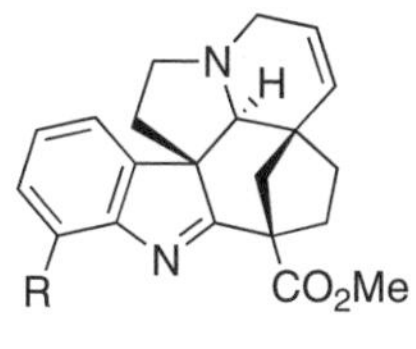

412 R = H

413 R = OMe

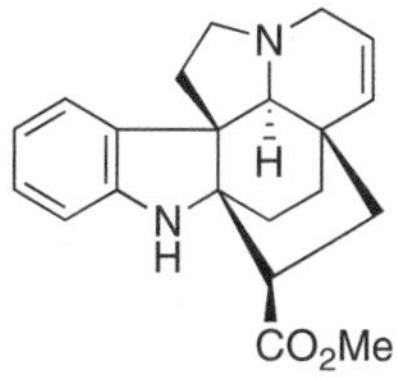

415

The presence of the hydroxyl group at position 2 was indicated by the downfield signal due to an oxygenated quaternary carbon at δ 97.1, the significant downfield shift due to its being linked to both a nitrogen and an oxygen atom. This assignment was also supported by the following correlations observed in the HMBC spectrum, namely, from H(6), H(17), H(18), and H(21) to C(2). The stereochemistry at C(21) is found to be similar to that in aspidofractinine alkaloids from the NOE enhancement of H(9) and H(19α) on irradiation of H(21). This result also confirmed the stereochemistry of the C(17) methylene bridge in the bicyclo[3.2.1]octane system constituting the C and F rings, as shown in **409**. The orientation of the C(2)-hydroxyl substituent was deduced to be α from the observed enhancement of H(17β) on irradiation of H(6β).

Kopsifolines B (**410**) and C (**411**) were characterized by hydroxy substitution in the piperidine D ring. The α-orientation of the C(15)–OH group in kopsifoline B was indicated by the observed H(15)/H(17α) NOE. In the case of kopsifoline C, the observed J_{14-15} value of 2 Hz indicated that the H(14) and H(15) were not *trans*-diaxially disposed. This, coupled with the observed H(15β)/H(17α) and H(14α)/H(3β) NOEs, is consistent with both H(14) and H(15) being oriented equatorially in a six-membered ring. The presence of an indolenine chromophore in kopsifolines D and E (**412** and **413**) was indicated by the UV spectrum and by the observed C(2) and C(7) resonances at δ 187.7 and 63.0, respectively, while kopsifoline F (**414**) was distinguished by aromatic methoxy substitution at C(11) instead of at C(12) in the other five alkaloids (193,194).

The kopsifolines represent a new family of indole alkaloids with an unprecedented carbon skeleton in which C(18) is not linked to C(2), but to C(16), unlike the case in aspidofractinine alkaloids. A kopsifoline-type structure was encountered earlier in one of the monomeric units constituting the bisindole alkaloid tenuiphylline, isolated from another *Kopsia* (*vide infra*). A possible biogenetic pathway to the kopsifolines (e.g., **412**) is via an intramolecular epoxide ring opening from an *Aspidosperma*-type precursor (such as **416**) as shown in Scheme 31 (194).

Padwa has reported an approach to the hexacyclic framework of the kopsifolines (as exemplified by **420**) from the piperidone **417** (196,197). The approach was based on a Rh(II)-catalyzed cyclization–cycloaddition cascade which yielded the cycloadduct **419** from the diazoimide **418**. This was, in turn,

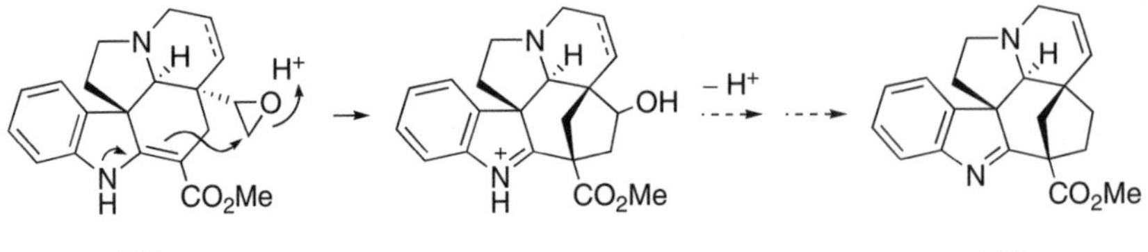

Scheme 31

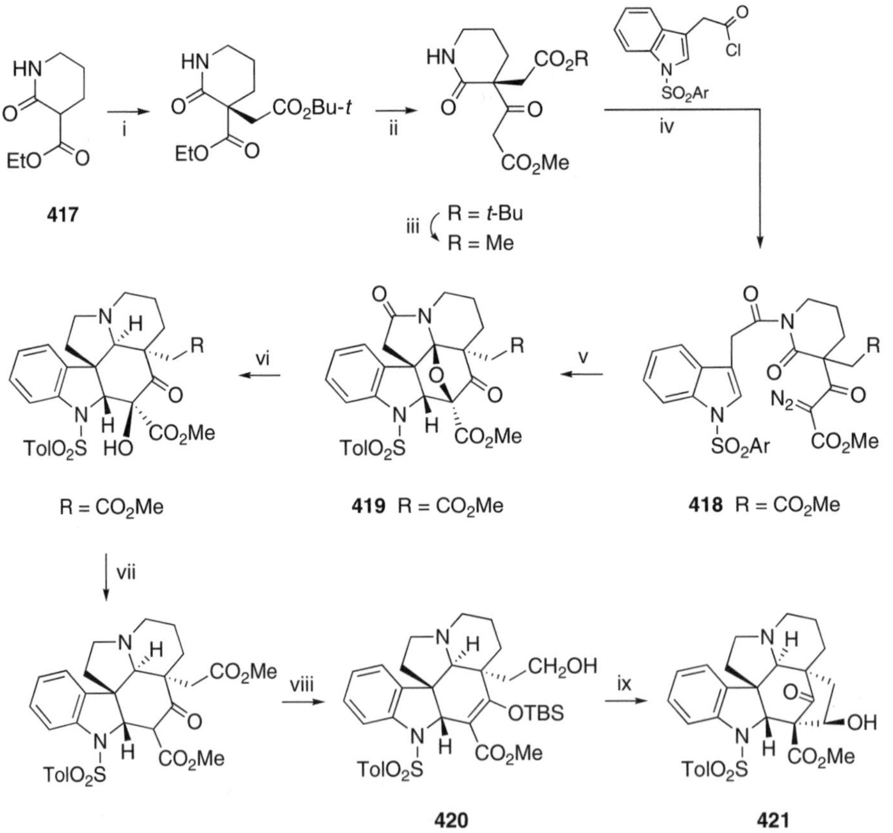

Scheme 32   Reagents: (i) *n*-BuLi, BrCH$_2$CO$_2$Bu-*t*; (ii) LiOH; CDI, MgCl$_2$, MeO$_2$CCH$_2$CO$_2$K; (iii) (MeO)$_3$CH, MeOH, *p*-TsOH; (iv) (indol-3-yl)-acetyl chloride, MsN$_3$, Et$_3$N; (v) Rh$_2$OAc$_4$, 80°C; (vi) P$_2$S$_5$; Raney-Ni; PtO$_2$, H$_2$, HCl; (vii) SmI$_2$, HMPA; (viii) TBDMSOTf; LAH; (ix) TPAP, NMO; CsF, MeCN.

transformed to the TBS enol ether **420**. Oxidation of the primary alcohol functionality, followed by reaction with CsF, gave compound **421** (Scheme 32).

I. Mersinine Alkaloids

The alkaloids of the mersinine group exemplify yet another new subclass of the monoterpenoid indole alkaloids. These alkaloids are characterized by a novel pentacyclic skeleton incorporating a quinolinic chromophore, and were also encountered exclusively and for the first time from the same plant (198,199). As in the case of the kopsifolines, the plant material was initially identified as *K. fruticosa* by Middleton, but was subsequently amended to *K. singapurensis* upon completion of his revision of the genus (1,195). From a biogenetic viewpoint, these alkaloids may be envisaged to have arisen from an aspido-fractinine precursor via formation of an aziridinium intermediate, followed in succession by aziridinium ring opening and reduction as shown in Scheme 33 (in analogy to the route used by Levy for a biomimetic entry into the *Melodinus* alkaloids, scandine and meloscine (200,201)). The proposed pathway leads to the numbering system adopted for the alkaloids of this new subclass.

Mersinines A (**422**) and B (**423**) have essentially the same structure, and share many spectral features. Both showed the same molecular ion in EIMS at m/z 500 indicating that they are isomers. The UV spectra of both were similar with absorption maxima characteristic of a tetrahydroquinoline chromophore. Both alkaloids possess OH, carbamate, and two methyl ester functions, as indicated by the IR, ^{1}H, and ^{13}C NMR data. Analysis of the NMR data (COSY, HMQC, HMBC) of both alkaloids led to the pentacyclic quinolinic structures as shown in **422** and **423**. The location of the methyl ester groups at C(16) and C(20) was indicated by the three-bond correlations from H(2) to C(16), and from H(15), H(19) to C(20), respectively, while the observed H(9)/H(21), H(21)/H(2) NOEs fixed the orienta-tions of H(21) and H(2). Other NOEs observed included those from H(19α) to H(2) and H(21). The main difference in the structure of **422** and **423** lies in the configuration of the quaternary center, C(16). The first indication of such an epimeric relationship was provided by the observation that mersinine A (**422**) was smoothly and irreversibly converted to mersinine B (**423**) by the action of dilute alcoholic NaOH, indicating that alkaloid **423** is the more stable epimer. Examination of models confirmed that the less sterically congested structure has a β-oriented hydroxyl substituent. This difference in the configuration at C(16) was also reflected in

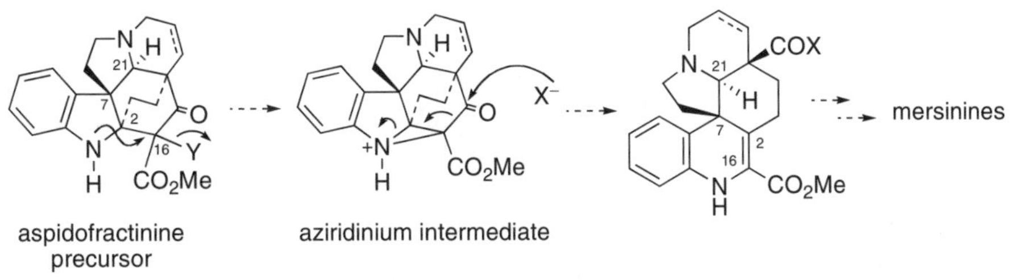

aspidofractinine   aziridinium intermediate
precursor

Scheme 33

differences in the NMR spectral data between **422** and **423**, which involve the resonances of H(6), C(16)–OH, and C(2). Alkaloid **423** and others with similar 16(*S*) configuration in the mersinine series showed strikingly similar C(16)–OH shifts in ^{1}H NMR (ca. δ 5.1), which are distinct from the C(16)–OH shifts in alkaloid **422** and others with the opposite, i.e., 16(*R*) configuration (ca. δ 4.6). A similar trend was indicated in the carbon shifts of C(2) (ca. δ 49 in **423** cf. δ 53 in **422**).

Additional support for this assignment was also provided by the paramagnetic deshielding of H(18) (α or β) by the proximate C(16)–OH group. The stereochemistry of the C(18) hydrogens can be independently assigned from consideration of the vicinal coupling constants and from NOE experiments. In alkaloid **422**, in which the C(16)–OH group is α, H(18α) experiences deshielding due to spatial proximity of the α-oriented OH (H-18α δ 1.78, H-18β δ 1.34), whereas in alkaloid **423**, the same effect operates on H-18β instead, which was deshielded to ca. δ 1.8 (H-18α ca. δ 1.3). It transpired that these trends are general in all the mersinine alkaloids (total of 18 in all) and provide a useful diagnostic tool for determining the relative configuration at C(16). The relative configuration at the other quaternary center, C(20), was assigned from the observed NOEs, as well as from its presumed relationship to the mersiloscines (*vide infra*). These conclusions have been subsequently vindicated by an X-ray analysis of mersinine A, which also confirmed the assignment of the relative configurations at the various stereocenters in **422** and **423** based on NOE data.

422 R^1 = CO$_2$Me, R^2 = OH   **424**   **425** R^1 = OH, R^2 = CO$_2$Me, R^3 = α-OH

423 R^1 = OH, R^2 = CO$_2$Me   **426** R^1 = CO$_2$Me, R^2 = OH, R^3 = α-OH

                                                             427 R^1 = OH, R^2 = CO$_2$Me, R^3 = β-OH

Mersinine C (**424**) was an isomer of mersinines A and B as indicated by the mass spectrum. The UV spectrum was similar to those of **422** and **423**, as was the IR spectrum which showed bands due to OH, ester, and carbamate functions. The IR spectrum, however, showed two additional (Wenkert–Bohlmann) bands at 2782 and 2874 cm^{-1}, which were not seen in the case of **422** and **423**. The ^{1}H and ^{13}C NMR spectral data were also generally similar, except for changes in several of the chemical shifts, as were the COSY and HMBC data, indicating affinity with the two previous alkaloids. In common with both **422** and **423**, the C(7) and C(21) configurations were defined by the observed reciprocal NOEs for H(21) and H(9), while the observed C(16)–OH shift at δ 5.13, as well as the deshielding of H(18β) relative to H(18α), were diagnostic of a 16(*S*) configuration as described for mersinine B (**423**) (*vide supra*). In contrast to **422** and **423**, however, the H(21)/H(2) NOE was not observed, although the observed H(18β)/H(6β) NOE in this case allowed the orientation of H(2) to be assigned as α. The C(2) resonance at δ 44.1, however, was not consistent with the trend observed for the other two alkaloids, and in addition, several other NOE interactions, such as those between H(21)/H(19α) and H(2)/H(19α), which were observed for **422** and **423**, were conspicuously not observed in mersinine C (**424**). These departures can be rationalized by the change in the D/E ring junction stereochemistry from *cis*, as in mersinines A and B, to *trans*, as in mersinine C, with H(21) and the N(4) lone pair oriented *anti* to each other as indicated by the observed Wenkert–Bohlmann bands in the IR spectrum of **424**, and by the change in the configuration at C(20) from R to S. The observed NOE data become intelligible on the basis of this structure. Examination of models showed that H(2) and H(19α) are no longer proximate to H(21) in mersinine C. Mersinine C (**424**) represents the first member of this subgroup, of which there are three others (*vide infra*), which are characterized by the presence of a *trans* D/E ring junction and C(20)–αCO$_2$Me.

The mersiloscines (**425–427**), which are characterized by incorporation of an additional lactone ring, represent another subgroup of the mersinines. The ^{1}H and ^{13}C NMR data of mersiloscine (**425**) showed a basic similarity with those of alkaloids **422** and **423**, except for some notable differences, which are consistent with the formation of a γ-lactone unit incorporating C(14), C(15), C(20), and C(17), and the presence of an OH substituent at C(15). This proposal was also supported by the observed three-bond correlations from H(14) and H(21) to the lactone carbonyl (C(17)) in the HMBC spectrum of **425**. Another significant difference was the downfield shift of H(18β) from δ 1.85 in **423** to δ 3.02 in **425**, which is caused by the anisotropic effect of the proximate lactone carbonyl function. This observation, in fact, allowed the assignment of the C(20) to C(17) bond as β, since the alternative structure (in which the lactone unit is α with respect to the general plane defined by the molecule) would result in the placement of H(19α) instead, within the anisotropic influence of the lactone carbonyl function. On the assumption that the lactone **425** was formed from a hypothetical hydroxy ester/acid precursor such as **428**, which was, in turn, derived from oxidation of the olefin **423**, it would be reasonable to conclude that the orientation of the ester group at C(20) in these closely related alkaloids (**422–423**, **425–428**) is also β. The orientation of the C(15)–OH group was established as α from the observed H(21)/H(19α) and H(19β)/H(15) reciprocal NOEs. The relative configuration of C(16) was deduced to be similar to those in mersinines B (**423**) and C (**424**), i.e., 16(*S*), from the diagnostic C(16)–OH and C(2) chemical shifts of ca. δ_H 5.1 and δ_C 48, respectively. The β-oriented C(16)–OH also caused deshielding of H(18β) relative to H(18α) (δ 3.0 vs. 1.3, respectively), but in the case of mersiloscine (**425**), this paramagnetic deshielding due to OH was reinforced by additional deshielding from anisotropy of the lactone carbonyl. Mersiloscine A (**426**) was readily deduced to be the C(16) epimer of mersiloscine (**425**) from the spectral data, which were in all respects similar to those of **425**, except for the diagnostic C(16)–OH and C(2) chemical shifts, which were indicative of a 16(*R*) configuration. Likewise, the NMR data for mersiloscine B

(**427**) indicated similarity with mersiloscine (**425**), except for the orientation of the C(15)–OH substituent, which was deduced to be β from the observed reciprocal H(15)/H(21) NOE.

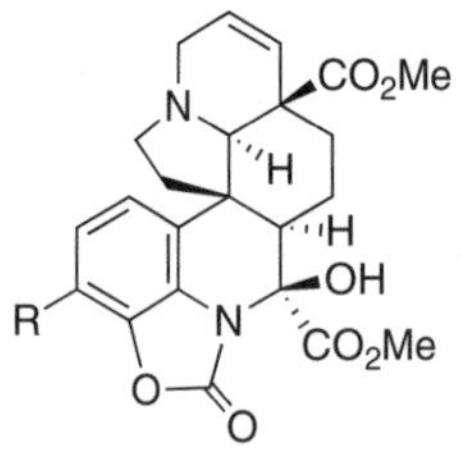

429 R = H
430 R = OMe

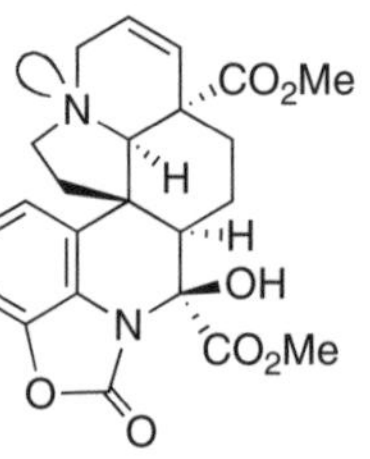

431

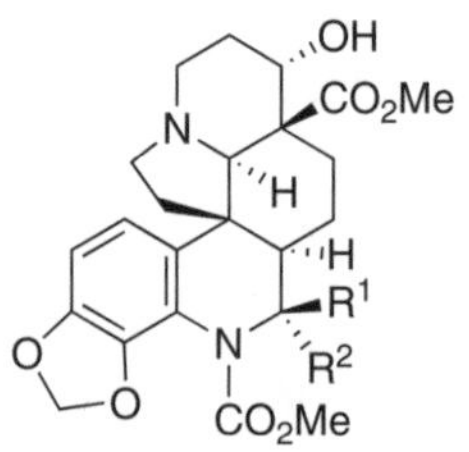

432 $R^1 = CO_2Me$, $R^2 = OH$
433 $R^1 = OH$, $R^2 = CO_2Me$

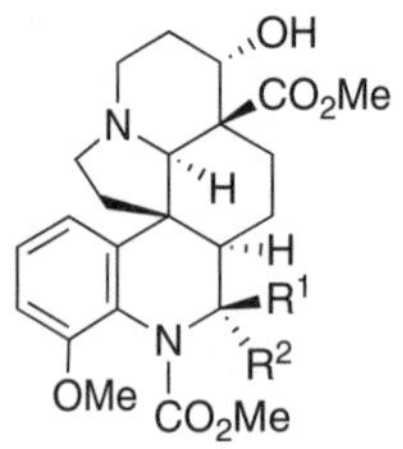

434 $R^1 = CO_2Me$, $R^2 = OH$
435 $R^1 = OH$, $R^2 = CO_2Me$

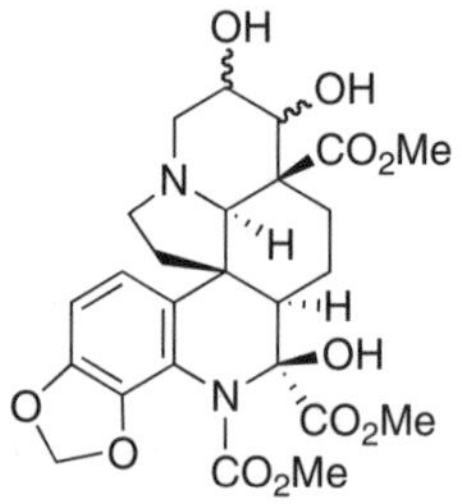

428 14β-OH, 15α-OH
436 14α-OH, 15β-OH

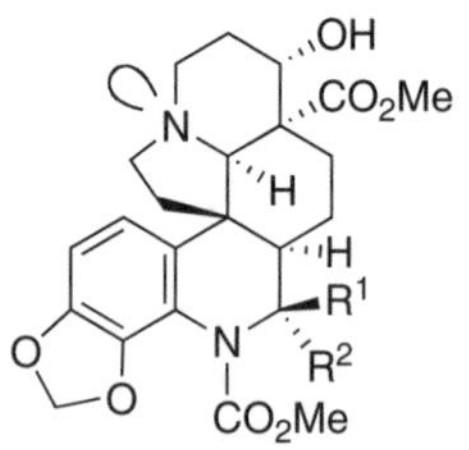

437 $R^1 = CO_2Me$, $R^2 = OH$
438 $R^1 = OH$, $R^2 = CO_2Me$

The mersifolines A–C (**429–431**) are characterized by incorporation of an unprecedented oxazoloquinoline chromophore. Mersifoline A (**429**) showed essentially similar UV, IR, and NMR data as those of the previous alkaloids in the mersinine series. The COSY spectrum showed the same partial structures that are present in the mersinines (**422–424**), and the similar HMBC data indicated that the carbon skeleton was essentially unchanged. The observation of characteristic reciprocal NOEs for H(9)/H(21), H(21)/H(19α), H(21)/H(2), and H(2)/H(19α), coupled with the similarity of the NMR spectral data, indicated that the configurations at C(7), C(21), C(2), and C(20) are similar to those of mersinines A and B. The H(9)/H(21) NOE indicated that substitution of the aromatic ring is at C(12) and the observed carbon resonance of C(12) at ca. δ 141 was suggestive of some form of oxygenation. However, although a carbamate function was indicated by the ^{13}C NMR spectrum (δ 152.1), the corresponding methyl resonance was conspicuously absent, and moreover, the molecular formula for mersifoline A (DBE 13) requires formation of an additional ring. These features together, indicated formation of a cyclic carbamate in which the oxygen of the carbamate function is linked to the aromatic C(12), thereby furnishing an oxazoloquinoline structure as shown in **429**. This conclusion receives additional support from the unusual IR band at $1784\,cm^{-1}$, which has been noted previously ($1780\,cm^{-1}$) for such oxazoloquinoline class of compounds (202). The mersifolines (**429–431**) represent the first examples of an indole or indole-derived alkaloid incorporating such a functionality. In the case of the mersifolines, the C(16)–OH resonance was observed at ca. δ 4.7 in all three alkaloids (**429–431**), suggesting a similar configuration in all three alkaloids. The shift of C(2) at δ 48.8 in mersifoline A (**429**) is diagnostic of a 16(*S*) configuration, and this conclusion was further supported by the observed deshielding of the H(18β) resonance compared to that of H(18α), due to the proximate, β-oriented C(16)–OH group. It would appear that the replacement of the usual carbamate substituent on the indolic nitrogen by a cyclic carbamate unit has resulted in a change in the C(16)–OH resonance, which has been shifted upfield from the normal value of ca. δ 5.1 for mersinine-type alkaloids with a 16(*S*) configuration. Mersifoline B (**430**) was in all respects similar to mersifoline A except for the presence of a methoxy substituent on the aromatic C(11).

Mersifoline C (**431**), on the other hand, was shown to belong to the mersinine C subgroup (*trans* D/E ring junction and C(20)–αCO_2Me), as indicated by the presence of Wenkert–Bohlmann bands in the IR spectrum. This was further confirmed on examination of the NMR spectral data, which showed departure of some of the ^1H and ^{13}C chemical shifts when compared with those of mersifoline A (**429**). The characteristic C(16)–OH at δ 4.70, as well as the deshielding of H(18β) relative to H(18α), are diagnostic of an *S* configuration at C(16), as is the case in mersifolines A (**429**) and B (**430**). However, unlike alkaloids **429** and **430**, the C(2) resonance in **431** was shifted upfield to ca. δ 42, which is similar to the situation in mersinine C (**424**), as a consequence of changes in both the D/E ring junction stereochemistry, as well as that of the C(20)-methyl ester group. The NOESY/NOE data of mersifoline C (**431**) were also similar to those of mersinine C (**424**), in that while the H(21)/H(19α), H(21)/H(2), and H(2)/H(19α) NOEs were not observed, the H(9)/H(21), H(18β)/H(6β), and H(3α)/H(21) NOEs were seen. These observations, coupled with the similarity of the NMR spectral data with those of mersinine C, are consistent with a structure with a *trans* D/E ring junction and an (*S*) configuration at C-20 (C(20)–αCO_2Me).

The mersidasines (**432–438**) are characterized by the presence of hydroxy substitution in the piperidine ring. Mersidasines A–E (**432–436**) belong to the mersinine A subgroup with *cis* D/E ring junction and C(20)–βCO_2Me, while mersidasines F and G (**437** and **438**) are characterized by the presence of a *trans* D/E ring junction and C(20)–αCO_2Me. The possibility that some of the isolated alkaloids could be artifacts formed as a result of epimerization at C(16) during isolation was discounted by control experiments.

The same plant yielded two additional alkaloids which are related to the mersinines, namely, the *seco*-mersinine alkaloid, mersirachine (**439**), and mersinaline (**440**), which represents a new variation of the mersinine series, having incorporated an additional carbon in the form of a formyl group, constituting part of a vinylogous amide unit associated with N(4) (203).

The ^{13}C NMR spectral data of mersirachine (**439**) showed, in addition to the carbamate and methyl ester functions, two other carbonyl signals, one due to a conjugated ester and the other due to a conjugated aldehyde. Furthermore, two double bonds were indicated, a 1,2-disubstituted double bond and a tetrasubstituted double bond. The characteristic quaternary C(16) associated with substitution by an OH and a methyl ester group was observed at δ 85.2. The ^{1}H NMR spectrum indicated the presence of a *trans*-double bond, an α,β-unsaturated aldehyde, a methyl ester group linked to C(16), and a conjugated methyl ester. The relative configuration at C(16) was deduced to be *R*, as indicated by the characteristic C(16)–OH and C(2) chemical shifts at δ_H 4.31 and δ_C 50.9, respectively (*vide supra*). The COSY spectrum showed the presence of NCH$_2$CH$_2$, CHCH$_2$CH$_2$, and NCH=CHCHO partial structures, while the isolated aminomethine normally associated with C(21) was not observed. The NMR data thus suggested that the A, B, C, and E rings of the mersinine-type skeleton were intact, with a double bond from C(20) to C(21) constituting part of the α,β-unsaturated methyl ester moiety. The α,β-unsaturated aldehyde fragment is therefore linked to N(4), constituting a vinylogous amide unit. These conclusions are supported by the HMBC data (H(5)/C(7); H(6)/C(2), C(21); H(18)/C(20); H(19)/C(2); H(3)/C(15)).

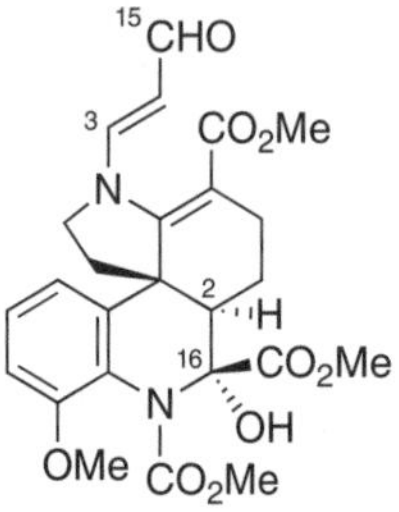

439 mersirachine 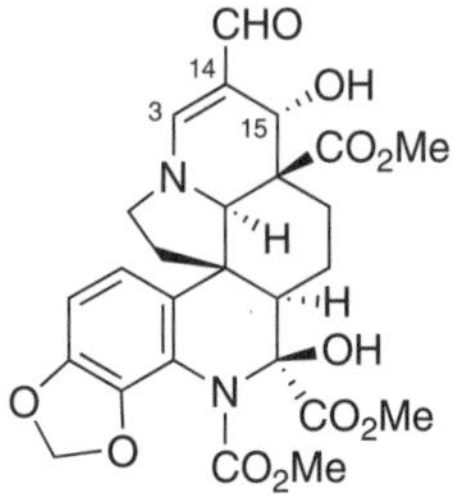   **440** mersinaline

The ^{13}C NMR spectral data of mersinaline (**440**) showed the presence of 26 carbon resonances (one more than mersinines A and B). The ^{13}C NMR spectrum revealed two methyl ester functions, a carbamate, a conjugated aldehyde, and the characteristic quaternary C(16) resonance at δ 86.2, associated with substitution by an OH and a methyl ester group. The ^{1}H NMR spectrum indicated the presence of a trisubstituted double bond from the lone vinylic signal at δ 7.27 (δ_C 153.7; the other quaternary olefinic resonance was observed at δ 115.5). An aldehyde-H signal was observed at δ 9.01 (δ_C 187.5), the upfield shift of both the ^{1}H and ^{13}C resonances suggesting conjugation with the trisubstituted olefinic moiety. The presence of an isolated oxymethine was indicated by the observed resonance at δ 64.2 (δ_H 4.87, s), although the OH signal itself was not observed.

The COSY and HMQC data indicated the presence of two fragments, namely, NCH_2CH_2 and $CHCH_2CH_2$, in addition to an isolated aminomethine corresponding to C(21), an isolated oxymethine, and a vinylogous amide unit incorporating a trisubstituted double bond (NCH=C(CHO)–). The former two fragments correspond to the usual C(5)–C(6) and C(2)–C(18)–C(19) fragments present in the mersinine-type alkaloids, and together with the NMR data suggested that while rings A, B, C, and E were intact, the piperidine ring D had undergone substantial changes. The presence of the vinylogous amide unit associated with N(4) was clearly indicated by the three-bond correlations from the vinylic H(3) to C(5) and C(21), and the observed CHO/H(3) NOE, while the H(3) to the oxymethine C(15) and H(15) to the aldehyde carbonyl correlations indicated that the oxymethine C(15) was linked to the olefinic C(14). Attachment of the oxymethine C(15) to the quaternary C(20) (from the H(21)/C(15), H(19)/C(17) HMBCs) completes the assembly of the molecule, which is in accord with the full HMBC data. The observed H(9)/H(21), H(2)/H(21), H(2)/H(19α), and H(21)/H(19α) reciprocal NOEs established the relative configurations at C(2), C(7), C(20), and C(21), while the 16(S) configuration was defined by the characteristic C(16)–OH and C(2) chemical shifts at δ_H 5.10 and δ_H 49.2, respectively. The orientation of the C(15)–OH was deduced to be α from the observed NOE between H(15) and H(19β).

It was suggested that both mersirachine (**439**) and mersinaline (**440**) share a common mersinine-type precursor **441**, as the novel tetracyclic indole, mersilongine (*vide infra*). Thus, the product of the Grob fragmentation, instead of undergoing protonation followed by elimination of acrolein in a retro-Michael reaction *en route* to mersilongine (*vide infra*), undergoes intramolecular conjugate addition to the α,β-unsaturated ester moiety, leading to the tetracyclic amino aldehyde **442**, which on subsequent oxidation or dehydrogenation yields the *seco*-mersinine alkaloid, the *E*-vinylogous amide **439**, corresponding to mersirachine. Alternatively, oxidation of **441** leads to the iminium ion intermediate **443**, which on deprotonation yields the enamine **444** (19). An intermolecular enamine-formaldehyde reaction then follows to provide the pentacyclic imino 1,3-diol **445**, which on subsequent deprotonation, followed by oxidation yields mersinaline (**440**) (Scheme 34) (203).

Mersinines A (**422**), B (**423**), C (**424**), and mersifolines B (**430**) and C (**431**) were found to reverse drug resistance in drug-resistant KB cells (IC_{50} 3–10 μg/mL in the presence of 0.1 μg/mL of vincristine), while the mersiloscines (**425–427**) and the mersidasines (**432–439**), which are characterized by the presence of a C(15)–OH substituent (as well as an additional lactone ring in the case of the mersiloscines), were found to be ineffective ($IC_{50} > 25$ μg/mL). Among the active alkaloids, the stereoisomers with a *trans* D/E ring junction and C(20)–αCO_2Me appear to be less effective compared to those with a *cis* D/E ring junction and C(20)–βCO_2Me (199).

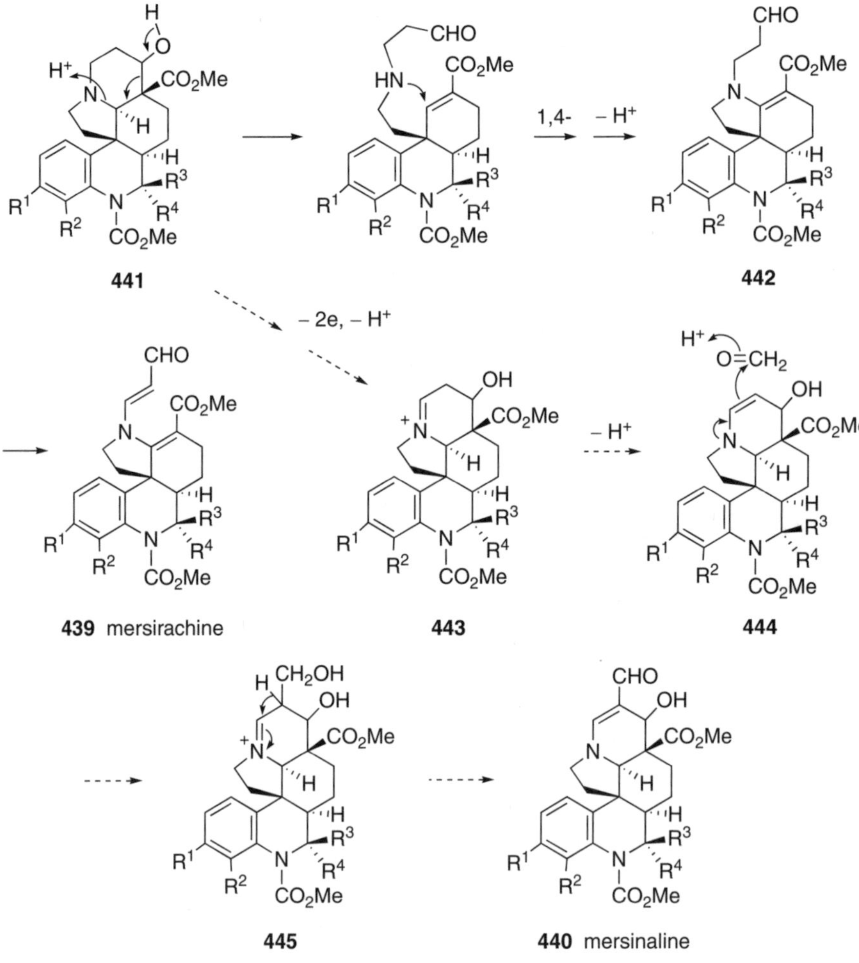

441

442

1,4-   − H⁺

− 2e, − H⁺

439 mersirachine

443

− H⁺

444

445

440 mersinaline

Scheme 34

VIII. MISCELLANEOUS NOVEL ALKALOIDS

Several *Kopsia* species provided minor alkaloids, which are characterized by novel skeletons derived from known monoterpenoid indole precursors through deep-seated rearrangements and/or elimination of key fragments. The same sample of *K. singapurensis* that provided the mersinines gave two such alkaloids, mersilongine (**446**) from the leaf and mersicarpine (**447**) from the stem-bark extract.

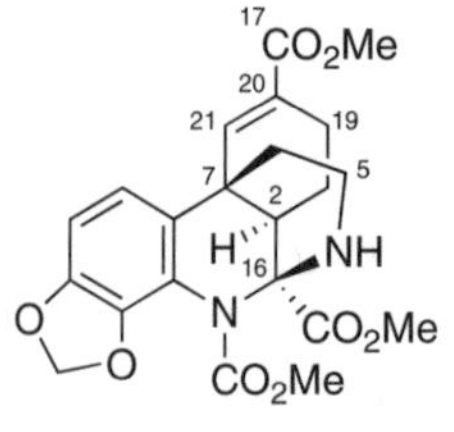

446 mersilongine

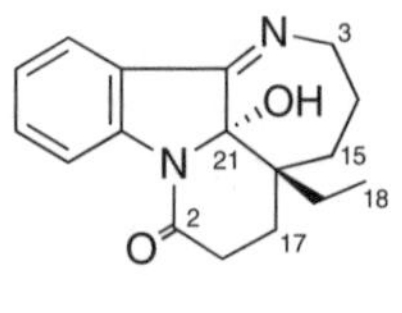

447 mersicarpine

Mersilongine (**446**) ($C_{22}H_{24}N_2O_8$) had a UV spectrum which was reminiscent of the mersinines, which suggested the presence of a similar tetrahydroquinoline chromophore (204). The IR spectrum showed bands due to NH, ester, and carbamate/conjugated carbonyl functionalities. The ^1H NMR spectrum showed the presence of three methoxy groups and a vinylic H (δ 7.40). The three methoxy groups were deduced to be associated with conjugated ester, carbamate, and methyl ester functions from their respective carbon signals at δ 167.6, 153.9, and 170.9, respectively. Since an NH function was indicated by the IR spectrum, this must due to an N(4)–H. In addition, two low-field signals at δ 128.4 and 141.9, corresponding to a quaternary and a methine carbon, respectively, were assigned to a trisubstituted double bond constituting part of the α,β-unsaturated ester moiety.

With these functionalities accounted for, a tetracyclic molecule was indicated from the molecular formula. The COSY spectrum showed the presence of two main fragments, namely, $NHCH_2CH_2$ and $CHCH_2CH_2C{=}CH$. In the latter fragment, long-range allylic coupling was clearly seen in the COSY spectrum, although the vinylic H appeared as a slightly broadened singlet. In the mersinine-type alkaloids (*vide supra*), the quaternary C(16) resonance was found at ca. δ 87, due to it being adjacent to both a nitrogen and an oxygen atom, as well as being substituted by a carbomethoxy group. In mersilongine, the quaternary C(16) resonance was observed at δ 77.3, shifted upfield by about 10 ppm, suggesting replacement of an adjacent oxygen by a nitrogen atom, compared to the mersinines. The HMBC spectrum showed the same correlations which establis-hed the six-membered ring B in the mersinines and thus confirmed the location of the methine C(2) within the quinolinic moiety. The ^2J and ^3J correlations from the lone vinylic H to C(7) and C(2), and the three-bond correlations from H(18) to C(7), C(16), and C(20), indicated that the $CHCH_2CH_2C{=}CH$ fragment forms part of the six-membered ring linked to ring B via C(2) and C(7). It only remains to link the CH_2CH_2NH fragment to complete the tetracyclic ring system to reveal the structure of mersilongine (**446**).

The structure is characterized by a six-membered ring C contiguously fused to the quinolinic portion via C(7) and C(2), and bridged by an aminoethylene unit from C(7) to C(16). The proposed structure is in complete accord with all of the spectral data, including the COSY, HMBC, and NOE/NOESY data. For instance, the three-bond correlations from H(5) to C(16) and C(7), and from H(6) to C(21) are consistent with the branching of the aminoethylene bridge from C(7) to C(16). NOESY, as well as NOE difference, experiments showed reciprocal NOE interactions between H(9)/H(21) and H(6R)/H(21), which are in agreement with the proposed structure. The stereochemistry of the aminoethylene bridge was deduced to be β on the basis of the observed NOE interactions, and this, in turn, fixed the relative configurations at C(7) and C(16), as well as that of C(2), which follow that of all the mersinine-type alkaloids discovered to date. In addition, it is also consistent with the proposed origin of **446** from a mersinine-type precursor (*vide infra*).

The structure of mersilongine (**446**) represents a departure from the mersinine group of alkaloids which occur exclusively in this plant. It appears to have lost an entire ring (corresponding to ring D of the mersinines), and in addition, a rearrangement seems to have occurred, resulting in cleavage of the N(4)–C(21)

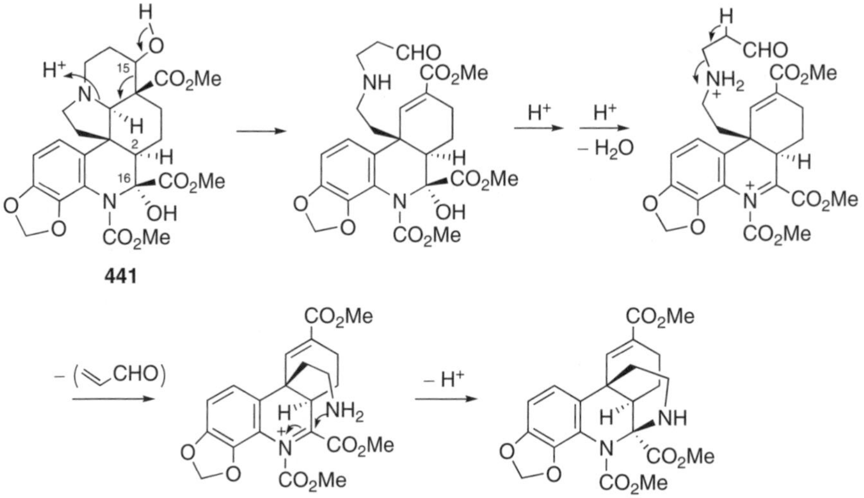

Scheme 35

bond, and formation of a new bond linking N(4) to C(16). A possible pathway to **446** is shown in Scheme 35 from a mersinine-type precursor **441**, involving a concerted Grob-like fragmentation, initiated by the C(15) alkoxy group, followed by a retro-Michael elimination of an acrolein fragment. Finally, intramolecular capture of the iminium ion **448** via 1,2-addition by the appositely oriented NH_2 function of the aminoethane side chain completes the formation of the aminoethylene bridge, yielding the novel tetracyclic ring system of mersilongine (**446**) (204). The same mersinine-type precursor **441** also accounts for the formation of the *seco*-mersinine derivative mersirachine (**439**), as well as the formyl-mersinine derivative mersinaline (**440**) (*vide supra*).

Mersicarpine (**447**) was a minor alkaloid isolated from the leaf extract of *K. singapurensis* (205). It was also subsequently found in *K. arborea* (54). The UV spectrum resembled those of an *N*-acyl dihydroindole chromophore (e.g., leuconoxine, **87**), but with additional bands, possibly due to additional conjugation with an imine function. The IR spectrum showed strong, broad bands at 3318 and $1654\,cm^{-1}$, suggesting the presence of OH and imine/lactam functionalities, respectively. The quaternary carbon resonances at δ 169.6 and 168.9 in the ^{13}C NMR spectrum were assigned to lactam carbonyl and imine functionalities, respectively, in agreement with the IR spectrum. In addition, another quaternary carbon resonance, which was observed downfield at δ 93.8 was attributed to a carbon adjacent to both a nitrogen and an oxygen atom. The ^1H NMR spectrum indicated an unsubstituted aromatic ring, a broad OH peak at δ 5.92, and an ethyl side chain.

The COSY and HMQC spectral data revealed two further partial structures, namely, $NCH_2CH_2CH_2$ and CH_2CH_2, in addition to the four contiguous aromatic hydrogens, and the ethyl side chain noted previously. The absence of an indole NH suggested that N(1) was associated with the lactam group, an inference which was further supported by the observed downfield shift of H(12) due to deshielding by the proximate amide/lactam carbonyl linked to the indolic nitrogen. In addition, the aromatic carbon and hydrogen shifts bore a striking similarity to those of *epi*-leuconolam, suggesting a close structural affinity in that part of the molecule with **447**.

The observed three-bond correlation from H(9) to the imine C(7) in the HMBC spectrum confirmed the assignment of H(9), and defined the location of the imine function. In addition, attachment of the propyl fragment to N(4) was supported by the H(3) to C(7) correlation, while the H(14) and H(18) correlations to the quaternary C(20) confirmed the branching of the propyl and ethyl fragments from this carbon. The three-bond correlations from H(15), as well as H(19) to the oxygenated C(21), permitted the assembly of the seven-membered imine-containing ring C. Since only a lactam carbonyl (which must be attached to the indolic nitrogen as noted above) remained as the source of unsaturation, another ring remained to be assembled from the molecular formula ($C_{17}H_{20}N_2O_2$) established by HREIMS. This was readily accomplished from the observed three-bond correlations from H(17) to C(2) and from H(16) to C(20). Other correlations from the HMBC spectrum were in complete accord with the proposed structure. In addition, the large geminal coupling observed for the C(16) hydrogens ($J = 18.5\,Hz$) is diagnostic of methylene hydrogens α to a carbonyl function, which provided further support for structure **447** (205).

The stereochemistry of the ethyl side chain was assumed to be similar to that in leuconolam (**86**), rhazinilam (**84**), and their congeners, which also occur in the stem-bark extract, on the grounds of a presumed close biogenetic relationship (*vide infra*). The NOESY and NOE difference experiments showed reciprocal NOEs between H(18) and H(15β), which, in turn, allowed the more deshielded signal to be assigned to H(15α). This was in agreement with the proposed α-face attachment of the hydroxyl substituent at C(21) (*vide infra*), as H(15α) would then be expected to experience deshielding due to spatial proximity of the α-oriented OH.

The structure of mersicarpine (**447**) represents a departure from the rhazinilam–leuconolam group of alkaloids, which coexist with **447** in the stem-bark extract of this plant. It appears to have lost the two-carbon tryptamine bridge corresponding to C(5) and C(6), normally present in the other monoterpenoid indole alkaloids. In addition, the presence of the lactam-containing ring D suggested an affinity to leuconoxine, although a further rearrangement appears to have occurred leading to loss of the two-carbon chain and formation of the seven-membered imine-containing ring C. A possible biogenetic pathway to **447** is shown in Scheme 36, from leuconolam (**86**) (or *epi*-leuconolam), involving formation of a dehydroleuconoxine derivative **93**, from the transannular attack by the indolic nitrogen on the C(21) iminium ion of **86**. Protonation of **93** yields the tertiary benzylic carbocation **449**, which then undergoes a 1,2-alkyl shift resulting in the iminium ion **451** (alternatively, **451** could also arise via a 1,2-alkyl shift from the protonated ketone **450**). Nucleophilic attack by a molecule of water on **451**, which would be anticipated to occur from the less hindered α-face, gives the tertiary alcohol **452**, characterized by the presence of a β-lactam unit associated with N(4). Subsequent cleavage of the

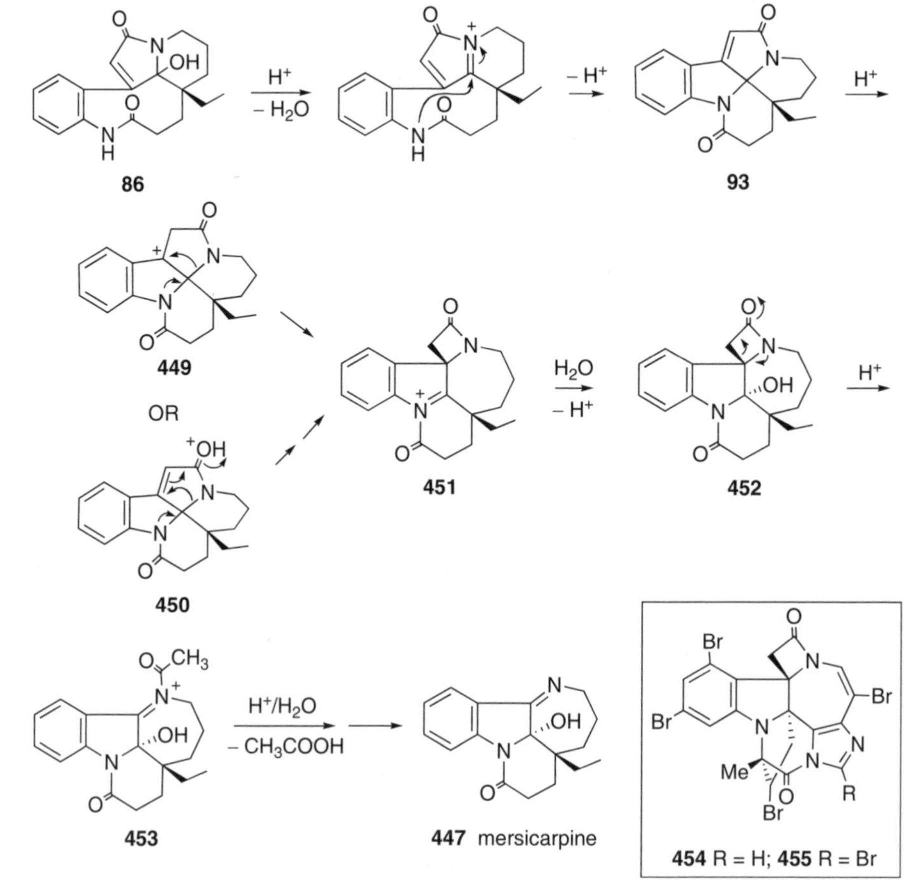

Scheme 36

β-lactam unit, followed by hydrolysis of the resultant acyl imine salt **453**, yields the novel ring system of mersicarpine (**447**) (205). Additional support for such a pathway was provided from an unexpected source. The halogenated marine alkaloids, chartellamides A (**454**) and B (**455**), from the marine bryozoan *Chartelle papyracea*, possess a structure displaying a remarkable resemblance to the 6-5-7 ring system of mersicarpine (**447**), and incorporate a β-lactam unit, corresponding to that present in the proposed intermediate, **452** (206).

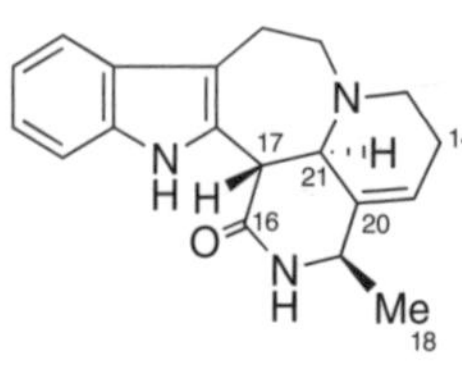

456 arboflorine

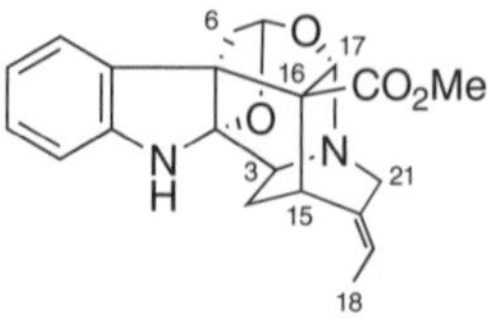

457 arbophylline

Two unusual alkaloids were obtained from *K. arborea*, namely, arboflorine (**456**) from the stem-bark and arbophylline (**457**) from the leaf extract. The UV spectrum of arboflorine (**456**) showed typical indole absorptions, while the IR spectrum suggested the presence of NH and lactam functionalities (207). The EIMS of **456** showed a molecular ion at m/z 307, the odd mass indicating the presence of a third nitrogen. This was confirmed by HREIMS, which revealed the molecular formula $C_{19}H_{21}N_3O$. The ^{13}C NMR spectrum showed a quaternary carbon resonance at δ 171.4, which was assigned to the lactam carbonyl, leaving two other downfield carbon signals (δ 133.6 and 120.8) which were attributed to a trisubstituted double bond. The ^1H NMR spectrum showed two broad NH peaks at δ 9.74 and 6.11, the former due to the indolic nitrogen from the observed three-bond correlations to C(7) and C(8) in the HMBC spectrum, a vinylic H seen as a broad doublet at δ 5.92, and a methyl doublet at δ 1.37. In addition, a pair of AB doublets was observed at δ 4.30 and 4.12 with $J = 9.4\,Hz$, the observed carbon shift associated with the lower-field doublet at δ 59.2, indicating that this two-carbon CHCH fragment was branched from a nitrogen. This was supported by the COSY and HMQC spectra, which also revealed the presence of NCH_2CH_2, $NCH_2CH_2CH=C$, and $CHCH_3$ fragments, in addition to the four contiguous aromatic hydrogens.

Analysis of the ^1H and ^{13}C shift values allowed recognition of the aminoethylene fragment as corresponding to the C(5)–C(6) unit, which was supported by the observed HMBC correlations from H(5) to C(7), and from H(6) to C(2), C(8). In a similar manner, the $NCH_2CH_2CH=C$ and NCHCH fragments correspond to the C(3)–C(14)–C(15)–C(20) and C(21)–C(16) units, from the observed correlations from H(3) to C(5), and from H(5) to C(21), respectively. The observed correlations from H(21) to C(2), and from H(16) to C(7) allowed assembly of the seven-membered ring incorporating N(4), which was fused to the dihydroindole moiety. The attachment of the $CHCH_3$ fragment to the quaternary C(20) was indicated by the correlations from the methyl H(18) to C(20), and from H(15) to C(19). This left the lactam function to be inserted to complete the formation of the fifth ring as required by the molecular formula. The linkage of the lactam carbonyl to C(16) was shown by the two-bond correlation to this carbon from H(16) and the three-bond correlations from the lactam NH to C(16) and C(20).

The relative stereochemistry at the various stereogenic centers was established by NOE experiments. The observed NOE between H(21) and H(19) required both these hydrogens to be *syn* with respect to each other. NOE was not observed between H(16) and H(21), but was seen for H(16)/H(5β) and H(16)/H(3β). The orientations of the C(3) hydrogens can be independently established from the observed *trans*-diaxial $J_{3\beta-14\alpha}$ coupling of 12 Hz and the NOE between H(14β) and the olefinic H(15). In addition, the observed NOEs between the lactam NH and both H(19) and H(21) allowed the stereochemistry of the nitrogen lone pair to be inferred.

A plausible precursor to the novel pentacyclic skeleton of **456** is the diester derivative of preakuammicine, **458**, which undergoes a Grob-like fragmentation to the iminium ion, **459** (Scheme 37). Isomerization leads to a conjugated iminium ion, which on conjugate addition by ammonia gives the tetracyclic amine, **460**. A retro-vinylogous Mannich reaction then follows to give the enolate, **461**, which on subsequent ring closure via an intramolecular Mannich reaction gives the

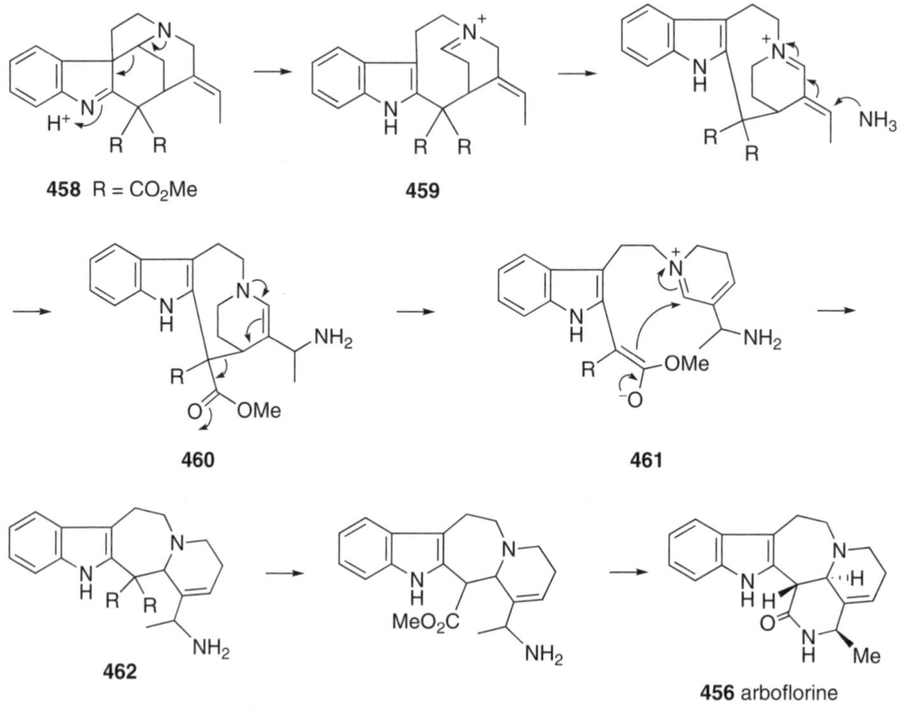

Scheme 37

tetracyclic indole, **462**, with an amine-containing side chain. Decarboxylation, followed by aminolysis, leads to the novel ring system of arboflorine (**456**) (207).

Arbophylline (**457**) ($C_{21}H_{22}N_2O_4$) showed a UV spectrum characteristic of a dihydroindole chromophore, while the IR spectrum indicated the presence of NH and ester functionalities (208). The presence of NH, ester carbonyl, and methoxy resonances in the NMR spectrum confirmed the presence of indolic NH and methyl ester functions. The ^{1}H NMR spectrum also indicated an ethylidene side chain, a deshielded methine doublet (δ 5.52) attributed to an acetal function, another isolated methine (δ 4.74) observed downfield as a singlet due to it being adjacent to both a nitrogen and an oxygen atom, and an isolated aminomethylene seen as an unresolved multiplet at δ 3.46 (δ_C 53.9). The observed shift of C(2) was noticeably downfield for a dihydroindole, and indicated oxygenation at this position.

The COSY and HMQC spectra revealed, in addition to the aromatic and ethylidene hydrogens, and the isolated methine and methylene mentioned previously, the presence of two additional spin systems, namely, a CHCH$_2$ and an NCHCH$_2$CH. The former corresponds to the C(5)–C(6) fragment, from the three-bond correlations observed from H(5) to C(7) and from H(6) to C(8) in the HMBC spectrum. The observed downfield shift of C(5) at δ 98.5 is consistent with it being associated with an acetal function, for which the corresponding methine shift at δ 5.52 was alluded to earlier. Since there are a total of four oxygens in **457**, and two are due to the ester function, one of the oxygens of the acetal function must be linked to C(2), while the other to the methine at C(17) (δ_H 4.74, δ_C 88.8), which is, in turn, linked to N(4). This assignment was also supported by the observed three-bond correlations from H(5) to C(17) and from H(17) to C(5) and C(7) in the HMBC spectrum. The second fragment branching from N(4) is the CHCH$_2$CH unit, which is linked from the C(15) methine to the quaternary C(16) bearing the methyl ester function, as indicated by the correlations from H(14) to C(16), and from H(15) to the ester carbonyl in the HMBC spectrum. The other methine corresponding to C(3) was deduced to be linked to C(2) from the observed correlations from H(14) to C(2), and from H(3) to C(7). Completion of the assembly of the structure of **457** required insertion of the remaining fragment, comprising the ethylidene side chain linked to the quaternary C(20), which was, in turn, linked to C(15) and the isolated aminomethylene C(21). Determination of the structure of arbophylline also allowed assignment of the base peak (m/z 323) observed in the mass spectrum of **457**, which though initially unintelligible, could now be attributed to the fragment ion resulting from loss of CO$_2$ (M$-$CO$_2$+H), as a consequence of the acetal functionality residing in the molecule.

The structure is also consistent with the NOE/NOESY data, which also confirmed the relative configurations at all of the stereogenic centers. Thus, irradiation of the indolic NH caused enhancement of H(12) and H(3), while irradiation of H(3) resulted in enhancement of H(21α) and H(14α), in addition to NH. Irradiation of H(15) resulted in enhancement of H(18) and *vice versa*, confirming the geometry of the C(19),C(20)-double bond as *E*. NOEs were also observed between H(17) and H(21β), as well as between H(9) and the ester methyl. Although there are a total of seven stereogenic centers in **457**, the rigid cage architecture of the molecule in essence restricts the number of stereochemical possibilities to one enantiomeric pair, corresponding to the relative configuration shown.

The structure of arbophylline (**457**) constitutes a novel skeleton of the monoterpenoid indole alkaloids. Its unprecedented heptacyclic ring system incorporates a cage structure which is bounded by pyrrolidinyl, tetrahydrofuranyl, and tetrahydropyranyl rings. Another notable feature is the presence of an acetal functionality, with the acetal carbon being shared by the tetrahydrofuranyl and tetrahydropyranyl units.

The novel ring system of arbophylline (**457**) suggests that a rearrangement must have occurred from a known monoterpenoid alkaloid precursor, which in all probability involves cleavage of the C(5) to N(4) bond as a key step. A possible biogenetic route to **457** is shown in Scheme 38, from an akuammiline-type precursor such as rhazimal (**463**), which was also detected as a minor alkaloid in the plant. Cleavage of the C(5)–N(4) bond occurs via hydrolysis of the C(5)–N(4) iminium ion **464**, in turn derived from oxidation of **463**. An amine-aldehyde nucleophilic addition, involving attack of N(4) onto the C(16) aldehyde carbonyl leads to the pentacyclic hydroxy aldehyde **465**. A tandem intramolecular process then ensues to effect the formation of the remaining two rings. This cascade is initiated by hemiacetal formation involving attack by the secondary alcohol function on the aldehyde carbonyl. This is then followed by hemiacetal OH attack on the imine function to complete the assembly of the heptacyclic ring system of arbophylline (**457**) (208).

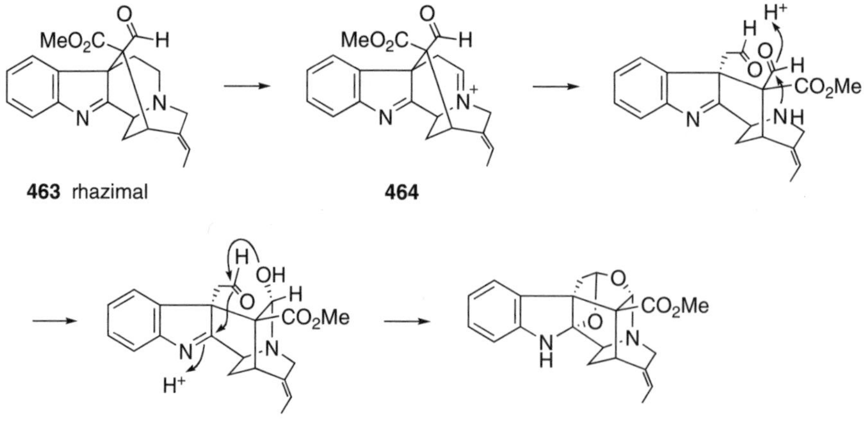

463 rhazimal   **464**

465   **457** arbophylline

Scheme 38

IX. BISINDOLE ALKALOIDS

Although known to be prolific producers of indole alkaloids, plants belonging to this genus provided only a modest number of bisindoles (209). Nitaphylline (**466**) was the only bisindole isolated from *K. teoi*, which otherwise yielded a large number of new indole alkaloids (140,210). The mass spectrum showed a molecular ion at m/z 910 with a significant peak at m/z 455 suggesting symmetrical cleavage of the parent ion along the bond bridging the monomeric moieties. This, together with the observation that the UV spectrum of nitaphylline was virtually superimposable with that of kopsingine (**4**), provided an early indication that nitaphylline is constituted from the union of two kopsingine units. This was further reinforced by the NMR spectral data. The ^{13}C NMR spectrum accounted for only 33 peaks, indicating overlap of 15 carbon resonances. In addition, four pairs of signals although distinguishable were only just so, differing in chemical shift by only 0.1 or 0.2 ppm. The bulk of the overlapped signals were readily assigned to the aromatic portions of the alkaloid, as well as the methoxy, carbamate, and ester groups, which were similar to those of kopsingine. The remaining signals could also be assigned based on 2D NMR data, and by comparison with kopsingine. The observation that the resonances of the piperidine ring carbons showed greater departure from kopsingine suggested that attachment of the monomeric entities involved the piperidine ring carbons.

The ^1H NMR spectrum showed six aromatic hydrogens, three olefinic hydrogens, two geminal hydrogens of an aminomethylene group, and only one H due to an aminomethine, indicating branching of the bisindole from C(3) of one kopsingine unit to the olefinic C(14′) or C(15′) of the other kopsingine unit. Since the geminal hydrogens of C(3′) were doublets ($J = 16$ Hz) with no evidence of coupling to any adjacent olefinic hydrogen, the bisindole was deduced to be bridged from C(3) of one kopsingine unit to C(14′) of the other. The stereochemistry of the point of branching could not be ascertained directly from the NMR spectrum due to overlap of the aminomethylene signal and poor resolution of the olefinic signals, but could be assigned by analogy to the bistabersonine alkaloids voafrine A (467) (β-branching) and voafrine B (468) (α-branching), which possess the same mode of attachment of the monomeric units (211). In nitaphylline (466), the carbon shifts of C(21) and C(21') were identical (overlapped) and were similar to the value in the monomer indicating that the substitution was β, with the pseudoequatorially oriented second kopsingine unit pointing away from the first by analogy to voafrine A. In the α-branched voafrine B, the resulting greater spatial proximity between the two monomeric units is reflected in the very different shifts observed for C(21). Nitaphylline (466) represents the first example of an aspidofractinine–aspidofracti-nine type dimeric alkaloid (140,210).

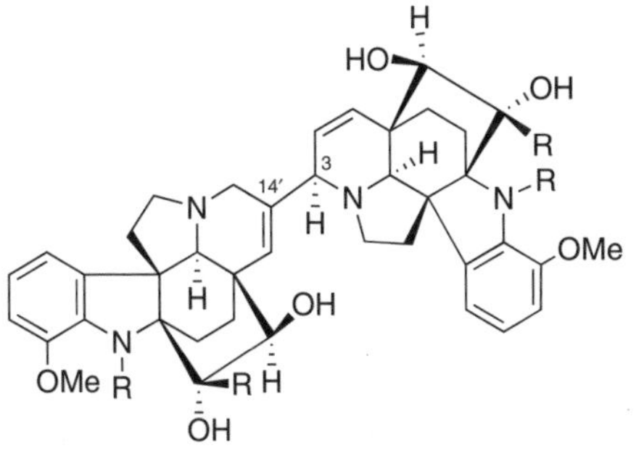

466 R = CO₂Me nitaphylline

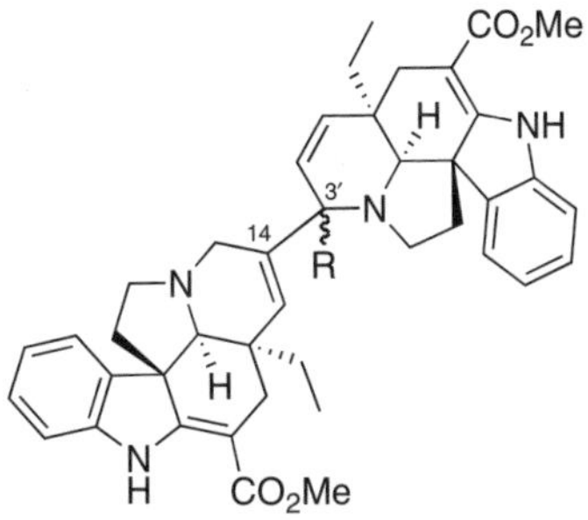

467 R = α-H voafrine A
468 R = β-H voafrine B

Tenuiphylline (469) was isolated in minute amounts, together with the tenuisines (*vide infra*), from *K. tenuis* occurring in Malaysian Borneo (188). The FABMS showed a MH⁺ peak at m/z 717 and high-resolution measurements provided the formula $C_{42}H_{44}N_4O_7$. The UV spectrum indicated the presence of dihydroindole chromophores and the IR spectrum indicated the presence of hydroxyl, urethane, ester, and lactone functions. In view of the small amount obtained, definitive structure elucidation of tenuiphylline required 600 MHz NMR spectra, since extensive overlap was encountered with lower-field instruments.

The NMR spectral data showed the presence of one carbamate and one methyl ester function. The aromatic region integrated for seven hydrogens indicating branching of the bisindole from an aromatic carbon of one monomeric unit. Absence of an NH signal and the presence of only one urethane function indicated the dihydroindole nitrogen of the other monomeric moiety as the other site of attachment. The NMR spectral data also showed the presence of four olefinic hydrogens, two associated with a six-membered ring and two with a five-membered ring. One unit of the bisindole was readily deduced to be identical to lapidilectine B (380) from the excellent correlation of the ¹³C shifts, in particular the nonaromatic shifts, with those of lapidilectine B. The other monomeric unit was considered as having a novel carbon skeleton or a rearranged venalstonine (415). The carbon shifts of this second unit generally resembled those of venalstonine (415), except for changes involving C(2), C(12), C(16), C(17), C(18), C(20), and C(21) (212). The COSY spectrum indicated the presence of the same groups as in venalstonine and the downfield shifts of C(20) and C(21) were reminiscent of those of the vindolinine derivatives (e.g., N-methylvindolinine (212)), suggesting a change in the connectivity involving the C(18)–C(19) fragment. The presence of a hydroxyl group on C(2) was supported by its downfield shift at δ 99.8 due to it being linked to an oxygen and a nitrogen, and by the observed three-bond correlations from H(6) and H(21) to C(2) in the HMBC spectrum. The presence of an OH on C(2) required that C(18) now be linked to C(16), which was consistent with the downfield shift of C(16) when compared with venalstonine (415), as well as with the observed correlations from H(17) to both C(18) and C(19) in the HMBC spectrum. From the ^{1}H NMR spectrum (COSY, NOE), it can be established that the aromatic portion associated with this unit was unsubstituted, and furthermore the upfield shift of H(12) when compared with venalstonine was consistent with the change from NH in venalstonine to N–C(11′) in tenuiphylline.

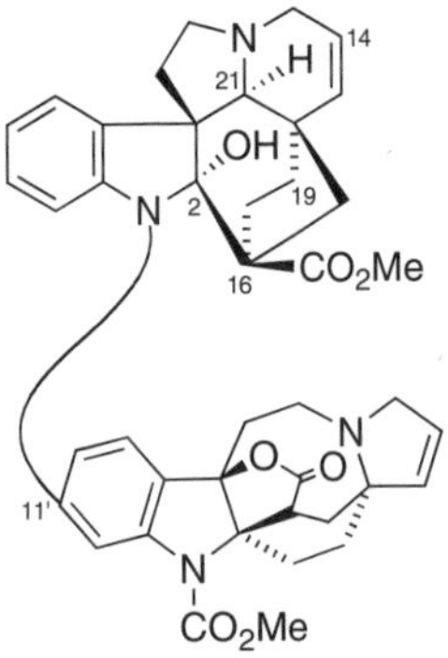

469 tenuiphylline

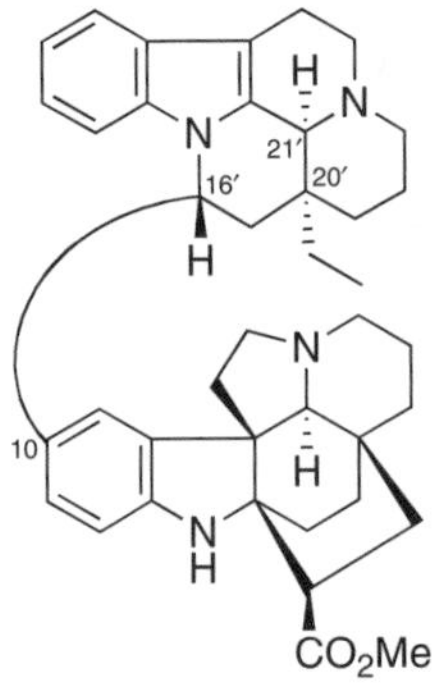

470 (+)-kopsoffine

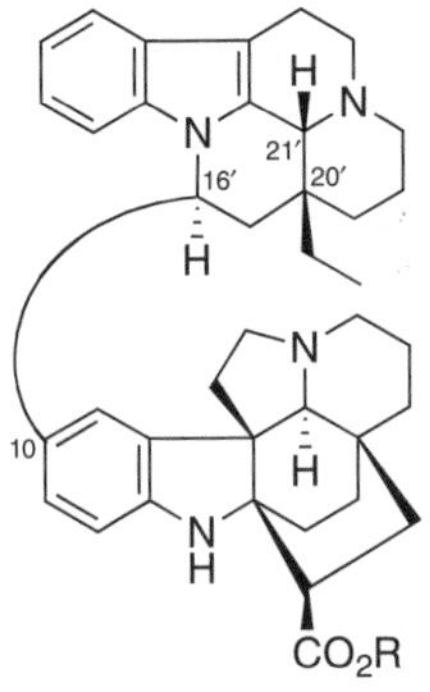

471 R = Me (−)-norpleiomutine
472 R = H

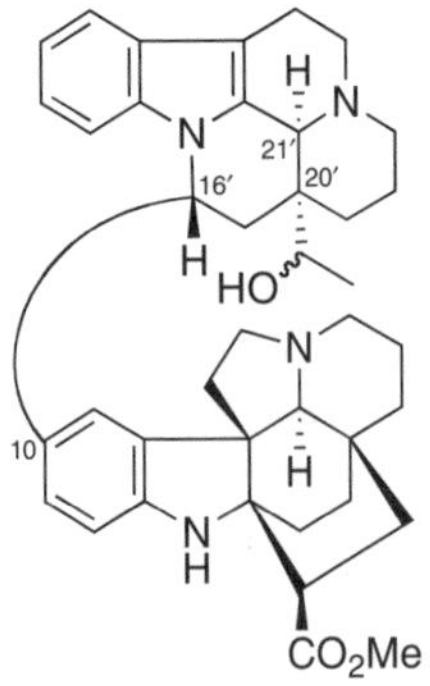

473 (+)-kopsoffinol

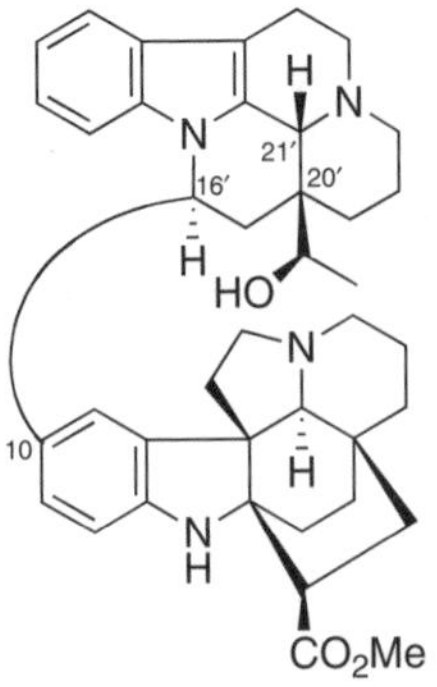

474 (+)-kopsoffinol

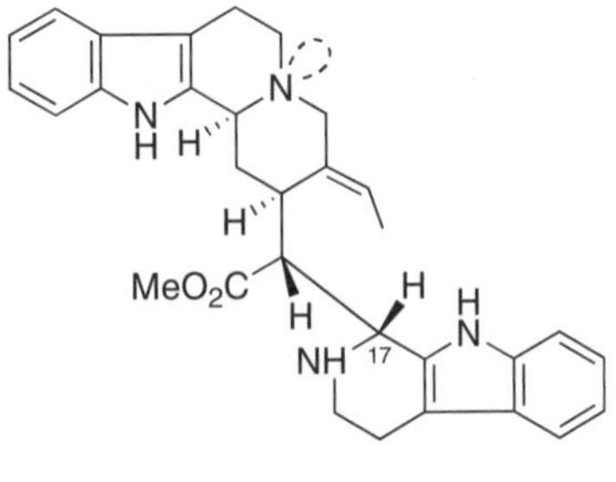

475 buchtienine

The attachment site on the lactone-containing monomer can be at either C(10′) or C(11′) from the coupling pattern of the aromatic hydrogens. The observed NOE between the aromatic doublet at δ 7.33 and H(6′β), and a H(9′)/C(7′) correlation in HMBC allowed assignment of this signal to H(9′), which must be coupled to H(10′), furnishing proof of C(11′) branching. The bisindole is thus linked from N(1) of one unit to C(11′) of the other. The stereochemistry of the C(2)–OH was assigned the α configuration since had it been β, the ester function would have been too far removed to experience NOE with H(9′) and H(10′), as well as the observed anisotropy from the other aromatic ring.

Six alkaloids (the kopsifolines, *vide supra*) possessing the carbon skeleton corresponding to the novel venalstonine-like monomer in tenuiphylline have since then been obtained from another Malaysian *Kopsia* (193,194).

K. pauciflora from Malaysian Borneo provided four bisindole alkaloids, namely, kopsoffine (**470**), norpleiomutine (**471**), (−)-demethylnorpleiomutine (**472**), and (+)-kopsoffinol (**473**) (80). Kopsoffine was also previously isolated from the Chinese species, *K. arborea* (*K. officinalis*) (79). These alkaloids are constituted from the union of eburnane and kopsinine units. They are remarkable in that while norpleiomutine (**471**) and demethylnorpleiomutine (**472**) incorporate eburnane units having the 20β,21β configuration, kopsoffine (**470**) and kopsoffinol (**473**) apparently incorporate eburnane units having the opposite (enantiomeric) configuration (20α,21α). The eburnane alkaloids present in the plant (including 19-hydroxyeburnamine (**81**), which corresponds to the eburnane half of kopsoffinol), however, belonged to the (−)-eburnamine series, with the 20β,21β configuration (4). It would thus appear from these results that *K. pauciflora* is exceptional in that both enantiomers of the eburnane moiety are involved in different bisindoles, even though the eburnane monomers existing in the plant belonged exclusively to one enantiomeric group. In another study of the same plant, only one bisindole, norpleiomutine (**471**), was isolated, while the eburnane alkaloids which were present were also all of the 20β,21β configuration (73). Norpleiomutine (**471**), demethylnorpleiomutine (**472**), and kopsoffinol (formulated as **474**) have been subsequently isolated from another Malaysian *Kopsia* from Borneo, *K. dasyrachis*, and complete high-field NMR spectral data were provided (85). In addition, (+)-19(R)-hydroxyeburnamine (**81b**), which constitutes the eburnane half in kopsoffinol, was also isolated, and its structure was established by X-ray analysis (74,85).

It was contended that all three bisindoles (**471**, **472**, and **474**) from *K. dasyrachis* incorporate in common, eburnane units having the 20β,21β configuration, for the following reasons (85,209). First, since the other two bisindoles present (**471** and **472**) have eburnane units with the 20β,21β configuration, it would be extremely unlikely that a third odd bisindole should have a eburnane unit of the opposite configuration. Moreover, since all four monomeric eburnane alkaloids present in the plant, namely, (+)-eburnamonine (**57**), (+)-19(R)-hydroxyeburnamine (**81b**), (+)-isoeburnamine (**61**), and (−)-19(R)-hydroxyisoeburnamine (**81a**), were of the 20β,21β configuration, it would be singularly unusual for kopsoffinol to incorporate a eburnane unit of the opposite configuration. Lastly, considering that the likely precursor of the eburnane half in kopsoffinol, (+)-19(R)-hydroxyeburnamine (81b), was also obtained from this plant, and its structure has been established by X-ray analysis, the most likely structure of kopsoffinol is 474, with the eburnane half corresponding to that of the monomeric precursor, 81b. Furthermore, the attachment of the kopsinine unit at C(16') is β (H(16'α)), as required by the observed coupling constants for the H(16') signal of 11 and 5 Hz for kopsoffinol (72).

Norpleiomutine (471) from *K. dasyrachis* showed strong cytotoxicity toward P388 cells (213), while the quasidimer buchtienine (475), first isolated from the Bolivian plant *Tabernaemontana buchtienii* (syn. *Peschiera buchtienii*) (214), but also subsequently found in *K. griffithii* from Malaysia (40), was found to show significant activity against *L. donovani* (41). The observed leishmanicidal activity of the alkaloid is consistent with the use of the Bolivian plant for the treatment of leishmaniasis (214).

X. ELECTROORGANIC TRANSFORMATIONS OF *KOPSIA* ALKALOIDS

The 3-to-17 oxo-bridged alkaloids kopsidines A–C (218–220) have been synthesized in high yields via the stable iminium salt obtained from electrochemical oxidation (Pt anode, 30% CH_2Cl_2–MeCN, 0.1 M Et_4NClO_4) of kopsingine (4) (Scheme 39) (215,216).

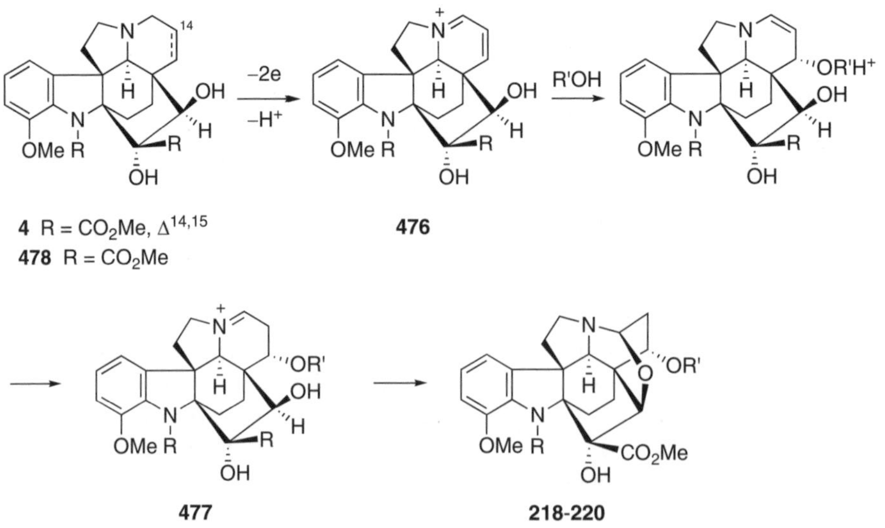

4 R = CO_2Me, $\Delta^{14,15}$
478 R = CO_2Me

476

477

218-220

Scheme 39

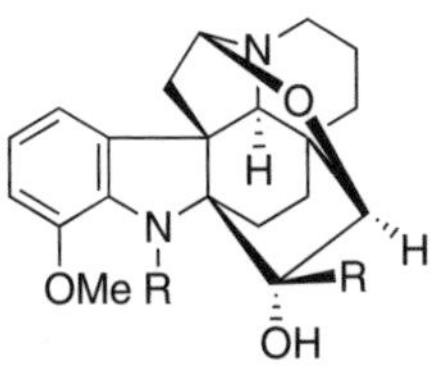

479 R = CO₂Me

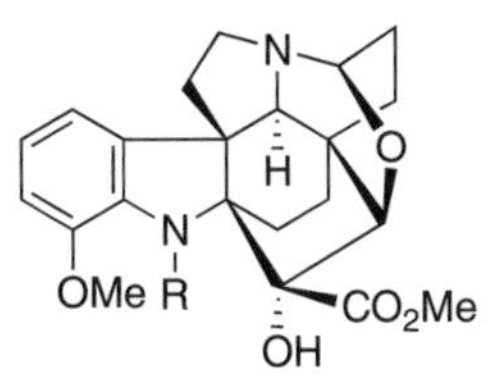

480 R = CO₂Me

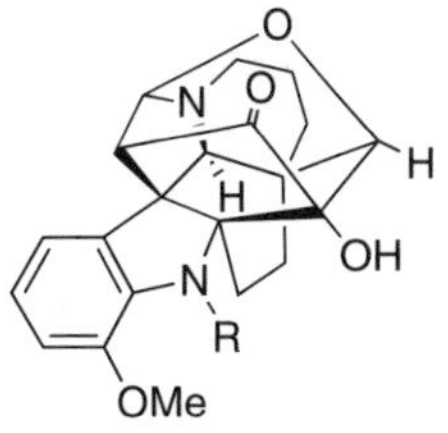

483 R = CO₂Me

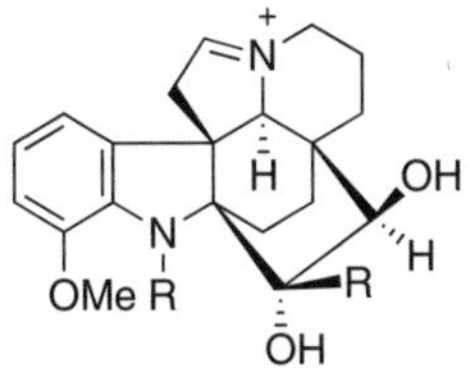

481 R = CO₂Me

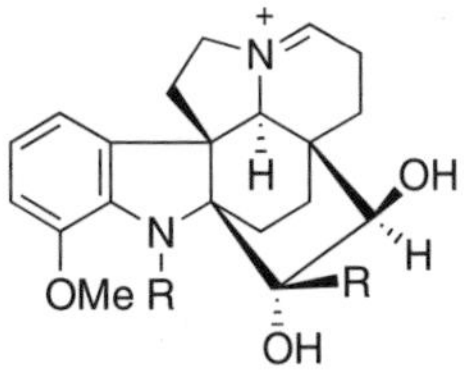

482 R = CO₂Me

Electrooxidation of kopsingine resulted in stepwise loss of an electron, deprotonation, followed by loss of another electron to yield the stable conjugated iminium salt **476** as the main product of the electrochemical process. Addition of methanol resulted in conjugate addition of the nucleophile to C(15), which was followed by an intramolecular 1,2-addition of the 17-β-OH function of the resulting iminium ion intermediate **477** yielding kopsidine A (**218**). Use of ethanol or water instead of methanol provided kopsidines B and C, respectively (215,216). A similar conversion from kopsingine via the Polonovski–Potier reaction has also been reported (217).

Electrooxidation of the semisynthetic derivative 14,15-dihydrokopsingine **478** on platinum in MeCN–CH₂Cl₂ gave the cyclized, oxo-bridged products **479** and **480** via direct intramolecular trapping of the presumably less stable iminium ions (**481** and **482**). However, if the same reaction was carried out in methanol, the cage product **483**, reminiscent of the kopsinitarines (*vide supra*), was also obtained in addition to **479** and **480**. The formation of the cage product **483** was attributed to two successive intramolecular cyclizations as shown in Scheme 40 (216).

It has been suggested that the ring-opened alkaloid danuphylline (**266**) originates from a methyl chanofruticosinate precursor **248** via hydrolysis of the iminium ion followed by subsequent retro-aldol-type reaction of the resultant unstable carbinol amine (*vide supra*). The proposal has been vindicated by the implementation of an electrochemically mediated semisynthesis as shown in Scheme 41 (165).

An interesting, electrochemically mediated transformation of the aspidofractinine alkaloid kopsamine (**166**) and related analogues has been reported (216,218). Anodic oxidation of kopsamine (**166**) on platinum in acetonitrile led to the cyclized products **487** and **488**, which have incorporated cyano and cyanomethyl groups, respectively. Electrooxidation on vitreous carbon in methanol, on the other hand, led to a dimerization product with C₂ symmetry, **490**. These results are rationalized

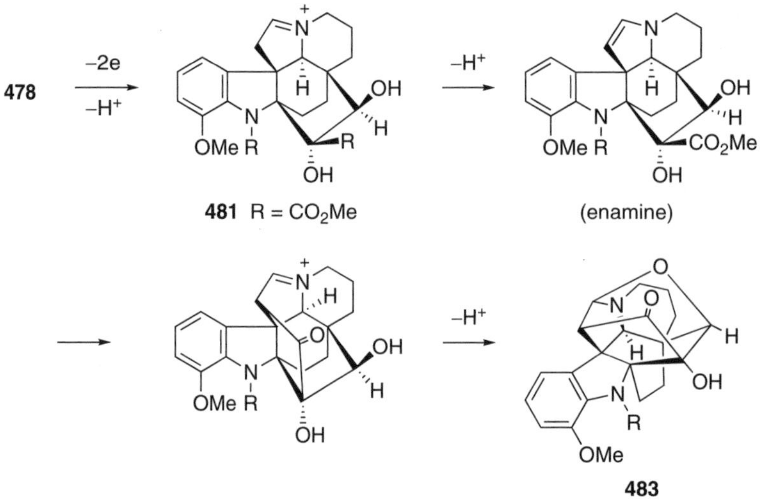

Scheme 40

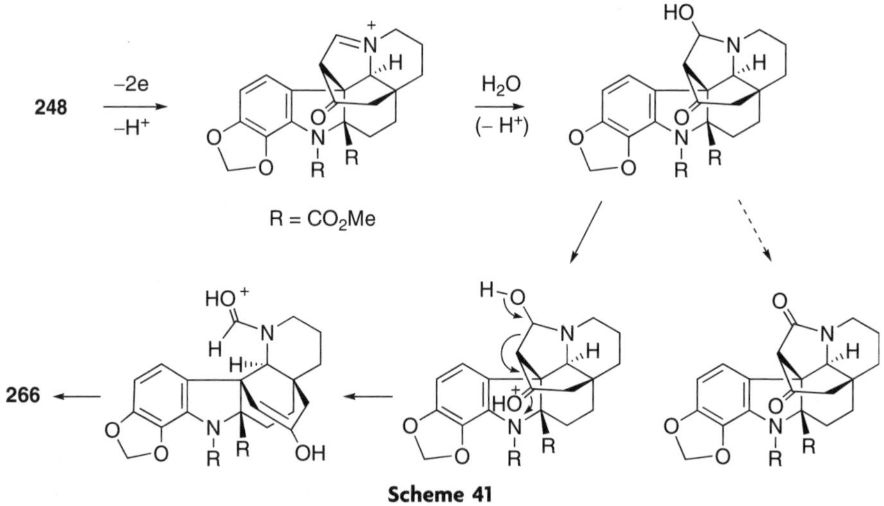

Scheme 41

in Schemes 42 and 43, respectively. Formation of the cyanated or cyanomethylated kopsine derivatives, **487** and **488**, respectively, were due to trapping of the intermediate iminium ion **486** by CN⁻ and cyanomethyl anions. The iminium ion **486** was formed from cyclization of the enamine **485**, which, in turn, resulted from deprotonation of the initially formed iminium ion **484** (Scheme 42). The source of the CN⁻ anion was probably from decomposition of the acetonitrile anion radical (formed by a side reaction in the counter electrode chamber), which can undergo the alternative decomposition to the cyanomethyl anion. The formation of the dimeric product **490** is rationalized in Scheme 43, involving as the key step, head-to-head coupling of the cation radical intermediate **489**, formed as a result of

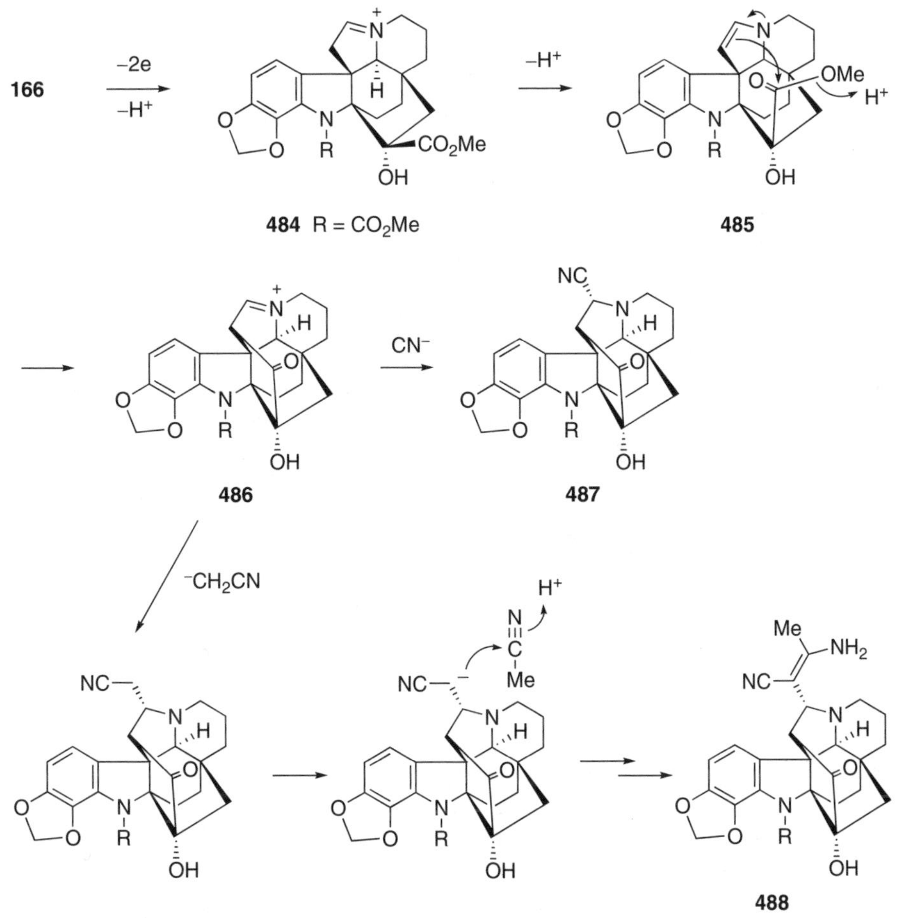

Scheme 42

a further one electron oxidation of the enamine **485**. The same behavior was reproduced in the related alkaloids, kopsiflorine (**164**) and kopsilongine (**165**) (216,218).

XI. ALKALOID DISTRIBUTION IN THE GENUS *KOPSIA*

The occurrence of alkaloids in the genus *Kopsia* is summarized in Table II. From the table, the predominance of the aspidofractinine skeleton is apparent, with kopsinine (**146**) and kopsamine (**166**) being the alkaloids most frequently encountered. Other common *Kopsia* alkaloids include tetrahydroalstonine, akuammiline and its derivatives, the rhazinilam–leuconolam alkaloids, and the eburnane alkaloids. In contrast, some alkaloids show more limited distribution, e.g., the simple indole alkaloids, harmane (**19**) and harmicine (**21**) in *K. griffithii*, vincophylline (**37**), mersilongine (**446**), the mersinines (**422–437, 439, 440**), and

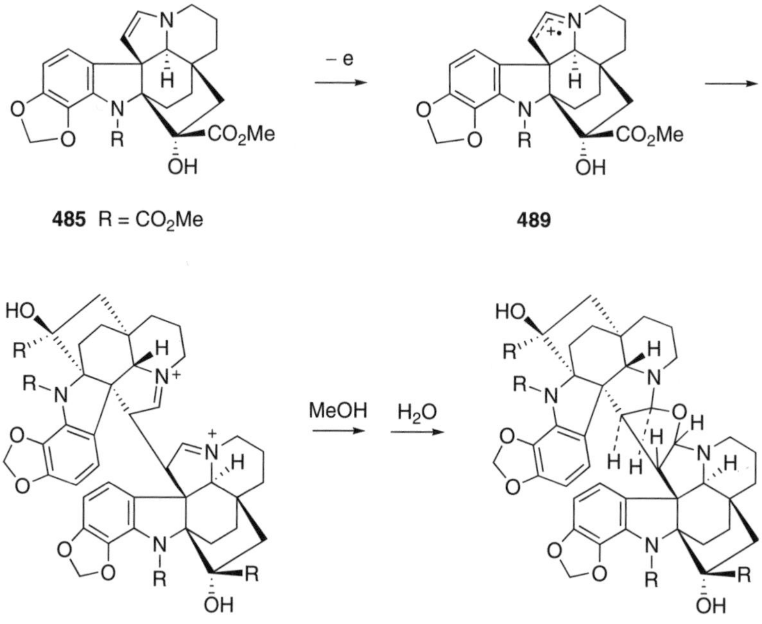

Scheme 43

kopsifolines (**409–414**) in *K. singapurensis*, the pericine alkaloids (**48, 50, 52**), arboricinine (**27**), arboflorine (**456**), and arbophylline (**457**) in *K. arborea*, the cyano-substituted lahadinines (**174, 175**) in *K. pauciflora*, the lundurines (**384–387**) in *K. tenuis*, lapidilectine A and its congeners (**375–379**) in *K. grandifolia*, the terengganensines (**82, 83**) in *K. profunda*, the quasidimeric buchtienine (**475**) in *K. griffithii*, and the bisindoles nitaphylline (**466**) and tenuiphylline (**469**) in *K. teoi* and *K. tenuis*, respectively. Other alkaloids with limited distribution include the methyl chanofruticosinate alkaloids (**247–263**) and the related danuphylline alkaloids (**266, 270**) (*K. arborea, K. dasyrachis*), fruticosine (**2**) and fruticosamine (**3**) (*K. fruticosa, K. arborea*), the kopsidines (**218–225**) and kopsinitarines (**237–241**) (*K. teoi, K. singapurensis*), lapidilectine B (**380**) and the tenuisines (**381–383**) (*K. grandifolia, K. tenuis*), the pauciflorine/kopsijasminilam alkaloids (**330–334**) (*K. pauciflora, K. arborea*), dasyrachine (**264**) (*K. dasyrachis, K. arborea*), and mersicarpine (**447**) (*K. singapurensis, K. arborea*).

There appears also to be a noticeable similarity between *K. singapurensis* and *K. teoi* from the viewpoint of their alkaloidal composition. Many alkaloids are common in both species, and in addition, some of these common alkaloids are restricted to these two species only, namely, kopsingine (**4**), kopsaporine (**190**), kopsinganol (**197**), 16-*epi*-kopsinine (**153**), 17-α-hydroxy-Δ^{14,15}-kopsinine (**178**), aspidodasycarpine (**38**), lonicerine, the oxo-bridged kopsidines (**218–220**) and singapurensines (**222–225**), and the cage kopsinitarines (**237–241**). Such a similarity from a chemotaxonomic viewpoint is also reflected in their

Table II   Occurrence of alkaloids in the genus *Kopsia*

Plant[a]	Plant part	Alkaloids	References
K. arborea Blume (Peninsular Malaysia)	Leaves	Arbophylline (**457**)	(208)
		Methyl *N*(1)-decarbomethoxychanofruticosinate (**249**)	(159)
		Methyl 11,12-methylenedioxy-*N*(1)-decarbomethoxychanofruticosinate (**252**)	(159)
		Methyl 11,12-methylenedioxychanofruticosinate (**248**)	(159)
		Methyl 11,12-methylenedioxy-*N*(1)-decarbomethoxy-$\Delta^{14,15}$-chanofruticosinate (**253**)	(159)
		Prunifoline A (**260**)	(160)
		Prunifoline B (**256**)	(160)
		Prunifoline C (**255**)	(160)
		Prunifoline D (**261**)	(160)
		Prunifoline E (**262**)	(160)
		Prunifoline F (**263**)	(160)
	Stem-bark	Akuammidine	(54)
		Arboflorine (**456**)	(54,207)
		Arboloscine (**92**)	(54,61)
		Arboricine (**26**)	(51,54)
		Arboricinine (**27**)	(51,54)
		Aspidofractinine	(54)
		Dasyrachine (**264**)	(54)
		N(1)-Decarbomethoxykopsamine (**172**)	(54)
		5,21-Dihydrorhazinilam (**85**)	(54)
		15α-Hydroxykopsinine (**149**)	(54)
		Kopsamidine A (**173**)	(54)
		Kopsamidine B (**148**)	(54)
		Kopsamine (**166**)	(54)
		Kopsamine *N*-oxide	(54)
		Kopsanone (**272**)	(54)
		Kopsifine (**229**)	(54)
		Kopsiflorine (**164**)	(54)
		Kopsilongine (**165**)	(54)
		Kopsinidine A (**234**)	(54)
		Kopsinidine B (**235**)	(54)
		Kopsinine (**146**)	(54)
		Leuconoxine (**87**)	(54)
		Mersicarpine (**447**)	(54,205)
		N(1)-Methoxycarbonyl-11,12-dimethoxykopsinaline (**167**)	(54)
		19(*S*)-Methoxytubotaiwine (**43**)	(54)
		19(*R*)-Methoxytubotaiwine (**44**)	(54)
		Methyl *N*(1)-decarbomethoxychanofruticosinate (**249**)	(54)
		11,12-Methylenedioxykopsine (**233**)	(54)
		O-Methylleuconolam	(54)
		Methyl 12-methoxychanofruticosinate (**250**)	(54)
		Methyl 11,12-methylenedioxychanofruticosinate (**248**)	(54)
		Methyl 11,12-methylenedioxy-*N*(1)-decarbomethoxychanofruticosinate (**252**)	(54)
		Methyl 11,12-methylenedioxy-*N*(1)-decarbomethoxy-$\Delta^{14,15}$-chanofruticosinate (**253**)	(54)
		Norflourocurarine	(54)
		Paucidactine B (**227**)	(54)

Table II (*Continued*)

Plant[a]	Plant part	Alkaloids	References
		Paucidactine C (**228**)	(54)
		Pericidine (**50**)	(54,61)
		Pericine (**48**)	(54)
		Pericine *N*-oxide (**49**)	(54)
		Pleiocarpamine (**41**)	(54)
		Prunifoline E (**262**)	(54)
		Rhazimal (**463**)	(54)
		Rhazinicine (**88**)	(54)
		Rhazinilam (**84**)	(54)
		Rhazinoline (**36**)	(54)
		Tetrahydroalstonine	(54)
		Valparicine (**52**)	(54,62)
		Venalstonidine	(54)
		Venalstonine (**415**)	(54)
		Vincadifformine	(54)
K. arborea Blume (*K. flavida*) (Peninsular Malaysia)	Leaves	Methyl 11,12-dimethoxychanofruticosinate (**251**)	(161)
		Methyl 11,12-methylenedioxy-*N*(1)-decarbomethoxychanofruticosinate (**252**)	(161)
		Methyl 12-methoxychanofruticosinate (**250**)	(161)
		Methyl 12-methoxy-*N*(1)-decarbomethoxychanofruticosinate (**254**)	(161)
		Methyl 3-oxo-11,12-methylenedioxy-*N*(1)-decarbomethoxy-14,15-didehydrochanofruticosinate (**258**)	(162)
		Methyl 3-oxo-12-methoxy-*N*(1)-decarbomethoxy-14,15-didehydrochanofruticosinate (**259**)	(162)
K. arborea Blume (*K. albiflora, K. pruniformis*) (India)	Leaves	Kopsine (**1**)	(9,13,219, 220)
K. arborea Blume (*K. jasminiflora*) (Thailand)	Leaves	14,15-Dehydrokopsijasminilam (**334**)	(145)
		10-Demethoxykopsidasinine (**329**)	(178)
		Deoxykopsijasminilam (**333**)	(145)
		Fruticosamine (**3**)	(145)
		Fruticosine (**2**)	(145)
		Jasminiflorine (**236**)	(145)
		Kopsijasmine (**208**)	(145)
		Kopsijasminilam (**332**)	(145)
K. arborea Blume (*K. longiflora*) (Queensland, Australia)	Leaves	Kopsamine (**166**)	(120,221)
		Kopsiflorine (**164**)	(120,221)
		Kopsilongine (**165**)	(120,221)
	Stem-bark	Kopsamine (**166**)	(120,221)
		Kopsilongine (**165**)	(120,221)
		Kopsinine (**146**)	(120,221)
K. arborea Blume (*K. pitardii*) (Vietnam)	Leaves	Methyl 11,12-methylenedioxy-*N*(1)-decarbomethoxy-$\Delta^{14,15}$-chanofruticosinate (**253**)	(223)
		Methyl *N*(1)-decarbomethoxychanofruticosinate (**249**)	(223)
		Methyl 11,12-methylenedioxychanofruticosinate (**248**)	(223)
K. arborea Blume (*K. officinalis*) (China)	Leaves	11,12-De(methylenedioxy)danuphylline (**270**)	(122)
		(+)-Eburnamonine (**57**)	(122)
		16β-Hydroxyaspidofractinine (**162**)	(122)

Table II (*Continued*)

Plant[a]	Plant part	Alkaloids	References
		(−)-19(*R*)-Hydroxyisoeburnamine (**81a**)	(122)
		(+)-19(*R*)-Hydroxyeburnamine (**81b**)	(122)
		(+)-Isoeburnamine (**61**)	(122)
		Kopsamine (**166**)	(122)
		Kopsiflorine (**164**)	(122)
		Kopsilongine (**165**)	(122)
		Kopsinine (**146**)	(122)
		Kopsinine *N*-oxide	(122)
		Kopsininic acid (**147**)	(122)
		Larutenine (= Larutensine) (**73**)	(122)
		11-Methoxykopsilongine (**167**)	(122)
		12-Methoxykopsinaline (**170**)	(122)
		Methyl 11,12-methylenedioxychanofruticosinate (**248**)	(122,158)
		Methyl chanofruticosinate (**247**)	(122,158)
		Methyl *N*(1)-decarbomethoxychanofruticosinate (**249**)	(158)
		Methyl *N*(1)-decarbomethoxy-$\Delta^{14,15}$-3-oxo-chanofruticosinate (**257**)	(122)
		Rhazinicine (**88**)	(122)
	Roots	5,22-Dioxokopsane (**271**)	(77)
		(−)-Isoeburnamine (**69**)	(77,79)
		Kopsinine (**146**)	(77,79)
		(+)-Kopsoffine (**470**)	(79)
		N(1)-Methoxycarbonyl-11,12-dimethoxykopsinaline (**167**)	(77)
		N(1)-Methoxycarbonyl-12-methoxykopsinaline (**165**)	(77)
		N(1)-Methoxycarbonyl-11,12-methylenedioxykopsinaline (**166**)	(77)
		12-Methoxykopsinaline (**170**)	(77)
		11,12-Methylenedioxykopsinaline (**172**)	(77)
		(−)-Quebrachamine	(77)
		Tetrahydroalstonine	(77)
	Fruits	*N*(1)-Carbomethoxy-11,12-dimethoxykopsinaline (**167**)	(133)
		N(1)-Carbomethoxy-11-hydroxy-12-methoxykopsinaline (**168**)	(133)
		N(1)-Carbomethoxy-12-hydroxy-11-methoxykopsinaline (**169**)	(133)
		N(1)-Carbomethoxy-12-methoxykopsinaline (**165**)	(133)
		5,22-Dioxokopsane (**271**)	(133)
		Eburnamenine[b]	(133)
		Kopsamine (**166**)	(133)
		Kopsamine *N*-oxide	(133)
		Kopsanone (**272**)	(133)
		Kopsinilam (**310**)	(133)
		Kopsinine (**146**)	(133)
		Pleiocarpine (**154**)	(133)
		Vincadifformine	(133)
	c	Eburnamine[b]	(222)
		Kopsinine (**146**)	(222)
		Kopsamine (**166**)	(222)
		Perivine	(222)

Table II (*Continued*)

Plant[a]	Plant part	Alkaloids	References
		19-Hydroxyeburnamine[b]	(222)
		19-Oxoeburnamenine[b]	(222)
		19-Oxoeburnamine[b]	(222)
K. dasyrachis Ridl.	Leaves	Danuphylline (**266**)	(35,164,165)
(Sabah,		11,12-Dimethoxykopsamine (**167**)	(35)
Malaysian		Kinabalurine G (**11**)	(35)
Borneo)		Kopsamine (**166**)	(35)
		Kopsamine *N*-oxide	(35)
		Kopsidasine (**211**)	(146)
		Kopsidasine *N*-oxide (**212**)	(146)
		Kopsidasinine (**326**)	(146)
		Kopsifine (**229**)	(35)
		Kopsirachine (**18**)	(35,36)
		12-Methoxypleiocarpine (**150**)	(35)
		Methyl chanofruticosinate (**247**)	(35)
		Methyl *N*(1)-decarbomethoxychanofruticosinate (**249**)	(35)
		Methyl 11,12-methylenedioxychanofruticosinate (**248**)	(35)
		Methyl 11,12-methylenedioxy-*N*(1)-decarbomethoxychanofruticosinate (**252**)	(35)
		Pleiocarpine (**154**)	(35)
	Stem-bark	*N*(1)-Carbomethoxy-5,22-dioxokopsane	(85)
		Dasyrachine (**264**)	(85)
		Decarbomethoxykopsifine (**230**)	(85)
		(−)-Demethylnorpleiomutine (**472**)	(85)
		(+)-Eburnamonine (**57**)	(85)
		(−)-19(*R*)-Hydroxyisoeburnamine (**81a**)	(74,85)
		(+)-19(*R*)-Hydroxyeburnamine (**81b**)	(74,85)
		16-Hydroxymethylpleiocarpamine (**42**)	(85)
		(+)-Isoeburnamine (**61**)	(85)
		Kopsamine (**166**)	(85)
		Kopsamine *N*-oxide	(85)
		Kopsifine (**229**)	(84,85)
		Kopsiflorine (**164**)	(85)
		Kopsiflorine *N*-oxide	(85)
		Kopsilongine (**165**)	(85)
		Kopsinarine (**231**)	(85)
		Kopsine (**1**)	(85)
		Kopsinine (**146**)	(85)
		Kopsinine *N*-oxide	(85)
		(+)-Kopsoffinol (**474**)	(85)
		Leuconoxine (**87**)	(85)
		11-Methoxykopsilongine (**167**)	(85)
		11-Methoxykopsilongine *N*-oxide	(85)
		12-Methoxypleiocarpine (**150**)	(85)
		11,12-Methylenedioxykopsinaline (**172**)	(85)
		11,12-Methylenedioxykopsine (**233**)	(85)
		(−)-Norpleiomutine (**471**)	(85)
		Paucidactine B (**227**)	(85)
		Pleiocarpamine (**41**)	(85)
		Pleiocarpine (**154**)	(85)
		Rhazinicine (**88**)	(84,85)
		Tetrahydroalstonine	(85)

Table II (*Continued*)

Plant[a]	Plant part	Alkaloids	References
K. deverrei	Leaves	14,15-Dihydro-10-methoxykopsinone (**158**)	(127)
L. Allorge		10-Methoxykopsinone (**156**)	(127)
(Peninsular		12-Methoxykopsinone (**157**)	(127)
Malaysia)	Stem-bark	N(1)-Carbomethoxy-17β-hydroxykopsinine (**151**)	(126)
		N(1)-Carbomethoxy-17β-hydroxy-$\Delta^{14,15}$-kopsinine (**152**)	(126)
		Deacetylakuammiline (= Rhazimol) (**31**)	(56)
		16-*epi*-Deacetylakuammiline (**33**)	(56)
		14α-Hydroxycondylocarpine (**46**)	(56)
		16-Hydroxymethylpleiocarpamine (**42**)	(56)
		N(1)-Methoxycarbonyl-11,12-methylenedioxykopsinaline (**166**)	(126)
		N(1)-Methoxycarbonyl-12-methoxykopsinaline (**165**)	(126)
		Kopsinone (**155**)	(126)
		Pleiocarpamine (**41**)	(56)
K. fruticosa	Leaves	Fruticosamine (**3**)	(14,22, 25,154)
(Roxb.) A. DC.			
(India,		Fruticosine (**2**)	(14,22, 25,154)
Indonesia,			
Malaysia)		Kopsine (**1**)	(7,8,10,14, 15,17,21, 22,154)
		Decarbomethoxykopsine	(21)
		Decarbomethoxyisokopsine	(21)
K. grandifolia D.J.	Stem-bark	*epi*-Lapidilectinol (**379**)	(186)
Middleton		Isolapidilectine A (**376**)	(186)
(*K. lapidilecta*)		Lapidilectam (**377**)	(186)
(Peninsular		Lapidilectine A (**375**)	(185,186)
Malaysia)		Lapidilectine B (**380**)	(186)
		Lapidilectinol (**378**)	(186)
		Venalstonine (**415**)	(186)
	Leaves	Lapidilectine B (**380**)	(185,186)
K. griffithii King	Leaves	Buchtienine (**475**)	(40)
& Gamble		N(1)-Carbomethoxy-11,12-dimethoxykopsinaline (**167**)	(40)
(Peninsular			
Malaysia)		N(1)-Carbomethoxy-11-hydroxy-12-methoxykopsinaline (**168**)	(40)
		(+)-Eburnamonine (**57**)	(40)
		Harmane (**19**)	(40)
		(+)-Harmicine (**21**)	(40)
		16(*R*)-19,20-*E*-Isositsirikine	(40)
		Kopsamine (**166**)	(40)
		Kopsamine N-oxide	(40)
		Kopsilongine (**165**)	(40)
		Kopsinine (**146**)	(40)
		Leuconolam (**86**)	(40)
		Leuconoxine (**87**)	(40)
		N(1)-Methoxycarbonyl-12-methoxy-$\Delta^{16,17}$-kopsinine (**214**)	(40)
		10-Demethoxy-12-methoxykopsidasinine (**328**)	(40)
		12-Methoxypleiocarpine (**150**)	(40)
		6-Oxoleuconoxine (**91**)	(90)

Table II (*Continued*)

Plant[a]	Plant part	Alkaloids	References
		Pleiocarpine (**154**)	(40)
		Rhazimol (= Deacetylakuammiline) (**31**)	(40)
		Tetrahydroalstonine	(40)
	Stem-bark	Akuammiline *N*-oxide (**34**)	(41)
		Buchtienine (**475**)	(41)
		Deacetylakuammiline (= Rhazimol) (**31**)	(41)
		(−)-Eburnamine (**58**)	(41)
		16-*epi*-Deacetylakuammiline (**33**)	(41)
		16-*epi*-Deacetylakuammiline *N*-oxide (**35**)	(41)
		Harmane (**19**)	(41)
		Kopsinine (**146**)	(41)
		Kopsinine *N*-oxide	(41)
		Leuconolam (**86**)	(41)
		Leuconoxine (**87**)	(41)
		11,12-Methylenedioxykopsinaline *N*-oxide	(41)
		Rhazinaline *N*-oxide	(41)
K. hainanensis	Stem-bark	5,22-Dioxokopsane (**271**)	(76)
Tsiang (China)		Eburnamenine[b]	(76)
		(+)-Eburnamine (**66**)	(76)
		(−)-Isoeburnamine (**69**)	(76)
		Kopsanone (**272**)	(76)
		Kopsinilam (**310**)	(76)
		Kopsinine (**146**)	(76)
		Kopsininic acid (**147**)	(76)
		Kopsinoline (= Kopsinine *N*-oxide)	(76)
		(+)-Kopsoffine (**470**)	(76)
		(+)-Tubotaiwine	(76)
K. larutensis King	Leaves	(−)-Eburnamine (**58**)	(69)
& Gamble		(+)-Eburnamonine (**57**)	(69)
(Peninsular		(+)-Eburnamonine *N*-oxide	(69)
Malaysia)		(+)-Isoeburnamine (**61**)	(69)
		(+)-Larutenine/larutensine (**73**)	(69)
	Stem-bark	(+)-Eburnamenine (**64**)	(72)
		(−)-Eburnamine (**58**)	(68,72)
		(−)-Eburnaminol (**74**)	(68)
		(+)-Eburnamonine (**57**)	(68,72)
		(+)-Eburnamonine *N*-oxide	(72)
		(−)-*O*-Ethyleburnamine (**59**)	(72)
		(+)-Isoeburnamine (**61**)	(68,72)
		Kopsinine (**146**)	(68,72)
		(+)-Larutensine/larutenine (**73**)	(68,72)
K. pauciflora	Leaves	Kinabalurine A (**5**)	(32,33)
Hook. f.		Kinabalurine B (**6**)	(33)
(Sabah,		Kinabalurine C (**7**)	(33)
Malaysian		Kinabalurine D (**8**)	(33)
Borneo)		Kinabalurine E (**9**)	(33)
		Kinabalurine F (**10**)	(33)
		Kopsamine (**166**)	(224)
		Kopsamine *N*-oxide	(224)
		Lahadinine A (**174**)	(136)
		Lahadinine B (**175**)	(136)

Table II (*Continued*)

Plant[a]	Plant part	Alkaloids	References
		11-Methoxykopsilongine (**167**)	(224)
		Paucidactine A (**226**)	(153)
		Paucidactine B (**227**)	(153)
		Paucifinine (**176**)	(136)
		Paucifinine *N*-oxide (**177**)	(136)
		Pauciflorine A (**330**)	(179)
		Pauciflorine B (**331**)	(179)
	Stem-bark	(−)-Demethylnorpleiomutine (**472**)	(80)
		(−)-Eburnamine (**58**)	(73)
		(+)-Eburnamonine (**57**)	(73)
		(+)-Isoeburnamine (**61**)	(73)
		Kopsamine *N*-oxide	(73)
		Kopsinine (**146**)	(73)
		(+)-Kopsoffine (**470**)	(80)
		(+)-Kopsoffinol (**474**)	(80)
		N(1)-Methoxycarbonyl-11,12-dimethoxykopsinaline (**167**)	(73)
		N(1)-Methoxycarbonyl-12-methoxy-$\Delta^{16,17}$-kopsinine (**214**)	(73)
		N(1)-Methoxycarbonyl-12-methoxykopsinaline (**165**)	(73)
		N(1)-Methoxycarbonyl-11,12-methylenedioxykopsinaline (**166**)	(73)
		10-Demethoxy-12-methoxykopsidasinine (**328**)	(73)
		(−)-Norpleiomutine (**471**)	(73,80)
		(+)-19-Oxoeburnamine (**80**)	(73)
		Pauciflorine A (**330**)	(224)
K. profunda (Peninsular Malaysia)	Leaves and stem-bark	Kopsinine (**146**)	(148)
		N(1)-Methoxycarbonyl-12-hydroxy-$\Delta^{16,17}$-kopsinine (**215**)	(148)
		N(1)-Methoxycarbonyl-12-methoxy-$\Delta^{16,17}$-kopsinine (**214**)	(147,148)
		N(1)-Methoxycarbonyl-12-methoxy-$\Delta^{16,17}$-kopsinine *N*-oxide (**217**)	(148)
		N(1)-Methoxycarbonyl-11,12-methylenedioxy-$\Delta^{16,17}$-kopsinine (**213**)	(147,148)
		N(1)-Methoxycarbonyl-11,12-methylenedioxy-$\Delta^{16,17}$-kopsinine *N*-oxide (**216**)	(148)
K. profunda Markgr. (*K. macrophylla*) (Peninsular Malaysia)	Leaves	Kopsilactone (**12**)	(39)
		Kopsone (**13**)	(39)
	Stem-bark	Akuammiline (**30**)	(39)
		Deacetylakuammiline (**31**)	(39)
		5,22-Dioxokopsane (**271**)	(39)
		Dregamine	(39)
		(+)-Kopsoffine (**470**)	(39)
		Norpleiomutine (**471**)	(39)
		Tabernaemontanine	(39)
K. profunda Markgr. (*K. terengganensis*) (Peninsular Malaysia)	Stem-bark	(−)-Eburnamine (**58**)	(75)
		(−)-Eburnaminol (**74**)	(75)
		(+)-Isoeburnamine (**61**)	(75)
		(+)-Larutensine (**74**)	(75)
		(+)-Quebrachamine	(75)

Table II (*Continued*)

Plant[a]	Plant part	Alkaloids	References
		Terengganensine A (**82**)	(75)
		Terengganensine B (**83**)	(75)
K. singapurensis	Leaves	16-*epi*-Akuammiline (**32**)	(55)
Ridl. (1)		16-*epi*-Deacetylakuammiline (**33**)	(55)
(Sample A)[d]		16-*epi*-Kopsinine (**153**)	(55)
(Peninsular		Kopsidine D (**221**)	(55)
Malaysia)		Kopsilongine (**165**)	(55)
		Kopsilongine *N*-oxide	(55)
		Kopsiloscine A (**180**)	(55)
		Kopsiloscine B (**181**)	(55)
		Kopsiloscine C (**182**)	(55)
		Kopsiloscine D (**183**)	(55)
		Kopsiloscine E (**184**)	(55)
		Kopsiloscine F (**185**)	(55)
		Kopsingine (**4**)	(55)
		Vincophylline (**37**)	(55)
	Stem-bark	Akuammidine	(55)
		Aspidodasycarpine (**38**)	(55)
		Aspidophylline A (**39**)	(55)
		16-*epi*-Akuammiline (**32**)	(55)
		16-*epi*-Deacetylakuammiline (**33**)	(55)
		17α-Hydroxy-$\Delta^{14,15}$-kopsinine (**178**)	(55)
		Kopsaporine (**190**)	(55)
		Kopsinganol (**197**)	(55)
		Kopsingine (**4**)	(55)
		Kopsinine (**146**)	(55)
		Leuconolam (**86**)	(55)
		Lonicerine	(55)
		Rhazinal (**90**)	(55,87)
		Rhazinilam (**84**)	(55)
		Tetrahydroalstonine	(55)
K. singapurensis	Leaves	Akuammidine	(58)
Ridl. (2)		16-*epi*-Akuammiline (**32**)	(58)
(Sample B)[d]		16-*epi*-Deacetylakuammiline (**33**)	(58)
(Peninsular		Kopsidarine (**199**)	(58)
Malaysia)		Kopsidine A (**218**)	(58)
		Kopsidine C (**220**)	(58)
		Kopsidine C *N*-oxide	(58)
		Kopsimaline F (**206**)	(58)
		Kopsinganol (**197**)	(58)
		Kopsingine (**4**)	(58)
		Kopsinitarine A (**237**)	(58)
		Kopsinitarine B (**238**)	(58)
		Mersingine A (**242**)	(58)
	Stem-bark	Akuammidine	(58)
		Aspidodasycarpine (**38**)	(58)
		Aspidophylline B (**40**)	(58)
		5,21-Dihydrorhazinilam (**85**)	(58)
		16-*epi*-Akuammiline (**32**)	(58).
		16-*epi*-Deacetylakuammiline (**33**)	(58)
		17α-Hydroxy-$\Delta^{14,15}$-kopsinine (**178**)	(58)

Table II (*Continued*)

Plant[a]	Plant part	Alkaloids	References
		Kopsiloscine C (**182**)	(58)
		Kopsiloscine G (**186**)	(58)
		Kopsinganol (**197**)	(58)
		Kopsingine (**4**)	(58)
		Kopsinine (**146**)	(58)
		Lonicerine	(58)
		Rhazinilam (**84**)	(58)
K. singapurensis	Leaves	Akuammidine	(128)
Ridl. (3) (*K.*		*N*(1)-Decarbomethoxykopsamine (**172**)	(128)
fruticosa) (195)		5,21-Dihydrorhazinilam (**85**)	(128)
(Peninsular		16-*epi*-Akuammiline (**32**)	(128)
Malaysia)		16-*epi*-Deacetylakuammiline (**33**)	(128)
		17α-Hydroxy-$\Delta^{14,15}$-kopsinine (**178**)	(128)
		Kopsiloscine G (**186**)	(128)
		Kopsiloscine J (**189**)	(128)
		Kopsimaline A (**201**)	(128)
		Kopsimaline B (**202**)	(128)
		Kopsimaline C (**203**)	(128)
		Kopsimaline D (**204**)	(128)
		Kopsimaline E (**205**)	(128)
		Kopsinicine (**196**)	(128)
		Kopsofinone (**161**)	(128)
		Mersinine A (**422**)	(198,199)
		Mersinine B (**423**)	(198,199)
		Mersinine C (**424**)	(199)
		Mersiloscine (**425**)	(198,199)
		Mersiloscine A (**426**)	(199)
		Mersiloscine B (**427**)	(199)
		Mersifoline A (**429**)	(199)
		Mersifoline B (**430**)	(199)
		Mersifoline C (**431**)	(199)
		Mersidasine A (**432**)	(199)
		Mersidasine B (**433**)	(199)
		Mersidasine C (**434**)	(199)
		Mersidasine D (**435**)	(199)
		Mersidasine E (**436**)	(199)
		Mersidasine F (**437**)	(199)
		Mersidasine G (**438**)	(199)
		Mersilongine (**446**)	(128,204)
		Mersirachine (**439**)	(128,203)
		Mersinaline (**440**)	(128,203)
		Rhazinilam (**84**)	(128)
	Stem-bark	Akuammidine	(128)
		Aspidodasycarpine (**38**)	(128)
		Burnamine	(128)
		Deacetylakuammiline (**31**)	(128)
		5,21-Dihydrorhazinilam (**85**)	(128)
		16-*epi*-Akuammiline (**32**)	(128)
		16-*epi*-Deacetylakuammiline (**33**)	(128)
		14α-Hydroxycondylocarpine (**46**)	(128)
		16-Hydroxymethylpleiocarpamine (**42**)	(128)

Table II (*Continued*)

Plant[a]	Plant part	Alkaloids	References
		Kopsiloscine H (**187**)	(128)
		Kopsiloscine I (**188**)	(128)
		Kopsinine (**146**)	(128)
		Leuconolam (**86**)	(128)
		Leuconoxine (**87**)	(128)
		Lonicerine	(128)
		Mersicarpine (**447**)	(128,205)
		Mossambine	(128)
		Picramicine	(128)
		Rhazinilam (**84**)	(128)
		Tetrahydroalstonine	(128)
K. singapurensis	Leaves	15α-Hydroxykopsinine (**149**)	(123)
Ridl. (4) (*K.*		Kopsifoline A (**409**)	(123,193, 194)
fruticosa) (195)			
(Peninsular		Kopsifoline B (**410**)	(123,193, 194)
Malaysia)			
		Kopsifoline C (**411**)	(123,193, 194)
		Kopsifoline D (**412**)	(123,194)
		Kopsifoline E (**413**)	(123,194)
		Kopsifoline F (**414**)	(123,194)
		Kopsorinine (**232**)	(123)
		Venacarpine A (**209**)	(123)
		Venacarpine B (**210**)	(123)
	Stem-bark	Akuammigine	(123)
		16-*epi*-Deacetylakuammiline (**33**)	(123)
		16-*epi*-Kopsinine (**153**)	(123)
		15α-Hydroxykopsinine (**149**)	(123)
		16-Hydroxymethylpleiocarpamine (**42**)	(123)
		Kopsanone (**272**)	(123)
		Kopsinine (**146**)	(123)
		Kopsorinine (**232**)	(123)
		Lonicerine	(123)
		Picramicine	(123)
		Pleiocarpamine (**41**)	(123)
		Venalstonine (**415**)	(123)
K. singapurensis	Leaves	5,21-Dihydrorhazinilam (**85**)	(102,152)
Ridl. (5)		11,12-Methylenedioxykopsaporine (**193**)	(102,152)
(Peninsular		Rhazinilam (**84**)	(102,152)
Malaysia)	Stem-bark	11,12-Methylenedioxykopsaporine (**193**)	(152)
		Singapurensine A (**222**)	(152)
		Singapurensine B (**223**)	(152)
		Singapurensine C (**224**)	(152)
		Singapurensine D (**225**)	(152)
K. singapurensis	Leaves	Kopsaporine (**190**)	(26,141)
Ridl. (6)		Kopsingine (**4**)	(26,141)
(Peninsular			
Malaysia)			
K. tenuis Leenh.	Leaves	Lundurine A (**384**)	(189,190)
& Steenis		Lundurine B (**385**)	(189,190)
(Sarawak,		Lundurine C (**386**)	(189,190)

Table II (*Continued*)

Plant[a]	Plant part	Alkaloids	References
Malaysian Borneo)		Lundurine D (**387**)	(189)
		Tenuiphylline (**469**)	(188)
		Tenuisine A (**381**)	(187,188, 189)
		Tenuisine B (**382**)	(187,188, 189)
		Tenuisine C (**383**)	(187,188, 189)
	Stem	Kopsinine (**146**)	(224)
		Kopsinine *N*-oxide	(224)
		Leuconoxine (**87**)	(224)
		Tetrahydroalstonine	(224)
K. teoi L. Allorge (1) (Peninsular Malaysia)	Leaves	11-Hydroxykopsingine (**191**)	(143)
		Kopsidine A (**218**)	(143,150, 151)
		Kopsidine B (**219**)	(143,150, 151)
		Kopsidine C (**220**)	(143,151)
		Kopsidine D (**221**)	(143,151)
		14,15-α-Epoxykopsingine (= Kopsimaline E) (**205**)	(143)
		Kopsinganol (**197**)	(143,150)
		Kopsingine (**4**)	(143,150)
		Kopsinitarine A (**237**)	(143,155, 156)
		Kopsinitarine B (**238**)	(143,155, 156)
		Kopsinitarine C (**239**)	(143,155, 156)
		Kopsinitarine D (**240**)	(143,156)
		Mersingine A (**242**)	(143,156, 157)
		Mersingine B (**243**)	(143,156, 157)
		12-Hydroxy-11-methoxykopsinol (**195**)	(143)
		11-Methoxykopsingine (**192**)	(143)
		Nitaphylline (**466**)	(140,143, 210)
	Stem-bark[d]	Akuammiline (**30**)	(132,143, 150)
		Aspidodasycarpine (**38**)	(143)
		16-*epi*-Deacetylakuammiline (**33**)	(143)
		17α-Hydroxy-$\Delta^{14,15}$-kopsinine (**178**)	(132,135, 140,143, 150)
		Kopsaporine (**190**)	(132,135, 143, 150)
		Kopsinganol (**197**)	(132,143, 150)
		Kopsingine (**4**)	(132,135, 143,150)

Table II (*Continued*)

Plant[a]	Plant part	Alkaloids	References
		Kopsinginine (**171**)	(132,135, 143, 150)
		Kopsinginol (**163**)	(132,143, 150)
		Kopsinol (**194**)	(132,143, 150)
		Lonicerine	(143)
		11,12-Methylenedioxykopsaporine (**193**)	(143)
		Rhazimol (= Deacetylakuammiline) (**31**)	(132,143, 150)
		Rhazinilam (**84**)	(132,143, 150)
K. teoi L. Allorge	Stem-bark[d]	Akuammiline (**30**)	(90)
(2) (Peninsular		Aspidodasycarpine (**38**)	(90)
Malaysia)		Deacetylakuammiline (= Rhazimol) (**31**)	(90)
		16-*epi*-Akuammiline (**32**)	(90)
		16-*epi*-Deacetylakuammiline (**33**)	(90)
		16-*epi*-Kopsinine (**153**)	(90)
		17α-Hydroxykopsinine (= kopsiloscine G) (**186**)	(90)
		16-Hydroxymethylpleiocarpamine (**42**)	(90)
		Kopsamine (**166**)	(90)
		Kopsidine A (**218**)	(90)
		Kopsijasminine (**207**)	(90)
		Kopsinganol (**197**)	(90)
		Kopsingine (**4**)	(90)
		Kopsinine (**146**)	(90)
		Kopsinitarine E (**241**)	(90)
		Kopsonoline (**160**)	(90)
		Leuconoxine (**87**)	(90)
		Lonicerine	(90)
		N(1)-Methoxycarbonyl-12-methoxy-$\Delta^{16,17}$-kopsinine (**214**)	(90)
		Pleiocarpamine (**41**)	(90)
		Tetrahydroalstonine	(90)
K. teoi L. Allorge	Leaves	Kopsingine (**4**)	(139)
(3) (Peninsular		11,12-Methylenedioxykopsaporine (**193**)	(139)
Malaysia)	Stem-bark	17α-Hydroxy-$\Delta^{14,15}$-kopsinine (**178**)	(139)
		Isoeburnamine	(139)
		Kopsingine (**4**)	(139)
		Lonicerine	(139)
		Rhazinilam (**84**)	(139)

[a]Classification according to Middleton (1) with original attribution in parenthesis.
[b]$[\alpha]_D$ not reported.
[c]Plant part not specified.
[d]Plant material collected at same location, but at different dates.

morphology. Middleton has acknowledged that the suggestion that *K. singapurensis* and *K. teoi* are close is indeed borne out by their morphology and has proposed further phylogenetic analysis, preferably incorporating the use of molecular data, to resolve the issue (1).

K. singapurensis (55,58) and *K. teoi* (90,132,135) were found to show a seasonal dependence of the alkaloidal composition from studies carried out on samples collected in the same location but at different times of the year.

K. singapurensis also presents another interesting feature with respect to the variation of alkaloid types within the species, in addition to the seasonal dependence mentioned. The predominant alkaloid skeleton in four separate studies of the species is aspidofractinine (e.g., kopsingine 4). However, in two other recent studies, both involving samples which were initially identified as *K. fruticosa* but subsequently amended to *K. singapurensis* by Middleton (1,195), the dominant alkaloid skeletons were very different, being mersinine in one sample (e.g., mersinine A, **422**) (198,199,203) and kopsifoline (e.g., kopsifoline A, **409**) in the other (193,194). Furthermore, both samples are characterized by the absence of kopsingine (4) (128,194), which was the major alkaloid present in previous studies of *K. singapurensis* (26,55,58,152). There appears therefore to be three morphologically similar *K. singapurensis* variants, which are nevertheless distinguishable by the occurrence of different dominant alkaloid structure types, namely, aspidofractinine, mersinine, and kopsifoline (58). The same appears to be also true for *K. profunda* where three separate studies of different samples, which are deemed to belong to the same species, revealed distinctly different alkaloidal composition. The dominant skeletons in two samples are aspidofractinine (147,148) and eburnane (75), while the third sample comprised a mixture of monoterpene, akuammiline, and bisindole alkaloids (39).

REFERENCES

[1] D. J. Middleton, *Harvard Pap. Bot.* **9**, 89 (2004).
[2] C. L. Blume, *Catalogus Batavia* (1823).
[3] F. Markgraf, *Blumea* **20**, 416 (1972).
[4] T. Sévenet, L. Allorge, B. David, K. Awang, A. H. A. Hadi, C. Kan-Fan, J. C. Quirion, F. Remy, H. Schaller, and L. E. Teo, *J. Ethnopharmacol.* **41**, 147 (1994).
[5] D. J. Middleton, *Adansonia* **27**, 287 (2005).
[6] M. Greshoff, *Ber. Deut. Chem. Ges.* **23**, 3537 (1890).
[7] A. Battacharya, A. Chatterjee, and P. K. Bose, *J. Am. Chem. Soc.* **71**, 3370 (1949).
[8] A. Battacharya, *J. Am. Chem. Soc.* **75**, 381 (1953).
[9] A. Battacharya, *Sci. Cult. (Calcutta)* **22**, 120 (1956).
[10] T. R. Govindachari, S. Rajappa, and N. Viswanathan, *J. Sci. Ind. Res. (India)* **20B**, 557 (1961).
[11] G. Spiteller, A. Chatterjee, A. Battacharya, and A. Deb, *Naturwissenschaften* **49**, 279 (1962).
[12] G. Spiteller, A. Chatterjee, A. Battacharya, and A. Deb, *Monatsh. Chem.* **93**, 1220 (1962).
[13] A. Chatterjee and A. Deb, *Sci. Cult. (Calcutta)* **28**, 195 (1962).
[14] A. R. Battersby and H. Gregory, *J. Chem. Soc.*, 22 (1963).
[15] T. R. Govindachari, K. Nagarajan, and H. Schmid, *Helv. Chim. Acta* **46**, 433 (1963).
[16] T. R. Govindachari, B. R. Pai, S. Rajappa, N. Viswanathan, W. G. Kump, K. Nagarajan, and H. Schmid, *Helv. Chim. Acta* **45**, 1146 (1962).

[17] T. R. Govindachari, B. R. Pai, S. Rajappa, N. Viswanathan, W. G. Kump, K. Nagarajan, and H. Schmid, *Helv. Chim. Acta* **46**, 572 (1963).

[18] A. Guggisberg, A. A. Gorman, B. W. Bycroft, and H. Schmid, *Helv. Chim. Acta* **52**, 76 (1969).

[19] B. M. Craven, B. Gilbert, and L. A. Paes Leme, *J. Chem. Soc., Chem. Commun.*, 955 (1968).

[20] B. M. Craven, *Acta Crystallogr., Sect. B* **B25**, 2131 (1969).

[21] A. Guggisberg, T. R. Govindachari, K. Nagarajan, and H. Schmid, *Helv. Chim. Acta* **46**, 679 (1963).

[22] A. Guggisberg, M. Hesse, W. Von Philipsborn, K. Nagarajan, and H. Schmid, *Helv. Chim. Acta* **49**, 2321 (1966).

[23] W. Klyne, R. J. Swan, A. A. Gorman, A. Guggisberg, and H. Schmid, *Helv. Chim. Acta* **51**, 1168 (1968).

[24] A. R. Battersby, J. C. Byrne, H. Gregory, and S. Popli, *J. Chem. Soc., Chem. Commun.*, 786 (1966).

[25] A. R. Battersby, J. C. Byrne, H. Gregory, and S. P. Popli, *J. Chem. Soc., C*, 813 (1967).

[26] D. W. Thomas, K. Biemann, A. K. Kiang, and R. D. Amarasingham, *J. Am. Chem. Soc.* **89**, 3235 (1967).

[27] B. Gilbert, *in: "The Alkaloids"* (R. H. F. Manske, ed.), vol. 8, p. 335. Academic Press, New York, 1965.

[28] B. Gilbert, *in: "The Alkaloids"* (R. H. F. Manske, ed.), vol. 11, p. 206. Academic Press, New York, 1968.

[29] G. A. Cordell, *in: "The Alkaloids"* (R. H. F. Manske and R. G. A. Rodrigo, eds.), vol. 17, p. 199. Academic Press, New York, 1979.

[30] A. Weissberger and E. C. Taylor, *in: "Indoles: The Monoterpenoid Indole Alkaloids"* (J. E. Saxton, ed.), p. 331. Wiley Interscience, New York, 1983.

[31] A. Weissberger and E. C. Taylor, *in: "Indoles: The Monoterpenoid Indole Alkaloids"* (J. E. Saxton, ed.), p. 357. Wiley Interscience, New York, 1994.

[32] T. S. Kam, K. Yoganathan, and W. Chen, *Nat. Prod. Lett.* **8**, 231 (1996).

[33] T. S. Kam, K. Yoganathan, and W. Chen, *J. Nat. Prod.* **60**, 673 (1997).

[34] Y. M. Chi, W. M. Yan, D. C. Chen, H. Noguchi, Y. Iitaka, and U. Sankawa, *Phytochemistry* **31**, 2930 (1992).

[35] T. S. Kam, Y. M. Choo, W. Chen, and J. X. Yao, *Phytochemistry* **52**, 959 (1999).

[36] K. Homberger and M. Hesse, *Helv. Chim. Acta* **67**, 237 (1984).

[37] E. M. Dickinson and G. Jones, *Tetrahedron* **25**, 1523 (1969).

[38] M. Streeter, G. Adolphen, and H. H. Appel, *Chem. Ind. (London)*, 1631 (1969).

[39] C. Kan-Fan, T. Sévenet, A. H. A. Hadi, M. Bonin, J. C. Quirion, and H. P. Husson, *Nat. Prod. Lett.* **7**, 283 (1995).

[40] T. S. Kam and K. M. Sim, *Phytochemistry* **47**, 145 (1998).

[41] T. S. Kam, K. M. Sim, T. Koyano, and K. Komiyama, *Phytochemistry* **50**, 75 (1999).

[42] S. Corsano and S. Algieri, *Ann. Chim. (Italy)* **50**, 75 (1960).

[43] W. R. Ashchroft, J. M. Silvio, and J. A. Joule, *Tetrahedron* **37**, 3005 (1981).

[44] A. I. Meyers and M. F. Loewe, *Tetrahedron Lett.* **25**, 2641 (1984).

[45] R. C. Bernotas and R. V. Cube, *Tetrahedron Lett.* **32**, 161 (1991).

[46] S. B. Mandal, V. S. Giri, M. S. Sabeena, and S. C. Pakrashi, *J. Org. Chem.* **53**, 4236 (1988).

[47] T. Itoh, M. Miyazaki, K. Nagata, M. Yokoya, S. Nakamura, and A. Ohsawa, *Heterocycles* **58**, 115 (2002).

[48] T. Itoh, M. Miyazaki, K. Nagata, S. Nakamura, and A. Ohsawa, *Heterocycles* **63**, 655 (2004).

[49] J. Szawkalo, S. J. Czarnocki, A. Zawadzka, K. Wojtasiewicz, A. Leniewski, J. K. Maurin, Z. Czarnocki, and J. Drabowicz, *Tetrahedron: Asymmetry* **18**, 406 (2007).

[50] H. J. Knölker and S. Agarwal, *Synlett*, 1767 (2004).

[51] K. H. Lim, K. Komiyama, and T. S. Kam, *Tetrahedron Lett.* **48**, 1143 (2007).

[52] R. Besselievre, B. P. Cosson, B. C. Das, and H. P. Husson, *Tetrahedron Lett.* **21**, 63 (1980).

[53] A. I. Meyers, T. Shoda, and M. F. Loewe, *J. Org. Chem.* **51**, 3108 (1986).

[54] K. H. Lim, O. Hiraku, K. Komiyama, T. Koyano, M. Hayashi, and T. S. Kam, *J. Nat. Prod.* **70**, 1302 (2007).

[55] G. Subramaniam, O. Hiraku, M. Hayashi, T. Koyano, K. Komiyama, and T. S. Kam, *J. Nat. Prod.* **70**, 1783 (2007).

[56] C. Kan, J. R. Deverre, T. Sévenet, J. C. Quirion, and H. P. Husson, *Nat. Prod. Lett.* **7**, 275 (1995).
[57] Y. Ahmad, K. Fatima, Atta-ur-Rahman, J. L. Occolowitz, B. A. Solheim, J. Clardy, R. L. Garnick, and P. W. Le Quesne, *J. Am. Chem. Soc.* **99**, 1943 (1977).
[58] G. Subramaniam, M. Hayashi, T. Koyano, O. Hiraku, K. Komiyama, and T. S. Kam, *Helv. Chim. Acta* **91**, 930 (2008).
[59] H. Arens, H. O. Borbe, B. Ulbrich, and J. Stockigt, *Planta Med.* **46**, 210 (1982).
[60] J. Kobayashi, M. Sekiguchi, S. Shimamoto, H. Shigemori, H. Ishiyama, and A. Ohsaki, *J. Org. Chem.* **67**, 6449 (2002).
[61] K. H. Lim and T. S. Kam, *Helv. Chim. Acta* **90**, 31 (2007).
[62] K. H. Lim, Y. Y. Low, and T. S. Kam, *Tetrahedron Lett.* **47**, 5037 (2006).
[63] J. P. Kutney, V. R. Nelson, and D. C. Wigfield, *J. Am. Chem. Soc.* **91**, 4278 (1969).
[64] J. P. Kutney, V. R. Nelson, and D. C. Wigfield, *J. Am. Chem. Soc.* **91**, 4279 (1969).
[65] A. Ahond, A. Cavé, C. Kan-Fan, Y. Langlois, and P. Potier, *J. Chem. Soc., Chem. Commun.*, 517 (1970).
[66] A. I. Scott, C. L. Yeh, and D. Greenslade, *J. Chem. Soc. Chem., Commun.*, 947 (1978).
[67] T. S. Kam, *in:* "Alkaloids: Chemical and Biological Perspectives" (S. W. Pelletier, ed.), vol. 14, p. 285. Pergamon, Amsterdam, 1999.
[68] K. Awang, M. Pais, T. Sévenet, H. Schaller, A. M. Nasir, and A. H. A. Hadi, *Phytochemistry* **30**, 3164 (1991).
[69] T. S. Kam, P. S. Tan, and C. H. Chuah, *Phytochemistry* **31**, 2936 (1992).
[70] M. Lounasmaa and A. Tolvanen, *in:* "The Alkaloids" (G. A. Cordell, ed.), vol. 42, p. 104. Academic Press, New York, 1992.
[71] M. Lounasmaa and E. Karvinen, *Heterocycles* **36**, 751 (1993).
[72] T. S. Kam, P. S. Tan, and W. Chen, *Phytochemistry* **33**, 921 (1993).
[73] T. S. Kam, L. Arasu, and K. Yoganathan, *Phytochemistry* **43**, 1385 (1996).
[74] T. S. Kam, G. Subramaniam, and W. Chen, *Nat. Prod. Lett.* **12**, 293 (1998).
[75] S. Uzir, A. M. Mustapha, A. H. A. Hadi, K. Awang, C. Wiart, J. F. Gallard, and M. Pais, *Tetrahedron Lett.* **38**, 1571 (1997).
[76] J. Zhu, A. Guggisberg, and M. Hesse, *Planta Med.*, 63 (1986).
[77] X. Z. Feng, C. Kan, P. Potier, S. K. Kan, and M. Lounasmaa, *Planta Med.* **48**, 280 (1983).
[78] Y. L. Zhou, Z. H. Huang, L. Y. Huang, J. P. Zhu, C. M. Li, and G. L. Wu, *Acta Chim. Sinica (English Edition)* **1**, 82 (1985).
[79] X. Z. Feng, C. Kan, H. P. Husson, P. Potier, S. K. Kan, and M. Lounasmaa, *J. Nat. Prod.* **47**, 117 (1984).
[80] C. Kan-Fan, T. Sévenet, H. P. Husson, and K. C. Chan, *J. Nat. Prod.* **48**, 124 (1985).
[81] H. H. A. Linde, *Helv. Chim. Acta* **48**, 1822 (1965).
[82] A. Banerji, P. L. Majumder, and A. Chatterjee, *Phytochemistry* **9**, 1491 (1970).
[83] K. T. De Silva, A. H. Ratcliffe, G. F. Smith, and G. N. Smith, *Tetrahedron Lett.*, 913 (1972).
[84] T. S. Kam and G. Subramaniam, *Nat. Prod. Lett.* **11**, 131 (1998).
[85] T. S. Kam, G. Subramaniam, and W. Chen, *Phytochemistry* **51**, 159 (1999).
[86] N. Aimi, N. Uchida, N. Ohya, H. Hosokawa, H. Takayama, S. Sakai, L. A. Mendonza, L. Polz, and J. Stöckigt, *Tetrahedron Lett.* **32**, 4949 (1991).
[87] T. S. Kam, Y. M. Tee, and G. Subramaniam, *Nat. Prod. Lett.* **12**, 307 (1998).
[88] B. David, T. Sévenet, O. Thoison, K. Awang, M. Pais, M. Wright, and D. Guenard, *Bioorg. Med. Chem. Lett.* **7**, 2155 (1997).
[89] F. Abe and T. Yamauchi, *Phytochemistry* **35**, 169 (1994).
[90] S. H. Lim, K. M. Sim, Z. Abdullah, O. Hiraku, M. Hayashi, K. Komiyama, and T. S. Kam, *J. Nat. Prod.* **70**, 1380 (2007).
[91] A. H. Ratcliffe, G. F. Smith, and G. N. Smith, *Tetrahedron Lett.*, 5179 (1973).
[92] J. A. Johnson and D. Sames, *J. Am. Chem. Soc.* **122**, 6321 (2000).
[93] J. A. Johnson, N. Li, and D. Sames, *J. Am. Chem. Soc.* **124**, 6900 (2002).
[94] A. E. Shilov and G. B. Shu'pin, *Chem. Rev.* **97**, 2879 (1997).
[95] P. Magnus and T. Rainey, *Tetrahedron* **57**, 8647 (2001).
[96] A. L. Bowie Jr., C. C. Hughes, and D. Trauner, *Org. Lett.* **7**, 5207 (2005).
[97] Z. Liu, A. S. Wasmuth, and S. G. Nelson, *J. Am. Chem. Soc.* **128**, 10352 (2006).

[98] M. G. Banwell, A. J. Edwards, K. A. Jolliffe, J. A. Smith, E. Hamel, and P. Verdier-Pinard, *Org. Biomol. Chem.* **1**, 296 (2003).

[99] M. Banwell, A. Edwards, J. Smith, E. Hamel, and P. Verdier-Pinard, *J. Chem. Soc., Perkin Trans. 1*, 1497 (2000).

[100] D. J. Abraham, R. D. Rosenstein, R. L. Lyon, and H. H. S. Fong, *Tetrahedron Lett.*, 909 (1972).

[101] O. Baudoin, F. Claveau, S. Thoret, A. Herrbach, D. Guenard, and F. Gueritte, *Bioorg. Med. Chem.* **10**, 3395 (2002).

[102] O. Thoison, D. Guenard, T. Sévenet, C. Kan-Fan, J. C. Quirion, H. P. Husson, J. R. Deverre, K. C. Chan, and P. Potier, *C. R. Acad. Sc. Paris II* **304**, 157 (1987).

[103] B. David, T. Sévenet, M. Morgat, D. Guenard, A. Moisand, Y. Tollon, O. Thoison, and M. Wright, *Cell Motil. Cytoskeleton* **28**, 317 (1994).

[104] O. Baudoin, D. Guenard, and F. Gueritte, *Mini-Rev. Org. Chem.* **1**, 333 (2004).

[105] O. Baudoin and F. Gueritte, *in:* "Studies in Natural Products Chemistry" (Atta-ur-Rahman, ed.), vol. 29, p. 355. Elsevier, Amsterdam, 2003.

[106] A. Décor, D. Bellocq, O. Thoison, N. Lekieffre, A. Chiaroni, J. Ouazzani, T. Cresteil, F. Gueritte, and O. Baudoin, *Bioorg. Med. Chem.* **14**, 1558 (2006).

[107] J. P. Alazard, C. Millet-Paillusson, O. Boye, D. Guenard, A. Chiaroni, C. Riche, and C. Thal, *Bioorg. Med. Chem. Lett.* **1**, 725 (1991).

[108] J. P. Alazard, C. Millet-Paillusson, D. Guenard, and C. Thal, *Bull. Soc. Chim. Fr.* **133**, 251 (1996).

[109] J. Lévy, M. Soufyane, C. Mirand, M. Doe de Maindreville, and D. Royer, *Tetrahedron: Asymmetry* **8**, 4127 (1997).

[110] C. Dupont, D. Guenard, L. Tchertanov, S. Thoret, and F. Gueritte, *Bioorg. Med. Chem.* **7**, 2961 (1999).

[111] C. Dupont, D. Guenard, C. Thal, S. Thoret, and F. Gueritte, *Tetrahedron Lett.* **41**, 5853 (2000).

[112] A. Décor, B. Monse, M. T. Martin, A. Chiaroni, S. Thoret, D. Guenard, F. Gueritte, and O. Baudoin, *Bioorg. Med. Chem.* **14**, 2314 (2006).

[113] E. Pasquinet, P. Rocca, S. Richalot, F. Gueritte, D. Guenard, A. Godard, F. Marsais, and G. Queguiner, *J. Org. Chem.* **66**, 2654 (2001).

[114] A. L. Bonneau, N. Robert, C. Hoarau, O. Baudoin, and F. Marsais, *Org. Biomol. Chem.* **5**, 175 (2007).

[115] C. Pascal, J. Dubois, D. Guenard, and F. Gueritte, *J. Org. Chem.* **63**, 6414 (1998).

[116] C. Pascal, J. Dubois, D. Guenard, L. Tchertanov, S. Thoret, and F. Gueritte, *Tetrahedron* **54**, 14737 (1998).

[117] O. Baudoin, D. Guenard, and F. Gueritte, *J. Org. Chem.* **65**, 9268 (2000).

[118] O. Baudoin, M. Cesario, D. Guenard, and F. Gueritte, *J. Org. Chem.* **67**, 1199 (2002).

[119] A. Herrbach, A. Marinetti, O. Baudoin, D. Guenard, and F. Gueritte, *J. Org. Chem.* **68**, 4897 (2003).

[120] W. D. Crow and M. Michael, *Aust. J. Chem.* **8**, 129 (1955).

[121] J. Zhu, A. Guggisberg, and M. Hesse, *Planta Med.* **1**, 63 (1986).

[122] H. Zhou, H. P. He, N. C. Kong, Y. H. Wang, X. D. Liu, and X. J. Hao, *Helv. Chim. Acta* **89**, 515 (2006).

[123] T. S. Kam and Y. M. Choo, *Phytochemistry* **65**, 2119 (2004).

[124] N. Langlois, L. Diatta, and R. T. Andriamialisoa, *Phytochemistry* **18**, 467 (1979).

[125] M. Zeches, J. Lounkokobi, B. Richard, M. Plat, L. Le Men-Olivier, T. Sévenet, and J. Pusset, *Phytochemistry* **23**, 171 (1984).

[126] C. Kan-Fan, S. K. Kan, J. R. Deverre, J. C. Quirion, H. P. Husson, Y. L. Zhou, and K. C. Chan, *J. Nat. Prod.* **51**, 703 (1988).

[127] M. Do Carmo Carreiras, C. Kan, J. R. Deverre, A. H. A. Hadi, J. C. Quirion, and H. P. Husson, *J. Nat. Prod.* **51**, 806 (1988).

[128] G. Subramaniam, O. Hiraku, M. Hayashi, T. Koyano, K. Komiyama, and T. S. Kam, *J. Nat. Prod.* **71**, 53 (2008).

[129] P. Le Menez, J. Sapi, and N. Kunesch, *J. Org. Chem.* **54**, 3216 (1989).

[130] M. Dufour, J. C. Gramain, H. P. Husson, M. E. Sinibaldi, and Y. Troin, *J. Org. Chem.* **55**, 5483 (1990).

[131] M. Dufour, J. C. Gramain, H. P. Husson, M. E. Sinibaldi, and Y. Troin, *Tetrahedron Lett.* **30**, 3429 (1989).

[132] T. S. Kam and K. Yoganathan, *Phytochemistry* **42**, 539 (1996).

[133] J. J. Zheng, Y. L. Zhou, and Z. H. Huang, *Acta Chim. Sinica*, 168 (1989).

[134] J. H. Wu, P. J. Zheng, J. J. Zhen, and Y. L. Zhou, *J. Struct. Chem.* **6**, 47 (1987).

[135] T. S. Kam, K. Yoganathan, C. H. Chuah, and W. Chen, *Phytochemistry* **32**, 1343 (1993).

[136] T. S. Kam and K. Yoganathan, *Phytochemistry* **46**, 785 (1997).

[137] M. Zeches, T. Ravao, B. Richard, G. Massiot, L. Le Men-Olivier, and R. Verpoorte, *J. Nat. Prod.* **50**, 714 (1987).

[138] T. S. Kam, G. Subramaniam, K. M. Sim, K. Yoganathan, T. Koyano, M. Toyoshima, M. C. Rho, M. Hayashi, and K. Komiyama, *Bioorg. Med. Chem. Lett.* **8**, 2769 (1998).

[139] T. Varea, C. Kan, F. Remy, T. Sévenet, J. C. Quirion, H. P. Husson, and A. H. A. Hadi, *J. Nat. Prod.* **56**, 2166 (1993).

[140] T. S. Kam, T. M. Lim, G. Subramaniam, Y. M. Tee, and K. Yoganathan, *Phytochemistry* **50**, 171 (1999).

[141] A. K. Kiang and R. D. Amarasingham, *Proceedings, Symposium on Phytochemistry*, p. 165. Kuala Lumpur, Malaysia (1957); *Chem. Abstr.* **53**, 14131 (1959).

[142] R. D. Amarasingham, M. Sc. Thesis, University of Malaya, 1961.

[143] T. S. Kam, K. Yoganathan, and S. L. Mok, *Phytochemistry* **46**, 789 (1997).

[144] S. L. Mok, K. Yoganathan, T. M. Lim, and T. S. Kam, *J. Nat. Prod.* **61**, 328 (1998).

[145] N. Ruangrungsi, K. Likhitwitayawuid, V. Jongbunprasert, D. Ponglux, N. Aimi, K. Ogata, M. Yasuoka, J. Haginiwa, and S. I. Sakai, *Tetrahedron Lett.* **28**, 3679 (1987).

[146] K. Homberger and M. Hesse, *Helv. Chim. Acta* **65**, 2548 (1982).

[147] T. S. Kam and P. S. Tan, *Phytochemistry* **29**, 2321 (1990).

[148] T. S. Kam and P. S. Tan, *Phytochemistry* **39**, 469 (1995).

[149] C. H. Chuah, T. S. Kam, and S. W. Ng, *Zeitschrift f. Krist.* **213**, 163 (1998).

[150] T. S. Kam, K. Yoganathan, and C. H. Chuah, *Tetrahedron Lett.* **34**, 1819 (1993).

[151] T. S. Kam, K. Yoganathan, and C. H. Chuah, *Phytochemistry* **45**, 623 (1997).

[152] K. Awang, O. Thoison, A. H. A. Hadi, M. Pais, and T. Sévenet, *Nat. Prod. Lett.* **3**, 283 (1993).

[153] T. S. Kam, K. Yoganathan, and W. Chen, *Tetrahedron Lett.* **37**, 3603 (1996).

[154] R. P. Glover, K. Yoganathan, and M. S. Butler, *Magn. Reson. Chem.* **43**, 483 (2005).

[155] T. S. Kam, K. Yoganathan, and C. H. Chuah, *Tetrahedron Lett.* **35**, 4457 (1994).

[156] T. S. Kam, K. Yoganathan, and W. Chen, *J. Nat. Prod.* **59**, 1109 (1996).

[157] K. Yoganathan, W. H. Wong, and T. S. Kam, *Nat. Prod. Lett.* **5**, 309 (1995).

[158] W. S. Chen, S. H. Li, A. Kirfel, G. Will, and E. Breitmaier, *Liebigs Ann. Chem.*, 1886 (1981).

[159] T. S. Kam, P. S. Tan, P. Y. Hoong, and C. H. Chuah, *Phytochemistry* **32**, 489 (1993).

[160] K. H. Lim and T. S. Kam, *Phytochemistry* **69**, 558 (2008).

[161] K. Husain, I. Jantan, N. Kamaruddin, I. M. Said, N. Aimi, and H. Takayama, *Phytochemistry* **57**, 603 (2001).

[162] K. Husain, I. Jantan, I. M. Said, N. Aimi, and H. Takayama, *J. Asian Nat. Prod. Res.* **5**, 63 (2003).

[163] T. R. Govindachary, K. Nagarajan, and H. Schmid, *Helv. Chim. Acta* **45**, 433 (1963).

[164] T. S. Kam, T. M. Lim, Y. M. Choo, and G. Subramaniam, *Tetrahedron Lett.* **39**, 5823 (1998).

[165] T. S. Kam, T. M. Lim, and Y. M. Choo, *Tetrahedron* **55**, 1457 (1999).

[166] C. Exon, T. Gallagher, and P. Magnus, *J. Am. Chem. Soc.* **105**, 4739 (1983).

[167] T. Gallagher, P. Magnus, and J. C. Huffman, *J. Am. Chem. Soc.* **105**, 4750 (1983).

[168] P. Magnus, T. Gallagher, P. Brown, and P. Pappalardo, *Acc. Chem. Res.* **17**, 35 (1984).

[169] T. Gallagher and P. Magnus, *J. Am. Chem. Soc.* **105**, 2086 (1983).

[170] P. Magnus, T. Gallagher, P. Brown, and J. C. Huffman, *J. Am. Chem. Soc.* **106**, 2105 (1984).

[171] P. Magnus, T. Katoh, I. R. Matthews, and J. C. Huffman, *J. Am. Chem. Soc.* **111**, 6707 (1989).

[172] P. Magnus, I. R. Matthews, J. Schultz, R. Waditschatka, and J. C. Huffman, *J. Org. Chem.* **53**, 5772 (1988).

[173] P. Magnus and P. Brown, *J. Chem. Soc., Chem. Commun.*, 184 (1985).

[174] M. E. Kuehne and P. J. Seaton, *J. Org. Chem.* **50**, 4790 (1985).

[175] E. Wenkert and M. J. Pestchanker, *J. Org. Chem.* **53**, 4875 (1988).

[176] E. Wenkert, K. Orito, and D. P. Simmons, *J. Org. Chem.* **48**, 5006 (1983).

[177] M. Ogawa, Y. Kitagawa, and M. Natsume, *Tetrahedron Lett.* **28**, 3985 (1987).

[178] M. O. Hamburger, G. A. Cordell, K. Likhitwitayawuid, and N. Ruangrungsi, *Phytochemistry* **27**, 2719 (1988).

[179] T. S. Kam, K. Yoganathan, T. Koyano, and K. Komiyama, *Tetrahedron Lett.* **37**, 5765 (1996).

[180] P. Magnus, L. Gazzard, L. Hobson, A. H. Payne, and V. Lynch, *Tetrahedron Lett.* **40**, 5135 (1999).

[181] P. Magnus, L. A. Hobson, N. Westlund, and V. Lynch, *Tetrahedron Lett.* **42**, 993 (2001).

[182] P. Magnus, L. Gazzard, L. Hobson, A. H. Payne, T. J. Rainey, N. Westlund, and V. Lynch, *Tetrahedron* **58**, 3423 (2002).

[183] M. E. Kuehne and Y. L. Li, *Org. Lett.* **1**, 1749 (1999).

[184] M. E. Kuehne, Y. L. Li, and C. Q. Wei, *J. Org. Chem.* **65**, 6434 (2000).

[185] K. Awang, T. Sévenet, A. H. A. Hadi, B. David, and M. Pais, *Tetrahedron Lett.* **33**, 2493 (1992).

[186] K. Awang, T. Sévenet, M. Pais, and A. H. A. Hadi, *J. Nat. Prod.* **56**, 1134 (1993).

[187] T. S. Kam, K. Yoganathan, and H. Y. Li, *Tetrahedron Lett.* **37**, 8811 (1996).

[188] T. S. Kam, K. Yoganathan, H. Y. Li, and N. Harada, *Tetrahedron* **53**, 12661 (1997).

[189] T. S. Kam, K. H. Lim, K. Yoganathan, M. Hayashi, and K. Komiyama, *Tetrahedron* **60**, 10739 (2004).

[190] T. S. Kam, K. Yoganathan, and C. H. Chuah, *Tetrahedron Lett.* **36**, 759 (1995).

[191] W. H. Pearson, Y. Mi, I. Y. Lee, and P. Stoy, *J. Am. Chem. Soc.* **123**, 6724 (2001).

[192] W. H. Pearson, I. Y. Lee, Y. Mi, and P. Stoy, *J. Org. Chem.* **69**, 9109 (2004).

[193] T. S. Kam and Y. M. Choo, *Tetrahedron Lett.* **44**, 1317 (2003).

[194] T. S. Kam and Y. M. Choo, *Helv. Chim. Acta* **87**, 991 (2004).

[195] Personal communication from D. J. Middleton on November 8, 2004. Although Dr. Middleton amended the identity of these specimens from the provisional *K. fruticosa* that he provided earlier to *K. singapurensis* upon completion of his review (1), he unfortunately did not update the identifications in his database. Consequently these specimens (K627, K649, K650, K651) appear in the review as *K. fruticosa* instead of the correct *K. singapurensis*.

[196] X. Hong, S. France, J. M. Mejia-Oneto, and A. Padwa, *Org. Lett.* **8**, 5141 (2006).

[197] X. Hong, S. France, and A. Padwa, *Tetrahedron* **63**, 5962 (2007).

[198] T. S. Kam, G. Subramaniam, and T. M. Lim, *Tetrahedron Lett.* **42**, 5977 (2001).

[199] G. Subramaniam, Y. M. Choo, O. Hiraku, K. Komiyama, and T. S. Kam, *Tetrahedron* **64**, 1397 (2008).

[200] G. Hugel and J. Levy, *J. Org. Chem.* **51**, 1594 (1986).

[201] G. Hugel and J. Levy, *J. Org. Chem.* **49**, 3275 (1984).

[202] A. Richardson Jr., *J. Org. Chem.* **28**, 2581 (1963).

[203] G. Subramaniam and T. S. Kam, *Tetrahedron Lett.* **48**, 6677 (2007).

[204] T. S. Kam and G. Subramaniam, *Tetrahedron Lett.* **45**, 3521 (2004).

[205] T. S. Kam, G. Subramaniam, K. H. Lim, and Y. M. Choo, *Tetrahedron Lett.* **45**, 5995 (2004).

[206] U. Anthoni, K. Bock, L. Chevolot, C. Larsen, P. H. Nielsen, and C. Christophersen, *J. Org. Chem.* **52**, 5638 (1987).

[207] K. H. Lim and T. S. Kam, *Org. Lett.* **8**, 1733 (2006).

[208] K. H. Lim and T. S. Kam, *Tetrahedron Lett.* **47**, 8653 (2006).

[209] T. S. Kam and Y. M. Choo, *in:* "The Alkaloids" (G. A. Cordell, ed.), vol. 63, p. 181. Academic Press, Amsterdam, 2006.

[210] T. S. Kam and K. Yoganathan, *Nat. Prod. Lett.* **10**, 69 (1997).

[211] J. Stöckigt, K. H. Pawelka, T. Tanahashi, B. Danieli, and W. E. Hull, *Helv. Chim. Acta* **66**, 2525 (1983).

[212] A. Ahond, J. Maurice-Marie, N. Langlois, G. Lukacs, P. Potier, P. Rasoanaivo, M. Sangare, N. Neuss, M. Plat, J. Le Men, E. W. Hagaman, and E. Wenkert, *J. Am. Chem. Soc.* **96**, 633 (1974).

[213] G. Subramaniam, Biologically Active Indole and Bisindole Alkaloids from *Kopsia*, Ph.D. Thesis, University of Malaya, Kuala Lumpur, 2004.

[214] M. Azoug, A. Loukaci, B. Richard, J. M. Nuzillard, C. Moreti, M. Zeches-Hanrot, and L. Le Men-Olivier, *Phytochemistry* **39**, 1223 (1995).

[215] G. H. Tan, T. M. Lim, and T. S. Kam, *Tetrahedron Lett.* **36**, 1327 (1995).

[216] T. S. Kam, T. M. Lim, and G. H. Tan, *J. Chem. Soc., Perkin Trans. 1*, 1594 (2001).

[217] C. Kan-Fan, J. C. Quirion, and H. P. Husson, *Nat. Prod. Lett.* **3**, 291 (1993).

[218] T. S. Kam, T. M. Lim, and G. H. Tan, *Heterocycles* **51**, 249 (1999).

[219] N. G. Bisset and A. Chatterjee, *Sci. Cult. (Calcutta)* **28**, 592 (1962).

[220] A. Bhattacharya, *Sci. Cult. (Calcutta)* **18**, 283 (1952).

[221] W. D. Crow and M. Michael, *Aust. J. Chem.* **15**, 130 (1962).

[222] Y. L. Zhou, Z. H. Huang, L. Y. Huang, J. P. Zhu, C. M. Li, and G. L. Wu, *Huaxue Xuebao* **42**, 1315 (1984); *Chem. Abstr.* **102**, 128817 (1984).

[223] T. T. H. Do, N. H. Nguyen, and Q. C. Nguyen, *Tap Chi Hoa Hoc* **45**, 152 (2007).

[224] K. Yoganathan, Alkaloids from Malaysian *Kopsia*. Chemistry and Bioactivity, Ph.D. Thesis, University of Malaya, Kuala Lumpur, 1997.

## CHAPTER 2

Alkaloids with Antiprotozoal Activity

Edison J. Osorio[1,*]**, Sara M. Robledo**[2] **and Jaume Bastida**[3]

I. INTRODUCTION

Protozoan parasites are among the most common chronic infections that occur primarily in rural and poor urban areas of tropical and subtropical regions

[1] Grupo de Investigación en Sustancias Bioactivas, Facultad de Química-Farmacéutica, Universidad de Antioquia, A. A. 1226, Medellín, Colombia

[2] Programa de Estudio y Control de Enfermedades Tropicales, Facultad de Medicina, Universidad de Antioquia, Medellín, Colombia

[3] Departament de Productes Naturals, Biologia Vegetal i Edafologia, Facultat de Farmàcia, Universitat de Barcelona, 08028 Barcelona, Spain

* Corresponding author.
E-mail address: josorio48@yahoo.com (E.J. Osorio).

The Alkaloids, Volume 66
ISSN: 1099-4831, DOI 10.1016/S1099-4831(08)00202-2

around the world. They are responsible for a large number of severe and widespread diseases including malaria, leishmaniasis, Chagas' disease, and sleeping sickness. These diseases, with the exception of malaria, belong to the group of Neglected Tropical Diseases (NTDs), since they are strongly linked with poverty and there is a lack of commercial markets for potential drugs. Like other NTDs, these diseases affect individuals throughout their lives, causing a high degree of morbidity and physical disability and, in certain cases, gross disfigurement. Patients can face social stigmatization and abuse as a result of contracting these diseases. The disability and the poverty associated with these diseases constitute large burdens on the health and economic development of low-income and middle-income countries in Africa, Asia, and the Americas (1). Strategies to control these diseases are based on surveillance, early diagnosis, vector control, and treatment (2). At present, there are only a few drugs on the market to treat these parasitic diseases, and they are not universally available in the affected areas. In addition, current drug treatments are unsatisfactory due to drug resistance, inefficiency, toxicity, prolonged treatment schedules, high cost, etc. Therefore, there is an urgent need for new treatments, which are safe, effective, cheap, and easy-to-administer, and for new lead compounds with novel mechanisms of action.

Living organisms are commonly used as natural sources of novel structures for the discovery and development of new drugs, since they contain countless molecules with a great variety of structures and pharmacological activities (3). Out of 1010 new active substances approved as drugs for medical conditions by regulatory agencies during the past 25 years, 490 (48.5%) were from a natural origin (4). The diversity of natural products with antiprotozoal activities has been illustrated in several reviews covering molecules that are mainly active against malaria, leishmaniasis, Chagas' disease, or sleeping sickness (5–11). This chapter deals exclusively and thoroughly with the alkaloid natural products that are particularly active against *Leishmania* sp., *Trypanosoma cruzi*, *Trypanosoma brucei*, and *Plasmodium falciparum*. The alkaloids with antiprotozoal activity are grouped according to their structures or origin in several categories: quinoline and isoquinoline alkaloids, indole alkaloids, steroidal alkaloids, alkaloids from marine organisms, and other alkaloids. A discussion on structure–antiprotozoal activity relationships is included, and the mechanism of action of several of these metabolites is described. Recent developments, as well as new experimental strategies in discovery and development of antiprotozoal compounds, are also discussed. Some of the *in vitro* antitrypanosomal or cytotoxic activities reported in the literature have been transformed into molar concentrations (mM, μM, or nM) to allow a better comparison, independent of their molecular weight. Nevertheless, direct comparison remains complex due to the different assay procedures used in various laboratories.

Alkaloids are one of the most important groups of natural products, already providing many drugs for human use (12). Although they can be seriously toxic for the host, alkaloid-containing plants and their biosynthesized alkaloids have a remarkable potential to provide pharmaceutical and biological agents contributing to the development of future antiparasitic drugs.

II. TROPICAL DISEASES CAUSED BY PROTOZOAN PARASITES

Leishmaniasis, Chagas' disease (or American trypanosomiasis), malaria, and African trypanosomiasis (or sleeping sickness) are vector-borne infectious diseases caused by protozoan parasites. Despite considerable control efforts, they are among the most prevalent parasitic diseases worldwide with a heavy social and economic burden (13). Those at greatest risk are populations that are poor or beyond the reach of adequate medical attention. Although these diseases are commonly associated with poverty, they are also a cause of hardship and a major hindrance to economic development (14). As a result of poor market incentives for companies to carry out research to develop new drugs, they are called "neglected diseases" (15). In most endemic countries, official Ministry of Health policy is to provide free treatment to all patients, but this is often unfeasible because the required drugs are in limited supply, especially in rural areas where the diseases mostly occur. Consequently, self-help patient organizations are often used to provide diagnosis and treatment. Thus, people afflicted by neglected diseases are vulnerable to violations of their basic human rights, such as access to health care and essential medicines (16).

The health impact of these diseases is measured by severe and permanent disabilities and deformities in affected people. In addition to the physical and psychological suffering they cause, these diseases inflict an enormous economic burden on affected communities owing to lost productivity and the high costs associated with long-term care, which in turn contributes to the entrenched cycle of poverty and ill-health in affected populations (13). Overall global prevalence is approximately 550 million cases, and close to 2.7 billion people living in endemic areas are at risk of contracting any of these diseases. In total, there are about 280 million new cases each year causing important health and socio-economic problems where these diseases are endemic. Chemotherapy remains one of the key measures used to control the intolerable burden of protozoan parasitic and other tropical diseases, but most of the available drugs are no longer effective due to drug resistance. Moreover, some of those that are still effective suffer from problems associated with toxicity, compliance, and high cost, resulting in an urgent need for new drugs.

A. Leishmaniasis

Leishmaniasis is a group of clinical diseases suffered by millions around the world, and affecting 88 countries in Africa, Asia, Europe, and America, 72 of which are developing countries and 13 are among the least developed. The spectrum of the disease is divided into three major syndromes: cutaneous (CL), mucocutaneous (MCL), and visceral leishmaniasis (VL). Annual incidence is estimated at 1.5 million cases of the cutaneous forms (CL and MCL) and 500,000 cases of the visceral form (VL), resulting in approximately 51,000 deaths per year. Overall prevalence is 12 million people and the population at risk is 350 million. However, this estimated global burden of disease is believed to be inaccurate partly due to the passive case detection data used to estimate the disease prevalence in many endemic countries. Apparently, for each symptomatic case, there are estimated to be 10 asymptomatic infections. The total burden of Disability Adjusted Life Years (DALY) is 2.09 million, with 840,000 for women and 1.25 million for men (13,17–19). The Special Program for Research and Training in Tropical Diseases (TDR) has classified Leishmaniasis as a group of emerging or uncontrolled diseases (Category I) (13,20).

Leishmaniasis is produced by at least 17 species of the protozoan *Leishmania* (order Kinetoplastida, family Trypanosomatidae). *L. donovani* and *L. infantum* are the causative agents of VL, while *L. major*, *L. tropica*, *L. aethiopica*, *L. braziliensis*, *L. panamensis*, *L. amazonensis*, and *L. mexicana* produce CL (17–19). *Leishmania* spp. are transmitted by sandflies of the genera *Phlebotomus* (Old World) and *Lutzomyia* (New World). The parasite exists in two morphological forms: the amastigote (aflagellated form) that proliferates intracellularly in the mammalian macrophages, and the promastigote (extracellular flagellated form) that proliferates in the gut of its sandfly vector and in the acellular culture medium. Infection occurs when an infected sandfly regurgitates infective promastigotes into the blood while feeding. The promastigotes are phagocytized by macrophages and transformed into amastigotes. The life cycle is continued when a sandfly vector feeding on the blood of an infected individual or an animal reservoir host ingests the macrophages infected with amastigotes (13,17–20).

Chemotherapy for leishmaniasis is still deficient. Most of the drugs have one or more limitations such as long-term administration, unaffordable cost, toxicity, or even worse, inefficacy due to development of resistance in the parasite (13,17–24). Pentavalent antimonials (Sb^V) have become the drug of choice for the treatment of all types of leishmaniasis. Pentostam® (sodium stibogluconate) (1) and Glucantime® (meglumine antimoniate) (2), are non-covalent chelates of Sb^V with improved solubility and uptake properties. However, they require long courses of treatment with parenteral administration, produce toxic side effects, and show variable efficacy. Increasing resistance to antimonials has been documented in several regions, but particularly in northeast India (25). Although these drugs constitute the main antileishmanial chemotherapy and have been used for over 50 years, information about their chemistry or precise mode of action and the identity of the biologically active components is uncertain. The second line drugs are pentamidine isethionate (3), and amphotericin B (4) (21,22). Pentamidine toxicity has mainly been associated with high cumulative doses (26). Toxic side effects of 3 include a sensation of burning, headache, tightness of the chest, dizziness, nausea, vomiting, and hypotension. Other serious side effects of pentamidine are hypoglycemia, hyperglycemia, and acute pancreatitis leading to diabetes (27). Common side effects of amphotericin B (4) include anaphylaxis, thrombocytopenia, flushing, generalized pain, chills, fever, phlebitis, anemia, convulsions, anorexia, decreased renal tubular and glomerular function, and hypokalemia. In order to increase the therapeutic index of amphotericin B, and to reduce toxicity, a lipid formulation of amphotericin B was developed (28,29).

Among the new drugs currently under clinical evaluation are miltefosine (5), an antitumor agent being used in India for the oral treatment of VL, and which is under clinical trials in different countries for the treatment of CL (21,30,31). Although miltefosine has proven effective, its long half-life could lead to the rapid emergence of resistance. In addition, miltefosine has demonstrated teratogenicity in animal studies leading to its contraindication in pregnancy, and recommended caution in women of childbearing potential (31). Two other drugs under clinical trials are paromomycin (6), and the 8-aminoquinoline derivative sitamaquine (7). Paromomycin is under clinical trial (phase III) for VL, and is currently used in combination with methyl benzethonium chloride or urea for CL. Sitamaquine is currently in phase II clinical trial for VL (21).

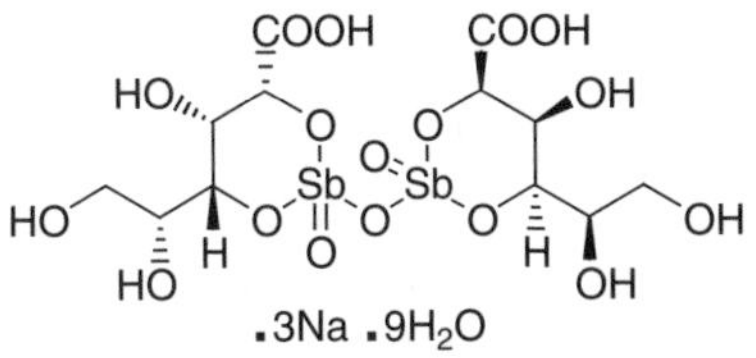

pentostam® 1
(sodium stibogluconate)

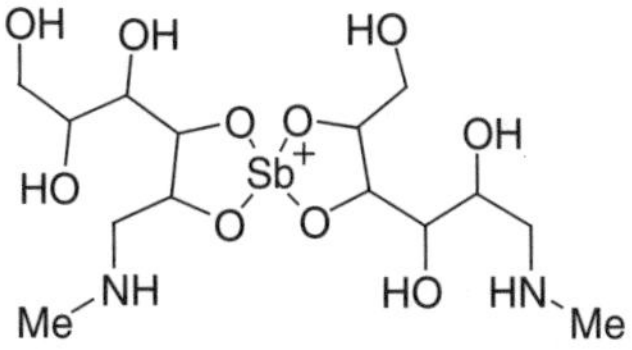

glucantime® 2
(meglumine antimoniate)

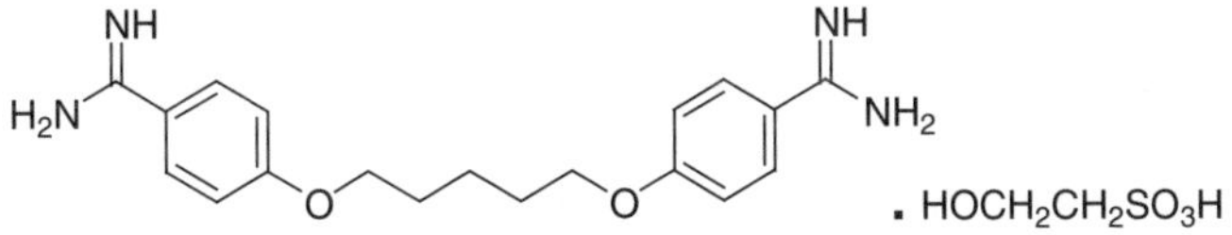

pentamidine isethionate 3

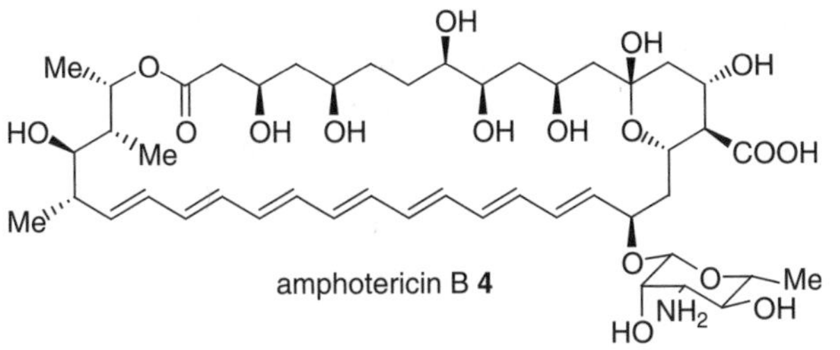

amphotericin B 4

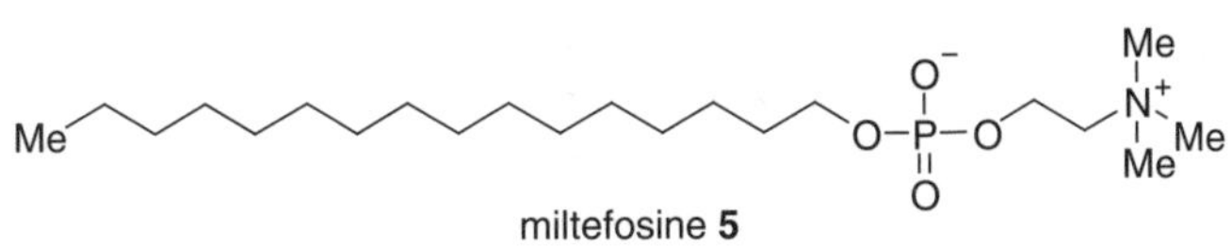

miltefosine 5

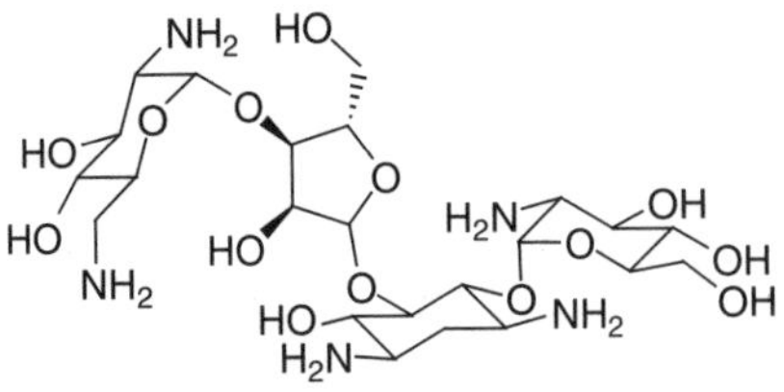

paromomycin **6** 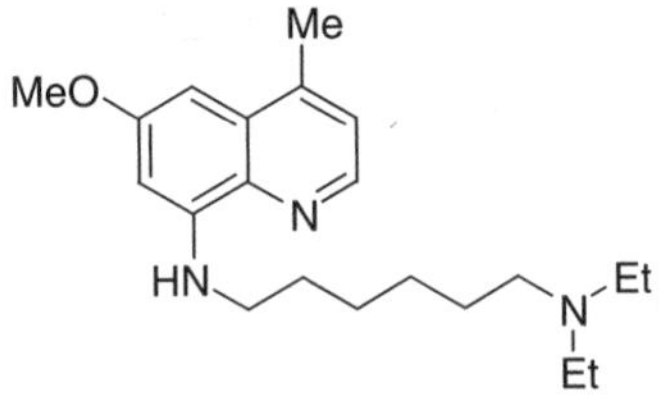   sitamaquine **7**

B. Chagas' Disease

Chagas' disease is endemic in Latin America, affecting 18 countries from northern Mexico to southern Argentina. There are about 20 million infected individuals, 120 million are at risk of acquiring the infection, and approximately 8 million are carriers of the disease (32–34). Control programs carried out during the past 15 years have reduced annual incidence of the disease from more than 500,000 new cases every year to around 50,000, resulting in about 14,000 deaths per year (35). Approximately 25 to 30% of those infected will progress to irreversible cardiac, esophageal, and colonic pathology, leading to considerable morbidity and mortality. The total DALY burden is 667,000, affecting men and women equally (32–34). TDR has classified Chagas' disease in category III, since the control strategy has proven effective and elimination is possible (1).

Chagas' disease is produced by the protozoan *T. cruzi* (order Kinetoplastida, family Trypanosomatidae), which is transmitted to humans and other mammals mostly by blood-sucking reduviid bugs of the subfamily Triatominae (family Reduviidae). The parasite exists in three morphological forms: amastigotes (aflagellated form) that proliferate intracellularly in the mammalian macrophages, epimastigotes (extracellular flagellated form) that proliferate in the insect vector and the acellular culture medium, and trypomastigotes (the infective flagellate form) found in the blood of the mammalian host and in the terminal part of the digestive and urinary tracts of vectors. Infection occurs when infected metacyclic trypomastigotes enter the body through wound openings or mucous membranes. The trypomastigotes enter various cells, differentiate into amastigotes, and multiply intracellularly. The amastigotes differentiate into trypomastigotes, which are then released back into the bloodstream. The life cycle is continued when a reduviid bug feeds on an infected person and ingests trypomastigotes in the blood meal. Transmission of Chagas' disease could also occur by blood transfusion or materno-fetal transmission (32–34).

Chemotherapy for Chagas' disease is based on two, empirically discovered, drugs, a nitrofuran derivative nifurtimox (**8**) and a nitroimidazole derivative benznidazole (**9**), but it is still insufficient (13,23,32,35). Both of these compounds are usually only effective when given during the acute stage of infection, being able to cure at least 50% of recent infections, as indicated by the disappearance of symptoms, and negativization of parasitemia and serology. Other important drawbacks include selective drug sensitivity in different *T. cruzi* strains, and a necessity for constant medical supervision due to the gastrointestinal and neurological side effects of nifurtimox, and the rash and gastrointestinal symptoms provoked by benznidazole. In addition, treatment is long (30 to 60 days), drug resistance develops, and these compounds are not effective in the chronic stage of the disease (23,32,34,36).

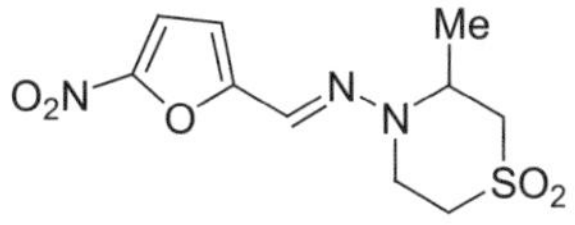

nifurtimox **8** 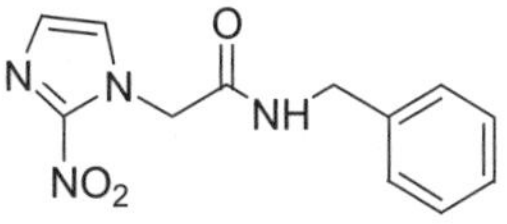   benznidazole **9**

C. African Trypanosomiasis (Sleeping Sickness)

African trypanosomiasis, also known as sleeping sickness, affects 36 countries in sub-Saharan Africa, and is fatal if left untreated. Although its precise prevalence is unknown, the population at risk is 60 million, while annual incidence is estimated at 300,000 to 500,000 cases, resulting in approximately 66,000 deaths per year. The total DALY burden is around 1.5 million, with 559,000 for women and 996,000 for men (37–39). TDR has classified African trypanosomiasis in Category I as an emerging or uncontrolled disease (13). In humans, it is caused by two kinetoplastid flagellates, *Trypanosoma brucei rhodesiense* and *T. b. gambiense* (order Kinetoplastida, family Trypanosomatidae), which are transmitted by several species of blood-feeding tsetse flies of the genus *Glossina. T. b. rhodesiense* occurs mainly in east and southern Africa, and *T. b. gambiense* mainly in west and central Africa. Like *T. cruzi*, the parasite exists in three morphological forms: amastigote, epimastigote, and trypomastigote. Infection occurs via the bite of the blood-sucking male and female tsetse flies that transfer the parasites from human to human. Following the bite of an infected tsetse, parasites multiply in the skin for 1–3 weeks before invading the hemolymphatic system (37–39).

The type of treatment depends on the stage of the disease (37,39). The drugs used in the early hemolymphatic stage are less toxic, easier to administer, and more effective, the current standard treatment being pentamidine (**3**) for *T. b. gambiense* or suramin (**10**) for *T. b. rhodesiense* (40). Successful treatment in the late stage, when the infection has spread to the central nervous system (CNS), requires a drug that can cross the blood-brain barrier to reach the parasite, but such drugs are quite toxic and complicated to administer. The late stage of the disease is treated by melarsoprol (**11**) or, alternatively, eflornithine (**12**) (40,41). All these drugs have to be administered intravenously and produce adverse side effects. Parasite resistance to melarsoprol has been reported (42).

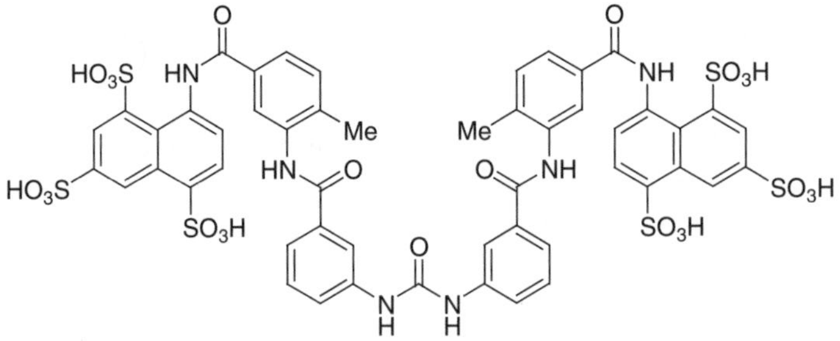

suramin **10**

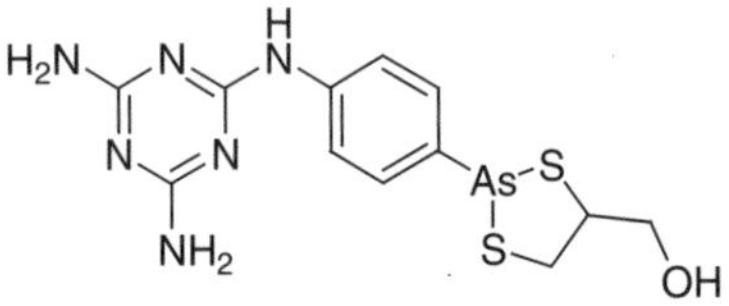

melarsoprol **11** 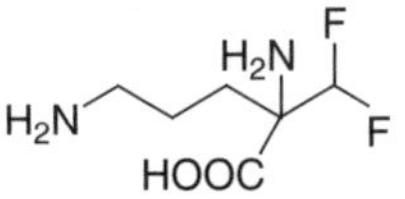   eflornithine **12**

Unfortunately, no vaccine is currently available for any of these diseases. The existing chemotherapy is unsatisfactory in terms of its lack of effectiveness, and also due to the toxicity associated with long-term treatments with empirically discovered drugs. Drug resistance and varying strain sensitivity to the available drugs is another drawback for clinically accessible chemotherapy. The search for new pharmacological alternatives is therefore a scientific priority in order to improve the health and quality of life of people suffering from these diseases.

D. Malaria

Malaria is a vector-borne infectious disease caused by protozoan parasites of the genus *Plasmodium* that are spread from person to person through the bites of several species of infected female mosquitoes of the genus *Anopheles*. Widespread in tropical and subtropical regions of America, Asia, and Africa, malaria is a public health problem in more than 90 countries inhabited by approximately 2400 million people, representing about 35% of the world population (13,43–46). Annual incidence of malaria is estimated at 300–500 million cases, resulting in approximately 1–2 million deaths per year, mostly in Sub-Saharan Africa, but also in Asia, Latin America, the Middle East, and parts of Europe. The total DALY burden is 46.5 million, spread equally between women and men (13,43–45). The economic burden of malaria is extremely high in countries where the disease is endemic, accounting for an estimated 1.3% reduction of the annual economic growth rate, and a higher than 50% reduction of gross national product (GNP), which has a long-term impact (47). The burden of malaria differs according to age and sex. In Africa, almost all deaths occur in children under 5 years of age, when the disease tends to be atypical and more severe (47). The disease burden associated with pregnancy has an additional impact due to the effect of malaria on the health of the fetus (48). TDR has classified malaria in Category II because, although control strategies are available, the disease persists (13).

The main causes of malaria are *P. falciparum* and *P. vivax*, but also *P. ovale* and *P. malaria* can be involved. *P. falciparum* is the species of parasite responsible for cerebral malaria. Parasites include mosquito stages (zygote, exflagellated gamete, and sporozoite), and human stages (sporozoites, liver schizonts, trophozoite, blood schizont, merozoite, and gametocyte). Infection in humans begins when infected female anopheline mosquitoes inject sporozoites subcutaneously into the human host during a blood meal (43–45). The sporozoites enter the bloodstream and invade hepatocytes where they differentiate and multiply as merozoites. In hepatocytes, *P. vivax* and *P. ovale* sporozoites can also develop into latent hypnozoites, which can lie dormant for months or years before differentiating into merozoites (49). The merozoites invade erythrocytes to establish a blood-stage infection, where they appear initially as a ring stage, followed by a growing trophozoite, which develops into a dividing asexual schizont. Some merozoites may also differentiate into sexual forms, male and female gametocytes, which are picked up by female *Anopheles* mosquitoes during a blood meal. Within the mosquito midgut, male gametocytes produce flagellated microgametes that fertilize the female macrogamete. The resulting ookinete traverses the mosquito gut wall and encysts on the exterior of the gut wall as an oocyst releasing sporozoites that migrate to the mosquito salivary gland (43–45).

Malaria infections are treated by several drugs. Currently available antimalarial drugs for therapy and prophylaxis include: the quinoline-containing drugs, chloroquine (13), quinine (14), amodiaquine (15), primaquine (16), and mefloquine (17); the antifolate drugs, pyrimethamine (18), used in combination with sulfadoxine (19) and proguanil (20); and two newer classes of drugs that have been introduced in the last years, artemisinin (21) and its derivatives artesunate (22), artemether (23), and arteether (24); and the hydroxynaphtho-quinone, atovaquone (25) (43–45,50).

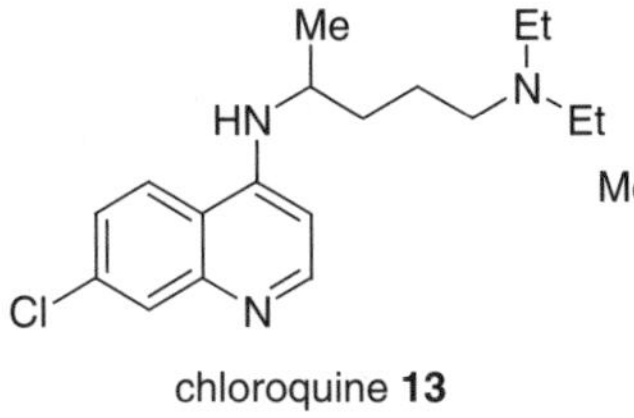

chloroquine **13** 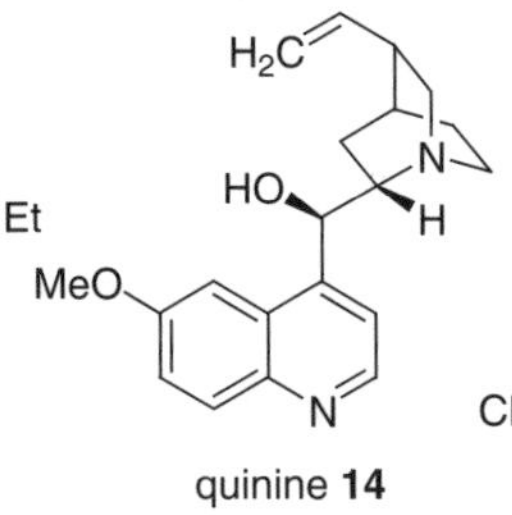   quinine **14** 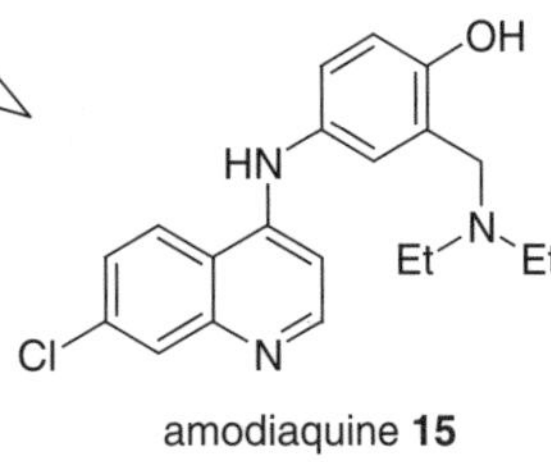   amodiaquine **15**

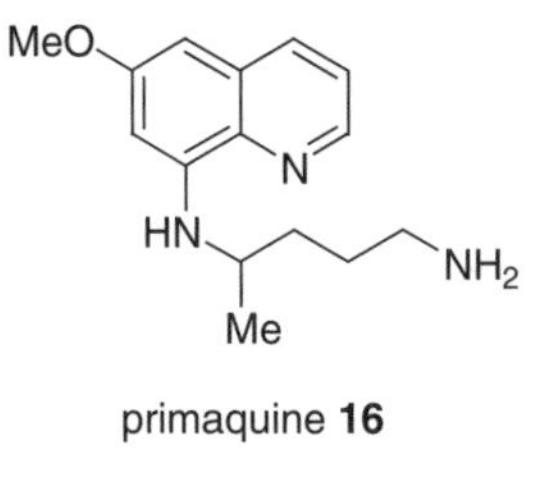

primaquine **16** 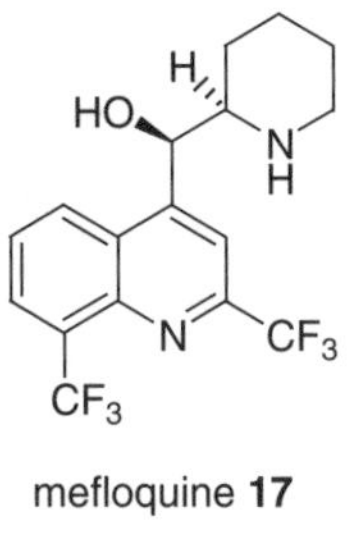   mefloquine **17** 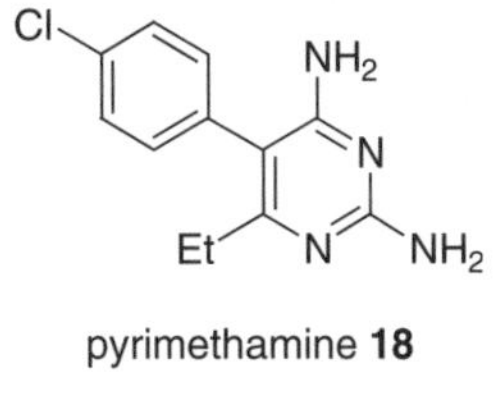   pyrimethamine **18**

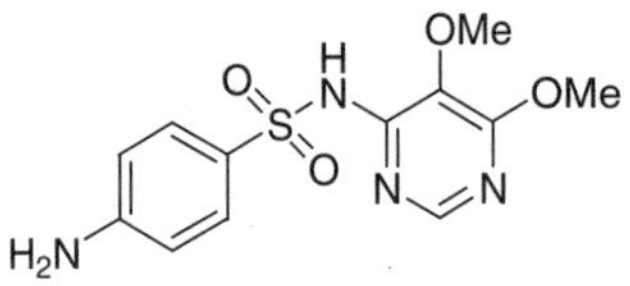

sulfadoxine **19**

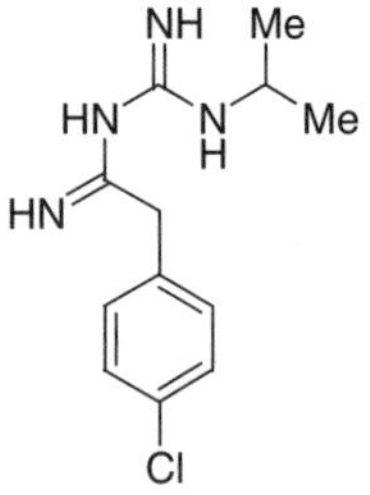

proguanil **20**

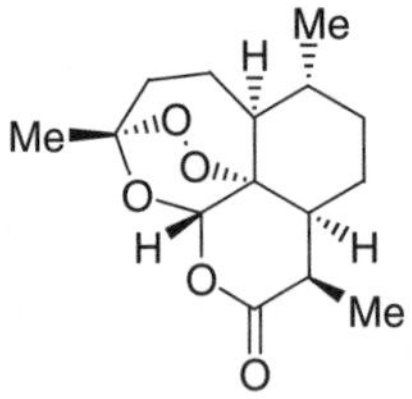

artemisinin **21**

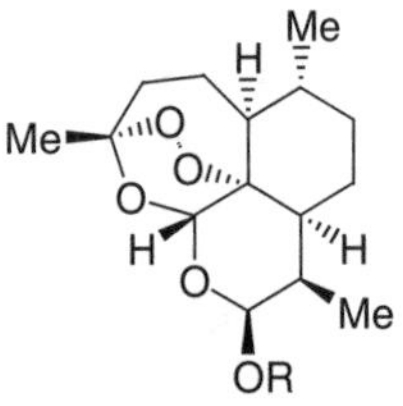

artesunate (**22**) R=CO(CH₂)₂COOH
artemether (**23**) R=Me
arteether (**24**) R=Et

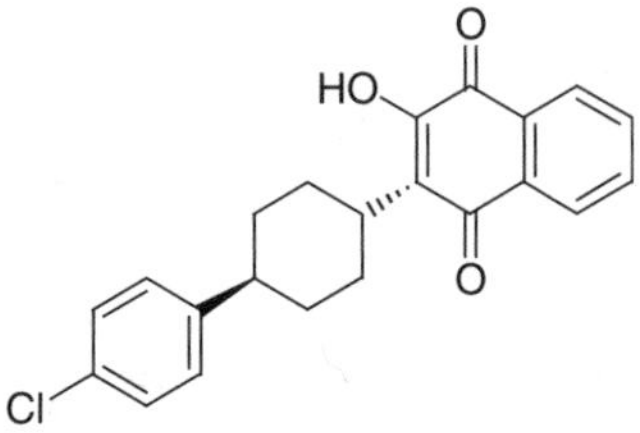

atovaquone **25**

Resistance to antimalarial drugs has been described for *P. falciparum* and *P. vivax*. *P. falciparum* has developed resistance to almost all antimalarial drugs currently used (51). In those areas where chloroquine is still effective it remains the first choice. Unfortunately, chloroquine-resistance is associated with reduced sensitivity to other drugs, such as quinine and amodiaquine. In some areas, *P. vivax* infection has become resistant to chloroquine and/or primaquine (44,50,52). Due to the global emergence of drug resistance, there is an urgent need for the development of new antimalarial drugs.

III. ALKALOIDS WITH ANTIPROTOZOAL ACTIVITY

A. Quinoline and Isoquinoline Alkaloids

1. Quinoline Alkaloids

The family Rutaceae is an important source of antiprotozoal quinoline alkaloids. Activity-guided fractionation of the extracts of *Galipea longiflora* K. Krause (Rutaceae), a Bolivian plant used locally by the Indian Chimaneses for the treatment of cutaneous leishmaniasis, has afforded active compounds identified as 2-substituted quinoline alkaloids, especially chimanine B (**26**) and chimanine D (**27**), with an IC₉₀ of around 0.14 mM, and 2-*n*-propylquinoline (**28**), with an IC₅₀ of around 0.29 mM, against the promastigote forms of

L. braziliensis and the epimastigote forms of *T. cruzi* (53). Results of *in vivo* experiments indicate that chimanine D (**27**) and 2-*n*-propylquinoline (**28**), at 100 mg/kg/day for 14 days, resisted infection with each New World cutaneous leishmaniasis-causing strain, including the virulent strains *L. amazonensis* PH8 and *L. venezuelensis* H-3 (54). In an experimental treatment of VL in infected BALB/c mice, subcutaneous chimanine D (**27**) at 0.54 mM/kg/day for 10 days resulted in 86.6% parasite suppression in the liver, while daily oral administration of 0.54 mM/kg of 2-*n*-propylquinoline (**28**) for 5 or 10 days to *L. donovani*-infected mice suppressed parasite burdens in the liver by 87.8 and 99.9%, respectively (55). Further biological and chemical studies of 2-substituted quinoline synthetic alkaloids have shown *in vitro* and/or *in vivo* antileishmanial properties in the *L. amazonensis* cutaneous infection murine model, as well as the *L. infantum* and *L. donovani* visceral infection murine models (56–58). Preliminary toxicological evaluations of 2-substituted quinoline alkaloids given to BALB/c mice indicated that the drugs had reasonable therapeutic indices (54), and they are currently being investigated in preclinical studies for the development of an oral treatment of VL (59). These studies suggest that the apparently poor *in vitro* properties and good *in vivo* efficiency of 2-*n*-propylquinoline (**28**) could be due to the pharmacological activity of its metabolites (59,60).

Other Rutaceae quinolines, like the 2-substituted tetrahydroquinoline alkaloids galipinine (**29**) and galipeine (**30**), isolated from *Galipea officinalis* Hancock, a plant native to Venezuela and used in folk medicine against fever, showed an *in vitro* antimalarial effect (IC$_{50}$: 0.24–6.12 and 0.33–13.78 µM, respectively) on chloroquine-resistant strains of *P. falciparum* (61). Bioassay-guided separations led to the isolation and structural elucidation of two 4-quinolinone alkaloids, dictyolomide A (**31**) and B (**32**), from the stem bark of *Dictyoloma peruvianum* Planch., a small tree used in folk medicine for the treatment of leishmaniasis. Antileishmanial activity could be attributed to these alkaloids, which induced complete and partial lysis in promastigote forms of *L. amazonensis* and *L. braziliensis*, respectively, at 0.35 mM (62). In the same way, seven alkaloids have been isolated from *Teclea trichocarpa* (Engl.) Engl. (a plant used in Kenyan traditional medicine for malaria treatment), including norme-licopicine (**33**) and arborinine (**34**), which displayed limited *in vitro* activities against both chloroquine-sensitive (HB3) and chloroquine-resistant (K1) strains of *P. falciparum*, with an IC$_{50}$ of 3.8–14.7 µM (63). Normelicopicine (**33**) was found to have some activity against *P. berghei* in mice with 32% suppression of parasitaemia at a dose of 25 mg/kg per day.

chimanine B (**26**) R=CHCHMe
2-*n*-propylquinoline (**28**) R=(CH$_2$)$_2$Me

chimanine D **27**

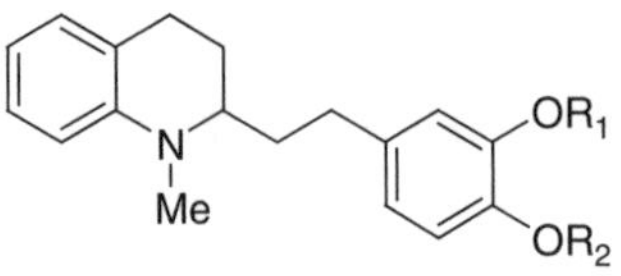

galipinine (**29**) R$_1$+R$_2$=CH$_2$
galipeine (**30**) R$_1$=H, R$_2$=Me

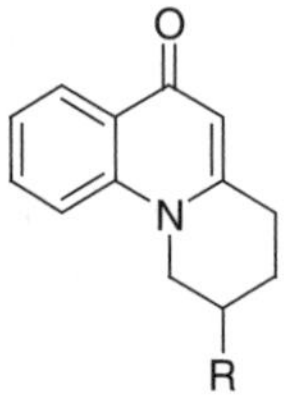

dictyolomide A (**31**) R=(CH$_2$)$_2$CH=CHCH$_2$Me
dictyolomide B (**32**) R=CHOH(CH$_2$)$_4$Me

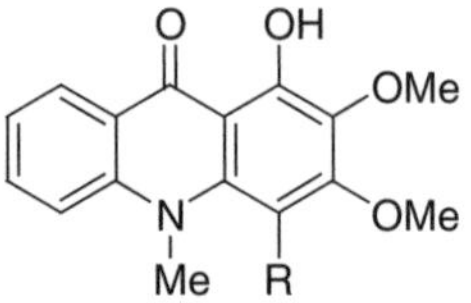

normelicopicine (**33**) R=OMe
arborinine (**34**) R=H

The decahydroquinolines (DHQs) are unusual quinoline alkaloid derivatives, characterized by the presence of a 2,3,5-trisubstituted *cis*-fused DHQ ring (64). Simple DHQs were first reported from the skins of dendrobatid frogs (65), and to date approximately 30 *cis-* and *trans*-DHQs have been isolated from amphibian sources (65,66). All of these alkaloids contain alkyl substituents attached at C-2 and C-5. The marine natural 2,5-dialkyl-DHQ alkaloids have also been isolated from ascidians (66,67), the flat worm *Prosthecergeus villatus* and its prey, the tunicate *Clavelina lepadiformis* (68), and a new tunicate species belonging to the genus *Didemnum* (69). Among plants, *Lycopodium* spp. are known to produce DHQs, but as parts of more complex alkaloidal molecules (70). DHQ alkaloids have also been isolated from myrmicine ants (71–73). Some DHQ alkaloids obtained from a tunicate species belonging to the genus *Didemnum* have shown significant and selective antiplasmodial and antitrypanosomal activity. The most pronounced biological activities were found for the alkaloids lepadin E (**35**) and F (**36**). Lepadin F has an IC$_{50}$ of 0.47 µM against the *P. falciparum* clone K1, while lepadin E showed an IC$_{50}$ of 0.94 µM. The low cytotoxicity of these molecules makes them suitable models for the development of potential therapeutic agents (69).

Among the quinoline alkaloids are the classic antimalarials such as quinine (**14**), the first important leading natural compound against malaria. The quinoline-containing drug derived from the bark of the Cinchona tree (Rubiaceae), native to Peru and the Andes, was used to treat fevers as early as the 17th century, although it was not until 1820 that the active ingredient of the bark was isolated and used in its purified form (74). Although quinine was replaced by synthetic compounds, it is again being applied against chloroquine-resistant strains, and as a treatment for uncomplicated and severe malaria in many different therapeutic regimens (75). Quinine was used as a template for the antimalarials chloroquine (**13**) and mefloquine (**17**), two fully aromatic quinoline type compounds (9). Quinimax®, which is a combination of quinine (14), quinidine (37) (dextrorotatory diastereomer of quinine), and cinchonine (38), all derived from cinchona bark, is also used (76). Quinine, quinidine, cinchonine, and cinchonidine (39) have significant trypanocidal activity with IC_{50} values of 4.9, 0.8, 1.2, and 7.1 µM, respectively, in *T. b. brucei*. For 37 and 38, the selectivity indices are greater than 200, indicating the potential of these alkaloids for further drug development (11). Likewise, quinine (14) completely inhibited the *in vitro* replication of *T. cruzi* at 15.4 µM (5).

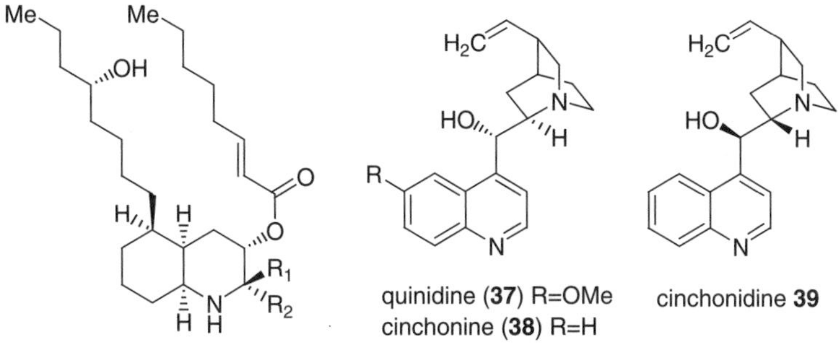

lepadin E (35) R_1=Me, R_2=H
lepadin F (36) R_1=H, R_2=Me

quinidine (37) R=OMe
cinchonine (38) R=H

cinchonidine 39

Although they are not natural products, the 8-aminoquinoline compounds deserve attention. Primaquine (16), a known 2-substituted 8-aminoquinoline, is the only tissue schizonticide (exoerythrocytic) drug available for radical treatment of *P. vivax* or *P. ovale* infections (77). Although primaquine has no clinical utility as a blood schizonticide, substantial efforts have been made to identify an 8-aminoquinoline with a better therapeutic index and activity against blood stages of malaria (77). On the other hand, sitamaquine (7) (originally WR6026), an 8-aminoquinoline in development as an antileishmanial agent (78), has undergone several small Phase 1/2 clinical trials with varying levels of success. For instance, 67% of patients were cured of *L. chagasi* in Brazil when treated with 2 mg/kg/day for 28 days (79); 92 and 100% of patients were cured of VL when treated with 1.75 mg/kg/day for 28 days in Kenya (80) and with 2 mg/kg/day for 28 days in India (81), respectively. Sitamaquine is rapidly metabolized, forming desethyl and 4-CH_2OH derivatives, which might be responsible for its activity (78), but little is known about its mechanism of action (82).

The mechanism of action of quinoline derivatives against *Leishmania* and *Trypanosoma* species remains unknown. Nevertheless, studies of antimalarial drugs against *Plasmodium* have revealed that quinoline compounds depend on interactions with heme (ferriprotoporphyrin IX) for their antimalarial action (83–85). The intraerythrocytic malaria parasite digests large quantities of host hemoglobin (86), a globular protein (globin) with an embedded heme group. The hemoglobin digestion process involves degradation of the protein component by proteolytic enzymes (87,88) and release of heme component (89,90). Because heme is toxic to *Plasmodium* parasites it is converted into a crystalline compound named malaria pigment or hemozoin, which is harmless to the parasite (91,92). Recent studies on cultured *P. falciparum* have shown that at least 95% of the hematin is incorporated into hemozoin (93). The antimalarial quinoline-type drugs act by binding to hemozoin crystal faces, which would inhibit their growth and result in a buildup of the toxic heme, thus leading to the death of the parasite (94–96). The mechanism of action of DHQ compounds is unknown and needs to be investigated to determine if it is similar to that of the fully aromatic quinoline-type compounds (69).

2. Aporphine and Oxoaporphine Alkaloids

The aporphine and oxoaporphine alkaloids are isoquinoline compounds known to have various pharmacological properties, including antiparasitic activity. Most aporphine derivatives occur in members of the Annonaceae family (97–101), one of the largest of the order Magnoliales, comprising about 120 genera and more than 2000 species (102). Among the active alkaloids, liriodenine (**40**) and anonaine (**41**), isolated from the roots and trunk bark of *Annona spinescens* Mart., have shown significant activity against promastigote forms of *L. braziliensis*, *L. amazonensis*, and *L. donovani* (103). However, while in this report liriodenine showed antileishmania activity with an IC_{100} of 0.36 mM against the *L. donovani* strain, in another study, liriodenine isolated from the bark of *Rollinia emarginata* Schltdl. (104) and *Unonopsis buchtienii* R.E.Fr. (105) presented IC_{100} values of 18.16 μM, and 11.33 μM, respectively, against the same promastigote forms of the parasite. Similarly, *O*-methylmoschatoline (**42**), isolated from the stem bark of *U. buchtienii* by activity-guided fractionation, showed an interesting *in vitro* activity against *T. brucei* with an IC_{100} of 19.45 μM, but without selectivity. On the other hand, crude extracts of *Uvaria klaineana* Engl. & Diels stems showed *in vitro* activity against *P. falciparum*. Three alkaloids were identified by bioassay-guided fractionation, the main alkaloid being crotsparine (1-hydroxy-2-methoxynor-proaporphine) (**43**), which showed IC_{50} values of 7.41, 11.29, and 12 μM against the chloroquine-resistant FcB1 and K1, and the chloroquine-sensitive Thai strains of *P. falciparum*, respectively (106).

Preliminary studies in the genus *Guatteria* (Annonaceae) have revealed interesting active compounds. Extracts from *G. foliosa* Benth., commonly known as "Sayakasi" among the Chimane Indians in Bolivia, have demonstrated antiparasitic activity against different strains of *Leishmania* and against *T. cruzi* (107). Fractionation of the alkaloid extract resulted in the isolation of several isoquinoline alkaloids, including isoguattouregidine (**44**), which showed significant activity against the promastigote forms of *L. donovani* and *L. amazonensis*, with total lysis of parasites at 0.29 mM. On the other hand, isoguattouregidine, argentinine (an aminoethylphenanthrene) (**45**), and 3-hydroxynornuciferine (**46**), evaluated at around 0.8 mM, presented significant trypanocidal activity against the bloodstream form (trypomastigote) of *T. cruzi* with partial lysis of 92, 81, and 68%, respectively (108). Aporphine alkaloids were also obtained in previous studies of *G. amplifolia* Triana & Planch. and *G. dumetorum* R.E.Fr. (109,110), whose extracts showed activity against *Leishmania* sp. and chloroquine-sensitive and -resistant strains of *P. falciparum* (111). In the search for compounds to treat leishmaniasis, the alkaloids xylopine (**47**) and nornuciferine (**48**) from *G. amplifolia* and cryptodorine (**49**) and nornantenine (**50**) from *G. dumetorum*, demonstrated significant activity against *L. mexicana* and *L. panamensis*. Xylopine (**47**) and cryptodorine (**49**) were among the most active compounds (IC$_{50}$ of 3 μM), and showed a 37- and 21-fold, respectively, higher toxicity towards *L. mexicana* than macrophages, the regular host cells of *Leishmania* sp. (110).

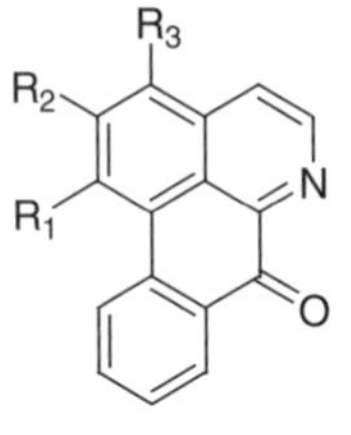

liriodenine (**40**) R$_1$+R$_2$=OCH$_2$O, R$_3$=H
methylmoschatoline (**42**) R$_1$=R$_2$=R$_3$=OMe

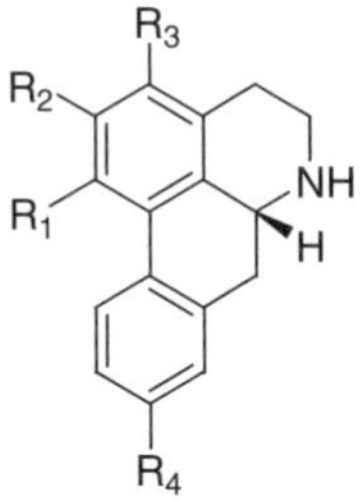

anonaine (**41**) R$_1$+R$_2$=OCH$_2$O, R$_3$=R$_4$=H
xylopine (**47**) R$_1$+R$_2$=OCH$_2$O, R$_3$=H, R$_4$=OMe

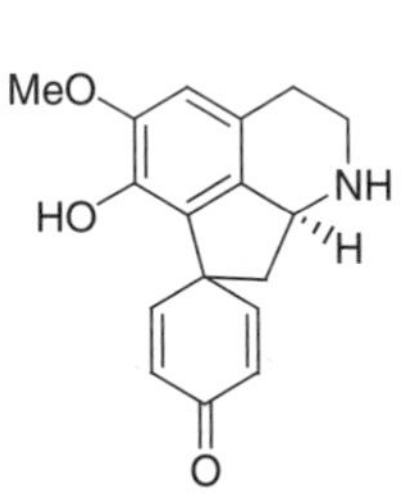

crotsparine **43** 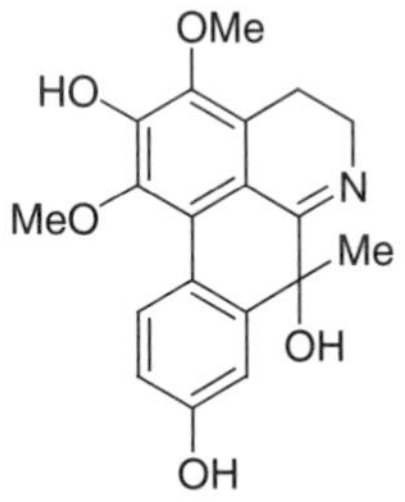   isoguattouregidine **44** 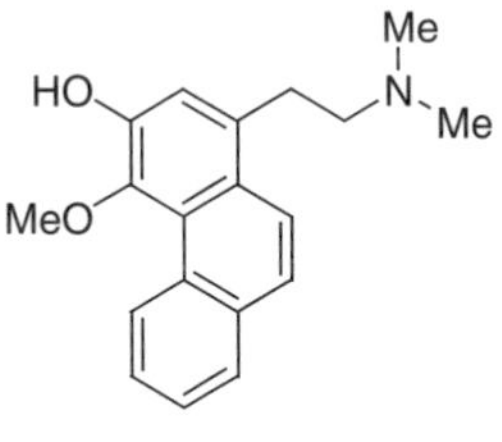   argentinine **45**

The aporphine derivatives that occur in members of the Annonaceae are also found in members of the Euphorbiaceae, Hernandiaceae, Lauraceae, Menisper-maceae, and Papaveraceae, among others. Bioactivity-guided fractionation of *Stephania dinklagei* Diels. (Menispermaceae) yielded liriodenine (**40**) and the zwitterionic oxoaporphine alkaloid *N*-methylliriodendronine (**51**). In agreement with previously reported results, *N*-methylliriodendronine was the most active against *L. donovani* amastigotes (IC$_{50}$ = 36.1 μM), while liriodenine showed the highest activity against *L. donovani* promastigotes and *P. falciparum*, with IC$_{50}$ values of 15.0 and 26.2 μM, respectively (105,112). Roemrefidine (**52**), an aporphine alkaloid isolated from the Bolivian vine *Sparattanthelium amazonum* Mart. (Hernandiaceae), as well as several members of the Papaveraceae, was also found to be active against both chloroquine-resistant and chloroquine-sensitive *P. falciparum* strains with IC$_{50}$ values of 0.58 and 0.71 μM, respectively (113). It has also been shown that roemrefidine acts on the parasite maturation, but has no effect on the erythrocytic reinvasion and no cumulative influence on the metabolic pathways of the parasite. In addition, the *in vitro* effect of a crude alkaloid extract of *Cassytha filiformis* L. (Lauraceae), a sprawling parasitic herb traditionally used to treat African trypanosomiasis, on bloodstream forms of *T. b. brucei* led to the evaluation of its three major aporphine alkaloids. Actinodaphnine (53), cassythine (54), and dicentrine (55) showed IC_{50} values of 3–15 μM against *T. b. brucei* (114). Potent antimalarial activity was also observed for stephanine (56) and two 6a,7-dehydroaporphine alkaloids, dehydrocrebanine (57) and dehydrostephanine (58), isolated from *Stephania venosa* Spreng. (Menispermaceae) (115). The IC_{50} values were 0.38, 0.22, and 0.12 μM, respectively, against the T9/94 strain of *P. falciparum*, thus the 6a,7-dehydro derivatives were about 2–3 times more potent than their parent compounds.

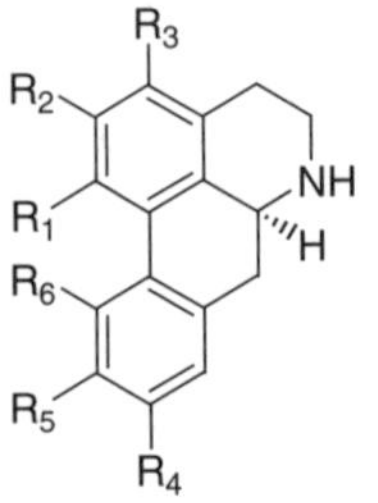

3-hydroxynornuciferine (46) $R_1=R_2=OMe$, $R_3=OH$, $R_4=R_5=R_6=H$
nornuciferine (48) $R_1=R_2=OMe$, $R_3=R_4=R_5=R_6=H$
cryptodorine (49) $R_1+R_2+R_4+R_5=OCH_2O$, $R_3=R_6=H$
nornantenine (50) $R_1=R_2=OMe$, $R_3=R_6=H$, $R_4+R_5=OCH_2O$
actinodaphnine (53) $R_1+R_2=OCH_2O$, $R_3=R_6=H$, $R_4=OH$, $R_5=OMe$
cassythine (54) $R_1+R_2=OCH_2O$, $R_3=R_5=OMe$, $R_4=OH$, $R_6=H$
norcorydine (62) $R_1=OH$, $R_2=R_5=R_6=OMe$, $R_3=R_4=H$

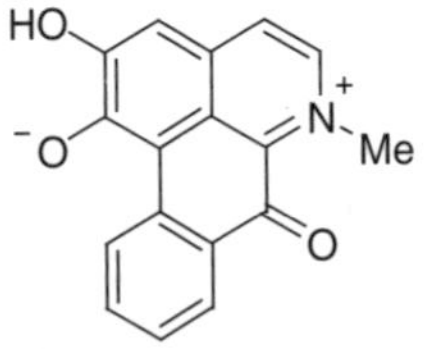

N-methylliriodendronine 51 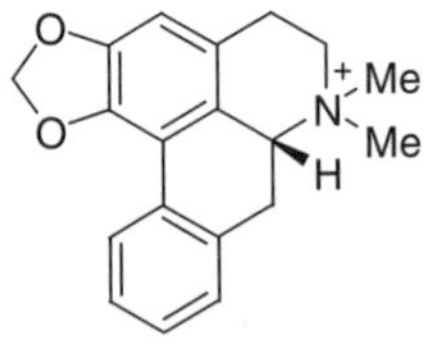   roemrefidine 52

Among the isoquinoline alkaloids with trypanocidal effect are the aporphine alkaloids predicentrine (59), glaucine (60), and boldine (61), which showed inhibition of the *in vitro* growth of *T. cruzi* epimastigotes with IC_{50} values of 0.08, 0.09, and 0.11 mM, respectively (116). In an attempt to discover new alkaloids with antiplasmodial properties, a number of monomeric isoquinoline alkaloids have also been evaluated. In the aporphine group, norcorydine (62) possessed the highest antiplasmodial activity ($IC_{50} = 3.08$ μM), while corydine, its *N*-methyl derivative, was seven-fold less active. The authors suggest that a secondary amino group and a phenolic substituent enhance the *in vitro* antiplasmodial activity of aporphines (117). Norcorydine was found to be non-toxic to KB cells ($IC_{50} = 733\,\mu M$), clearly showing selective toxicity against the K1 strain of *P. falciparum*. Antiplasmodial activity was also shown by ushinsunine (**63**) ($IC_{50} = 5.99\,\mu M$) and dehydroocoteine (**64**) ($IC_{50} = 5.78\,\mu M$); the latter alkaloid was six-fold more active than its C6a–C7 saturated derivative ocoteine (117).

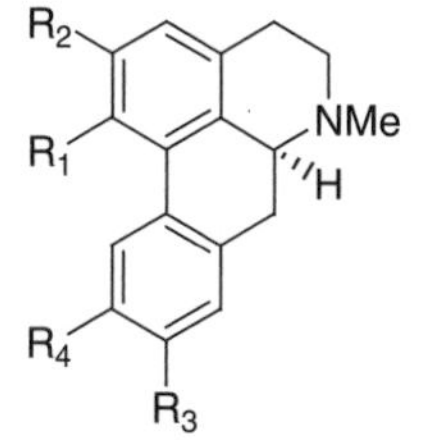

dicentrine (**55**) R_1+R_2=OCH$_2$O, R_3=R_4=OMe
predicentrine (**59**) R_1=R_3=R_4=OMe, R_2=OH
glaucine (**60**) R_1=R_2=R_3=R_4=OMe
boldine (**61**) R_1=R_4=OMe, R_2=R_3=OH

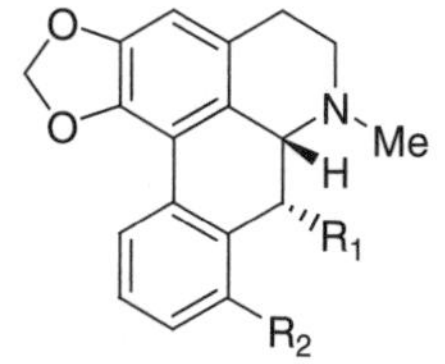

stephanine (**56**) R_1=H, R_2=OMe
ushinsunine (**63**) R_1=OH, R_2=H

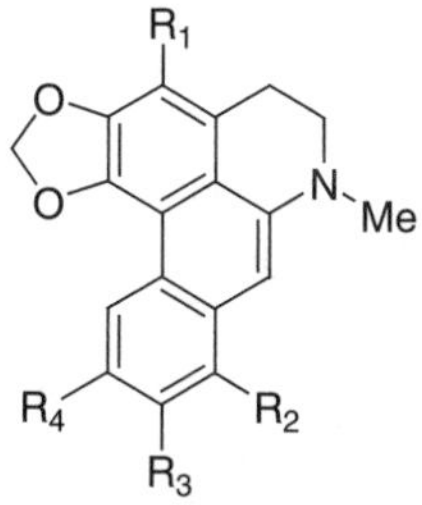

dehydrocrebanine (**57**) R_1=R_4=H, R_2=R_3=OMe
dehydrostephanine (**58**) R_1=R_3=R_4=H, R_2=OMe
dehydroocoteine (**64**) R_1=R_3=R_4=OMe, R_2=H

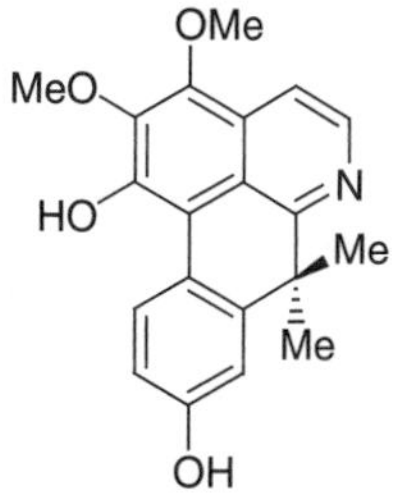

melosmine **65**

The mechanism of action of aporphine alkaloids seems to be related, at least in part, to the inhibition of topoisomerase II by DNA intercalation or minor groove binding (114,118). The identification of liriodenine (**40**) as a strong, topoisomerase II catalytic inhibitor and a topoisomerase II poison (119) has led to the search for other topoisomerase II inhibitors among the aporphine alkaloids. Liriodenine, like many strong, topoisomerase II inhibitors, is a very planar molecule, and thus a likely DNA intercalator. However, some derived aporphines lack the structural characteristics normally associated with conventional DNA intercalators, such as the presence of two or three fused aromatic rings. Instead, they only have two aromatic rings separated by saturated rings, making them non-planar molecules. In addition, these two aromatic rings are substituted with methylenedioxy and/or methoxy groups, which may hinder access of the molecule to the intercalation sites (114). On the basis of molecular modeling studies, it has been proposed that such non-planar molecules can be "adaptive intercalators," which undergo a conformational change upon binding to DNA to adopt a strained planar conformation (118). It has also been suggested that aporphine alkaloids with phenolic groups could participate as antioxidants and in this way inhibit cellular respiration in *Trypanosoma* (120). It is conceivable that the sterically hindered phenolic groups of the most active alkaloids may be acting as free radical chain breaking antioxidants and that this property may be related in some way to their antitrypanosomal effects (5). Our results would agree with this statement. As part of our study aimed at the isolation of antiprotozoal compounds from some medicinal plants of Colombia, we directed our attention toward the components of *Rollinia pittieri* Saff. (Annonaceae), obtaining a series of aporphine alkaloids with potent antiprotozoal activity (data not published). The most active alkaloid, melosmine (**65**), presented the greatest free radical scavenging activity (121).

3. Benzylisoquinoline, Protoberberine, and Related Alkaloids

Benzylisoquinoline, protoberberine, protopine, benzo[c]phenanthridine, and many other alkaloids all contain the basic building block of the isoquinoline skeleton (122). The benzylisoquinolinic alkaloids are widely distributed in Nature and they have been isolated from different plants commonly used in traditional medicine for the treatment of parasitic diseases (6). Among the active constituents are the phenolic benzylisoquinolines coclaurine (**66**) and norarmepavine (**67**), which have been shown to inhibit the growth of *T. cruzi* epimastigotes *in vitro* with an IC_{50} of around 0.30 mM (116).

The protoberberines are distributed in plant families such as Papaveraceae, Berberidaceae, Fumariaceae, Menispermaceae, Ranunculaceae, Rutaceae, and Annonaceae, with a few also found in the Magnoliaceae and Convolvulaceae (123–129). Most protoberberine alkaloids exist in plants either as tetrahydroprotoberberines or as quaternary protoberberine salts, although a few examples of dihydroprotoberberines have also been described. The quaternary protoberberine alkaloids (QPA) represent approximately 25% of all the currently known alkaloids with a protoberberine skeleton isolated from natural sources (122). Berberine (**68**), a QPA initially obtained by Buchner and Herberger in 1830 as a yellow extract from *Berberis vulgaris* L. (Berberidaceae), is probably the most widely distributed alkaloid of all (122), and is well known for its antiparasitic activity (130). This metabolite is the main constituent in various folk remedies used to treat cutaneous leishmaniasis, malaria, and amebiasis (131). Berberine (**68**) has been used clinically for the treatment of leishmaniasis for over 50 years, and has been shown to possess significant activity both *in vitro* and *in vivo* against several species of *Leishmania*. At a concentration of 29.7 µM, berberine effectively eliminates *L. major* parasites in mouse peritoneal macrophages, but shows minimum activity when applied topically on mouse cutaneous lesions caused by *L. major*. Vennerstrom *et al.* also tested berberine and several of its derivatives for antileishmanial activity against *L. donovani* and *L. panamensis in vivo* in golden hamsters (130). Even though berberine is effective against cutaneous ulcers caused by the New World pathogen *L. panamensis* in rats, it has been observed that in these cases viable amastigotes persist on the skin, resulting in the reappearance of the lesion (132,133). Similarly, although berberine has an ethnomedicinal history in India for cutaneous leishmaniasis, topical application was ineffective in *in vivo* tests (134). Canadine (**69**), a tetrahydroberberine, is controversially less toxic and more potent than berberine against *L. donovani*, but not as potent as meglumine antimonate (Glucantime®), a therapy standard (10).

Pessoine (**70**) and spinosine (**71**), members of the small group of catecholic isoquinoline alkaloids with a protoberberine skeleton, were isolated from the trunk bark of *Annona spinescens* (Annonaceae). Pessoine (**70**) was evaluated for its trypanocidal activity *in vitro* and induced a partial lysis (55%) in the trypomastigote forms of *T. cruzi* at 0.79 mM (103). On the other hand, burasaine (**72**), an alkaloid isolated from the roots of several species of the *Burasaia* genus, has shown *in vitro* antiplasmodial activities against the FcM29 strain of *P. falciparum*, being two times less potent than quinine (135). In an attempt to discover further alkaloids with antiplasmodial properties, a number of isoquinoline alkaloids were evaluated against *P. falciparum* (strain K1). The protoberberine group included the alkaloids with the highest antiplasmodial activities. Dehydrodiscretine (**73**) and berberine were the most active with IC_{50} values of less than 1 μM (0.64 and 0.97 μM, respectively), while three other QPA, columbamine (**74**), jatrorrhizine (**75**), and thalifendine (**76**) and the protopine-type alkaloid, allocryptopine (**77**), had values between 1 and 10 μM (117). The low antiplasmodial activity of the non-quaternary alkaloid canadine (**69**) found in this study, in contrast with the high activity of berberine, may indicate that a quaternary nitrogen is required for antiplasmodial activity in this series of alkaloids. Studies carried out to explain the relationship between structure and antimalarial activity in protoberberine alkaloids (136–138) have pointed out certain features. It appears that a quaternary nitrogen atom, especially in an isoquinolinium, rather than a dihydroisoquinolinium ion, contributes to increased antimalarial activity. This activity is also increased by the aromatization of ring C as a consequence of the quaternization of the ring-B nitrogen, and by the type of oxygen substituents on rings A, C, and D, and the position of the oxygen functions on ring D (136–138).

Quaternary benzo[c]phenanthridine alkaloids (QBA) occur mainly in plants of the Papaveraceae and Fumariaceae families, and can also be found in the Caprifoliaceae, Meliaceae, and Rutaceae (139). These elicitor-inducible secondary metabolites are called phytoallexins because of their antimicrobial and antifungal activities (140,141). Their main representatives are the 2,3,7,8-tetrasubstituted alkaloids chelerythrine (**78**) and sanguinarine (**79**). This latter alkaloid and berberine (**68**) have been reported as trypanocidal agents against *T. b. brucei* with IC_{50} values of 1.9 and 0.5 μM, respectively. No or little selectivity was observed for either alkaloid (selectivity index, SI = 0.7 and 51.0, respectively) (142). Bioassay-guided fractionation of *Toddalia asiatica* Lam. (Rutaceae), a plant used by the Pokot tribe of Kenya to treat fevers, resulted in the isolation of the QBA nitidine (**80**), a well-known cytotoxic agent that shows good activity against some chloroquine-sensitive and chloroquine-resistant strains of *P. falciparum* (IC_{50} around 0.12–0.47 μM) (143).

While the mechanism of action and pharmacological activities of berberine (68) have been extensively studied, very little is known about the other protoberberine alkaloids (144). It has been reported that berberine is a potent *in vitro* inhibitor of both nucleic acid and protein synthesis in *P. falciparum* (145). Despite its slightly buckled structure due to the partial saturation of the central ring, berberine has been previously characterized as a DNA intercalating agent, and as a cationic ligand, electrostatic forces playing an important role in its interaction with DNA (146–149). In a recent comprehensive study, the mode of binding of berberine to short oligonucleotide duplexes was examined by NMR and molecular modeling techniques. The authors concluded that the drug does not intercalate into DNA, but forms minor groove complexes by partial intercalation due to its buckled structure (150). Recent studies with other protoberberine alkaloids suggest that burasaine (72) forms intercalation complexes, showing a true intercalative mode for the binding with double-stranded DNA (144,151), although targets other than DNA may also be invoked (144). In relation to their mode of action, benzo[c]phenanthridine alkaloids have been shown to be topoisomerase inhibitors (152), so it is possible that their antimalarial action is mediated through the inhibition of the parasite enzyme. They also inhibit microtubule assembly and interact with DNA (153–155).

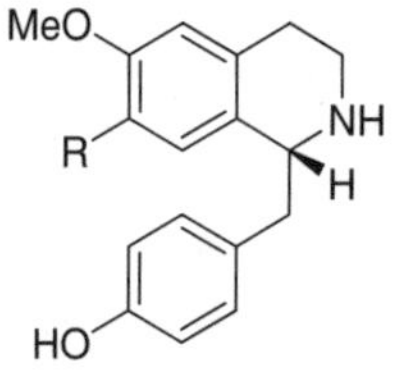

coclaurine (66) R=OH
norarmepavine (67) R=OMe

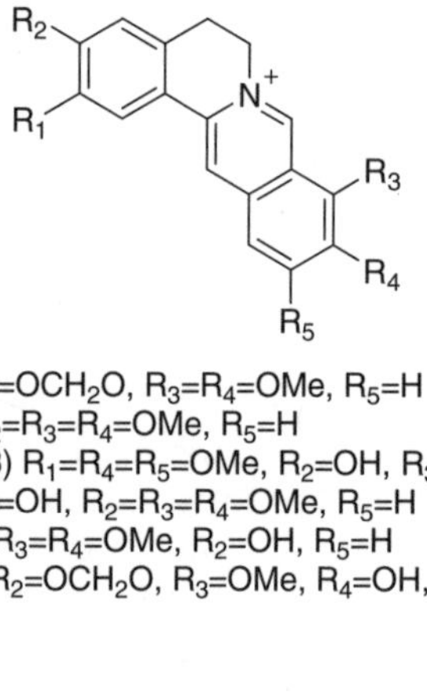

berberine (68) R_1+R_2=OCH$_2$O, R_3=R_4=OMe, R_5=H
burasaine (72) R_1=R_2=R_3=R_4=OMe, R_5=H
dehydrodiscretine (73) R_1=R_4=R_5=OMe, R_2=OH, R_3=H
columbamine (74) R_1=OH, R_2=R_3=R_4=OMe, R_5=H
jatrorrhizine (75) R_1=R_3=R_4=OMe, R_2=OH, R_5=H
thalifendine (76) R_1+R_2=OCH$_2$O, R_3=OMe, R_4=OH, R_5=H

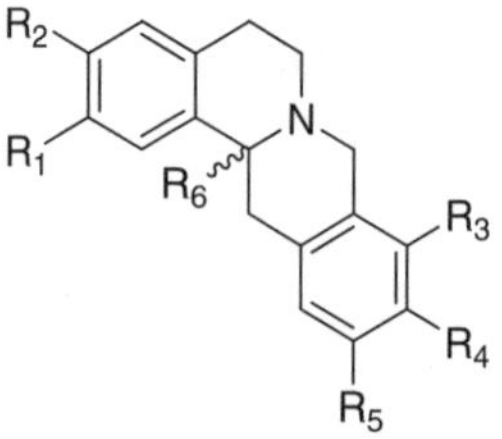

canadine (69) R_1+R_2=OCH$_2$O, R_3=R_4=OMe, R_5=H, R_6=Hβ
pessoine (70) R_1=R_4=R_5=OH, R_2=OMe, R_3=H, R_6=Hα
spinosine (71) R_1=R_2=OMe, R_3=H, R_4=R_5=OH, R_6=Hα

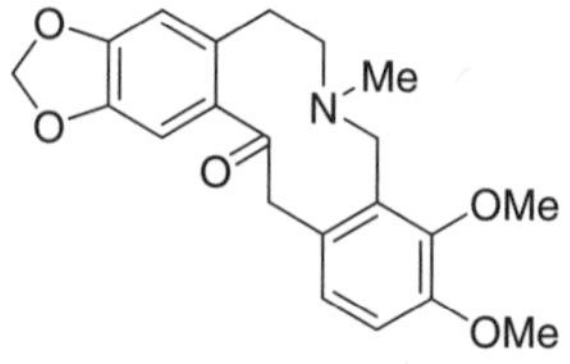

allocryptopine 77

chelerythrine (**78**) R$_1$=R$_2$=OMe, R$_3$=H
sanguinarine (**79**) R$_1$+R$_2$=OCH$_2$O, R$_3$=H
nitidine (**80**) R$_1$=H, R$_2$=R$_3$=OMe

4. Bisbenzylisoquinoline Alkaloids

The bisbenzylisoquinoline (BBIQ) alkaloids constitute a series of over 430 phenylalanine-derived phytometabolites widespread in nature. They are found in the following botanical families: Annonaceae, Aristolochiaceae, Berberidaceae, Hernandiaceae, Lauraceae, Menispermaceae, Monimiaceae, Nymphaeaceae, and Ranunculaceae (156–159). BBIQ alkaloids present a rich and varied chemistry and pharmacology (160), comprising two isoquinoline moieties (head portions) linked to two benzyl moieties (tail portions). They have been classified into 26 structural types (denoted by roman numerals) according to the nature, number, and attachment point of the bridges. The alkaloids within each group differ from one another by the nature of their oxygenated substituents, the degree of unsaturation of the heterocyclic rings, and the stereochemistry of their two chiral centers, C-1 and C-1' (156–159). A number of biological activities have been reported for BBIQ alkaloids including immunomodulatory effects (161,162), cardiovascular effects (163,164), antithrombosis (165), anti-HIV (166), and antiprotozoal activities, in particular against *Leishmania* sp. (160,167,168), *T. cruzi* (142,160,169–175), and *Plasmodium* sp. (176–192).

In an evaluation of 14 isoquinoline alkaloids, especially BBIQs extracted from Annonaceae, Berberidaceae, Hernandiaceae, and Menispermaceae, four are of particular interest. Daphnandrine (**81**) and limacine (**82**), isolated from *Albertisia papuana* Becc. and *Caryomene olivascens* Barneby & Krukoff, respectively (Menispermaceae), obaberine (**83**) obtained from *Pseudoxandra sclerocarpa* Maas (Annonaceae) and gyrocarpine (**84**), produced by *Gyrocarpus americanus* Jacq. (Hernandiaceae) showed strong activity against promastigotes of *L. braziliensis*, *L. amazonensis*, and *L. donovani*. Leishmanicidal activity of the first three alkaloids was at an IC$_{100}$ of nearly 84 μM, and gyrocarpine (**84**) was active *in vitro* at 16 μM (167). However, in an *in vivo* test against *L. amazonensis* (168), gyrocarpine (**84**) was less potent than Glucantime®. In the same study, isotetrandrine (**85**), a metabolite isolated from *Limaciopsis loangensis* Engl. (Menispermaceae), showed leishmanicidal activity at 16 μM against the promastigote forms of *L. braziliensis*, *L. amazonensis*, and *L. donovani*, and exhibited activity approximately equal to or greater than Glucantime® in BALB/c mice infected with *L. amazonensis* (168).

The structural requirements for the antiprotozoal effects of the BBIQ alkaloids were explored by determining the activity of several BBIQ alkaloids against *L. donovani* and *T. b. brucei* (160). The results revealed that 13 BBIQ alkaloids: isotetrandrine (**85**), fangchinoline (**86**), funiferine (**87**), tiliageine (**88**), oxyacanthine (**89**), aromoline (**90**), thalisopidine (**91**), obamegine (**92**), dinklacorine (**93**), isotrilobine (**94**), trilobine (**95**), gilletine (**96**), and insularine (**97**), had IC_{50} values of less than $5\,\mu M$ against *L. donovani* promastigotes, with fangchinoline ($IC_{50} = 0.39\,\mu M$) being as potent as the standard drug pentamidine ($IC_{50} = 0.4\,\mu M$). Of the 15 BBIQ alkaloids tested against *L. donovani* amastigote forms, phaeanthine (**98**) showed strong activity ($IC_{50} = 2.41\,\mu M$), but at this concentration was toxic to the infected macrophages. Cocsoline (**99**) was the only alkaloid that showed selective toxicity ($IC_{50} = 12.3\,\mu M$) towards *L. donovani* amastigotes in macrophages, being slightly less potent than the standard drug pentostam ($IC_{50} = 9.75\,\mu g\;Sb(V)/$ mL). Of the 12 BBIQ alkaloids tested for antitrypanosomal activity, eight: isotetrandrine (**86**), aromoline (**90**), thalisopidine (**91**), isotrilobine (**94**), trilobine (**95**), phaeanthine (**98**), daphnoline (**100**), and berbamine (**101**), had IC_{50} values of $1\text{--}2\,\mu M$ against *T. b. brucei* bloodstream trypomastigote forms. Thalisopidine (**91**) displayed the strongest trypanocidal activity ($IC_{50} = 1.14\,\mu M$), but none were as active as pentamidine ($IC_{50} = 0.3\,nM$) (160). Similarly, berbamine was active on *T. b. brucei* in the same concentration range as reported in the previous study, with an IC_{50} value of $2.6\,\mu M$ (SI = 7.0) (142). On the other hand, phaeanthine was shown to be inactive in mice infected with the same parasite (169). However, Camacho *et al.* suggest that the optimal structure for antileishmania and antitrypanosomal activity of the BBIQ alkaloids is not fully understood, and that further examination is necessary in order to define these relationships (160).

daphnandrine (**81**) R$_1$=H, R$_2$=R$_4$=OMe, R$_3$=OH (R,S)
obaberine (**83**) R$_1$=Me, R$_2$=R$_3$=R$_4$=OMe (R,S)
gyrocarpine (**84**) R$_1$=Me, R$_2$=R$_3$=OMe, R$_4$=OH (S,R)
oxyacanthine (**89**) R$_1$=Me, R$_2$=OH, R$_3$=R$_4$=OMe (R,S)
aromoline (**90**) R$_1$=Me, R$_2$=R$_3$=OH, R$_4$=OMe (R,S)
daphnoline (**100**) R$_1$=H, R$_2$=R$_3$=OH, R$_4$=OMe (R,S)
cepharanthine (**107**) R$_1$=Me, R$_2$=OMe, R$_3$+R$_4$=OCH$_2$O (R,S)
candicusine (**108**) R$_1$=Me, R$_2$=R$_3$=OH, R$_4$=OMe (R,R)
cycleapeltine (**110**) R$_1$=Me, R$_2$=R$_4$=OMe, R$_3$=OH (S,S)

limacine (**82**) R$_1$=Me, R$_2$=OH, R$_3$=R$_4$=OMe (R,R)
isotetrandrine (**85**) R$_1$=Me, R$_2$=R$_3$=R$_4$=OMe (R,S)
fangchinoline (**86**) R$_1$=Me, R$_2$=OH, R$_3$=R$_4$=OMe (S,S)
obamegine (**92**) R$_1$=Me, R$_2$=R$_3$=OH, R$_4$=OMe (R,S)
phaeanthine (**98**) R$_1$=Me, R$_2$=R$_3$=R$_4$=OMe (R,R)
berbamine (**101**) R$_1$=Me, R$_2$=R$_4$=OMe, R$_3$=OH (R,S)
thalrugosine (**113**) R$_1$=Me, R$_2$=OH, R$_3$=R$_4$=OMe (R,S)
2-norisotetrandine (**114**) R$_1$=H, R$_2$=R$_3$=R$_4$=OMe (R,S)
penduline (**118**) R$_1$=Me, R$_2$=R$_4$=OMe, R$_3$=OH (S,S)

The chemical and biological investigation of the stem bark alkaloidal extract of *Guatteria boliviana* H.Winkler (Annonaceae) led to the isolation of five new and four known BBIQ derivatives. The alkaloids were tested on trypomastigote forms of *T. cruzi* and all were active at 0.40 mM. Three of them, funiferine (**87**), antioquine (**102**), and guatteboline (**103**), with an imine function in the molecule, exhibited an IC_{90} lower than 0.16 mM (170). The substantial differences among the structures did not allow any structure–activity relationship to be elaborated.

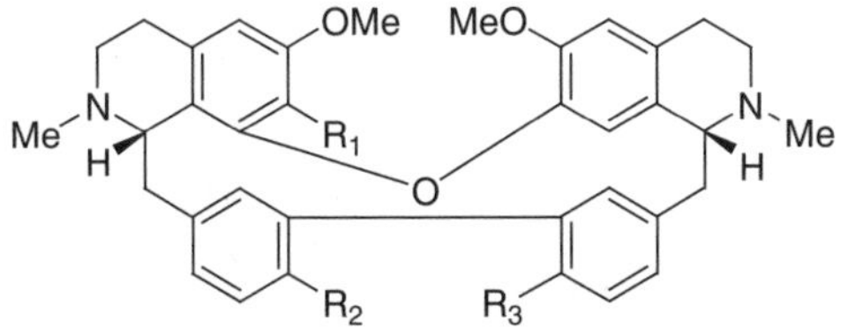

funiferine (**87**) $R_1=R_3=$OMe, $R_2=$OH (S,R)
tiliageine (**88**) $R_1=R_2=$OH, $R_3=$OMe (S,R)
antioquine (**102**) $R_1=R_3=$OH, $R_2=$OMe (S,R)

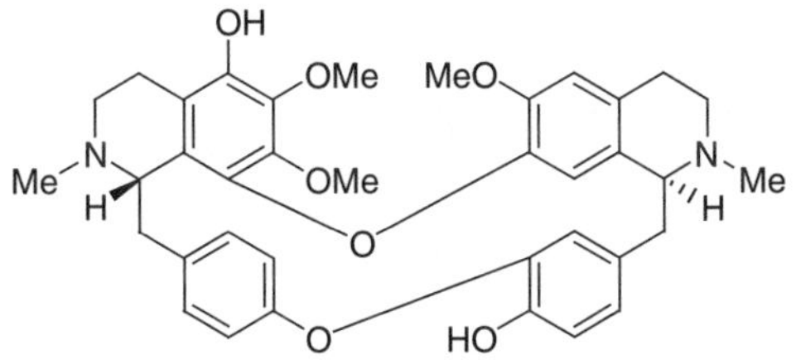

thalisopidine (**91**) (S,S)

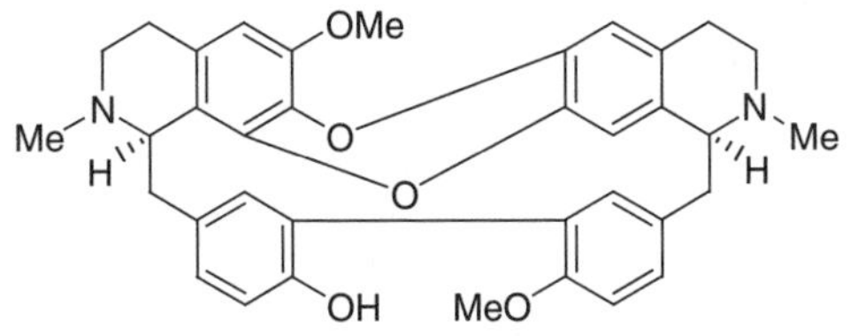

dinklacorine (**93**) (R,S)

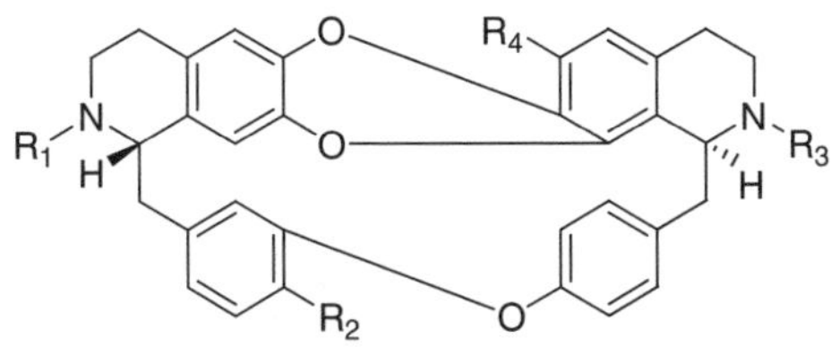

isotrilobine (**94**) $R_1=R_3=$Me, $R_2=R_4=$OMe (S,S)
trilobine (**95**) $R_1=$Me, $R_2=R_4=$OMe, $R_3=$H (S,S)
cocsoline (**99**) $R_1=$H, $R_2=$OH, $R_3=$Me, $R_4=$OMe (S,S)
12-*O*-methyltricordatine (**109**) $R_1=R_3=$Me, $R_2=$OMe, $R_4=$OH (S,S)

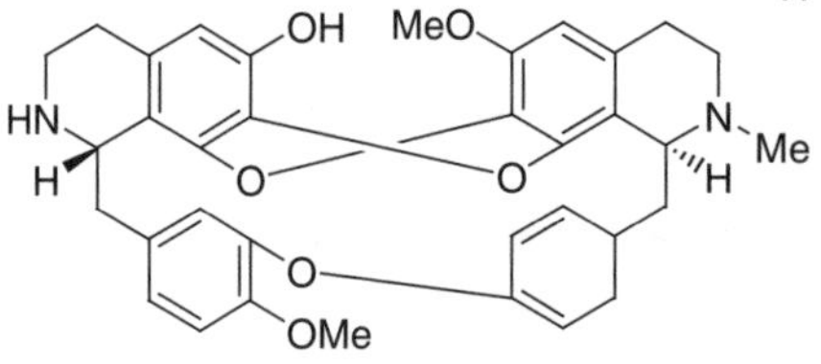

gilletine (**96**) (S,S) 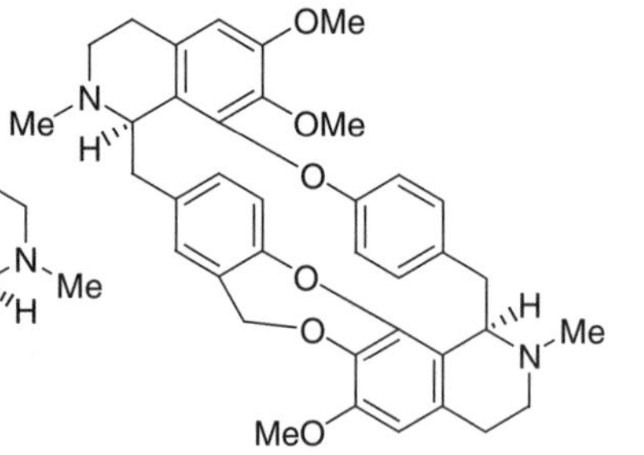   insularine (**97**) (R,R)

Fournet *et al.* have described the *in vitro* trypanocidal activity of some BBIQ alkaloids against the epimastigote forms (171) and bloodstream forms of *T. cruzi* (172). The alkaloids daphnandrine (**81**), gyrocarpine (**84**), phaeanthine (**98**), cocsoline (**99**), daphnoline (**100**), and isochondodendrine (**104**) completely lysed the trypomastigote forms of *T. cruzi* at a concentration of ca. 250 μg/mL (0.40–0.45 mM) (172). These authors also reported *in vivo* studies of some active BBIQ alkaloids (173,174). Five alkaloids, limacine (**82**), isotetrandrine (**85**), phaeanthine (**98**), curine (**105**), and cycleanine (**106**), were tested orally for their trypanocidal activity in *T. cruzi*-infected BALB/c mice in the acute phase; curine and cycleanine showed high efficacy and negative parasitemias 5–7 weeks after inoculation at 10 mg/kg (0.002 mM/kg) daily for 10 days of treatment. The other BBIQ alkaloids showed a relative efficacy against both strains (173). Interestingly, curine and cycleanine have two diphenylether linkages head to tail, while limacine, isotetrandrine and its isomer phaeanthine, which have two diphenyl-lether linkages head to head and tail to tail, did not show the same efficacy. In addition, the activity of daphnoline (**100**) and cepharanthine (**107**), as well as benznidazole, was determined in acute and chronic *T. cruzi*-infected mice (174). Compared to benznidazole, oral treatment with daphnoline was more parasitologically and serologically effective in the infected mice. In the case of chronic infection, the serological cure rate was similar to that of the standard drug. These results were not seen with cepharanthine, an excellent inhibitor of trypanothione reductase, a target for specific chemotherapy against Chagas' disease (175).

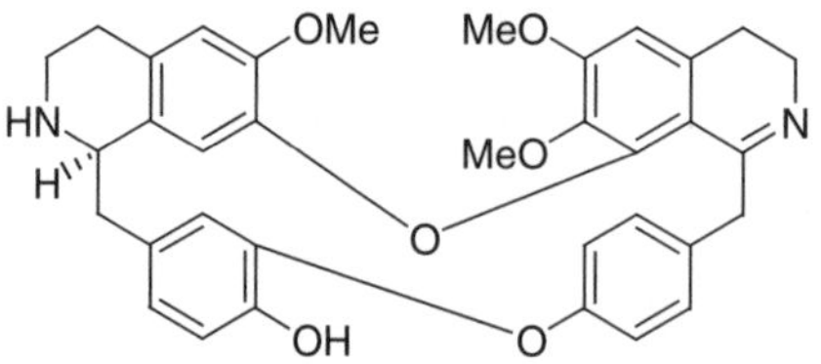

guatteboline (**103**) (R)

The antimalarial activity reported for a significant number of plants, mainly from the Annonaceae, Lauraceae, Menispermaceae, and Ranunculaceae families, has been attributed to the presence of a variety of BBIQ alkaloids (9). With a few exceptions, these compounds show reasonable antimalarial activity, but this is often coupled with cytotoxicity. The large number of BBIQ derivatives that occur in Nature prompted Angerhofer *et al.* (176) and Marshall *et al.* (177) to compare the antiplasmodial and cytotoxic activity of 53 and 24 alkaloids, respectively, isolated from different families. In the first study, the most active alkaloids against the chloroquine-sensitive clone D-6 of *P. falciparum* were candicusine (**108**) and 12-*O*-methyltricordatine (**109**), with IC_{50} values of 29 and 30 nM, respectively, whereas cepharanthine (**107**) and cycleapeltine (**110**), both with an IC_{50} of around 67 nM, were the most active against the chloroquine-resistant clone W-2. However, when the SI is considered, the most selective alkaloids were cycleanine (**106**) and malekulatine (**111**) (e.g., SI=>460 for cycleanine, as compared with >490 for malekulatine in clone D-6 of *P. falciparum*) (9,176). In the latter study, eight of the BBIQ alkaloids tested had IC_{50} values of less than 1 µM against the multidrug-resistant K1 strain of *P. falciparum in vitro*, thalisopidine (**91**) being the most potent antiplasmodial compound with an IC_{50} value of 90 nM. Under the same test conditions, the IC_{50} of chloroquine against the same multidrug-resistant strain of *P. falciparum* was 200 nM (177).

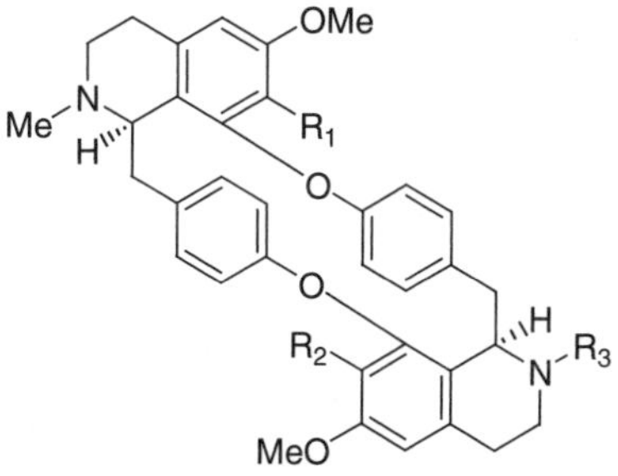

isochondodendrine (**104**) R₁=R₂=OH, R₃=Me (R,R)
cycleanine (**106**) R₁=R₂=OMe, R₃=Me (R,R)
N'-demethylcycleanine (**112**) R₁=R₂=OMe, R₃=H, (R,R)

A decoction of the root bark of the shrub *Albertisia villosa* (Exell) Forman of the Menispermaceae family is used in traditional medicine against malaria and many infectious diseases. In an attempt to explain this use, a root bark alkaloidal extract was analyzed and found to contain three BBIQ alkaloids, cocsoline (**99**), *N*-demethylcycleanine (**112**), and most abundantly, cycleanine (**106**), which has potent antiplasmodial activity. Two Menispermaceae plants, *Cyclea barbata* Miers and *Stephania rotunda* Lour., are used in traditional medicine to treat the fever associated with malaria (179,180). Through bioactivity-guided fractionation of *C. barbata*, five alkaloids were found to be responsible for the antimalarial activity. The most active compounds, cycleapeltine (**110**) and thalrugosine (**113**) were capable of inhibiting the growth of cultured *P. falciparum* strains D-6 and W-2 with IC_{50} values of 47.6–66.6 and 106.9–128.1 nM, respectively (179). Similarly, phytochemical studies of *S. rotunda* have demonstrated the presence of several alkaloids (180–183), cepharanthine (**107**) being the most active against the W-2 strain with an IC_{50} of 0.61 µM (180,184). Several BBIQ alkaloids have been isolated from *Stephania erecta* Craib (185). The antimalarial potential of each of these alkaloids was investigated with cultured *P. falciparum* strains D-6 and W-2, the most active alkaloid being 2-norisotetrandrine (**114**) ($IC_{50} = 108.6$ and 74.4 nM against D-6 and W-2, respectively) (185). Likewise, investigation of the active methanol fraction of the root of *Epinetrum villosum* (Exell) Troupin (Menisperma-ceae), a plant which is taken orally for the treatment of fever and malaria (186), led to the isolation of four BBIQ alkaloids (187). Among them, isochondodendrine (**104**) was found to have the most potent antiplasmodial activity ($IC_{50} = 0.20$ µM with SI 175), while cycleanine and cocsoline also acted against *P. falciparum* ($IC_{50} = 4.50$ and 0.54 µM, respectively). Qualitatively, these results for isochondo-dendrine correspond with those reported by Mambu *et al.* (188), who isolated and identified three BBIQ alkaloids from the stem bark of *Isolona ghesquierei* Cavaco & Keraudr. (Annonaceae). They found that curine (**105**) and isochondodendrine have strong *in vitro* antiplasmodial activity against the chloroquine-resistant strain FcM29 of *P. falciparum* with IC_{50} values of 0.35 and 0.89 µM, respectively.

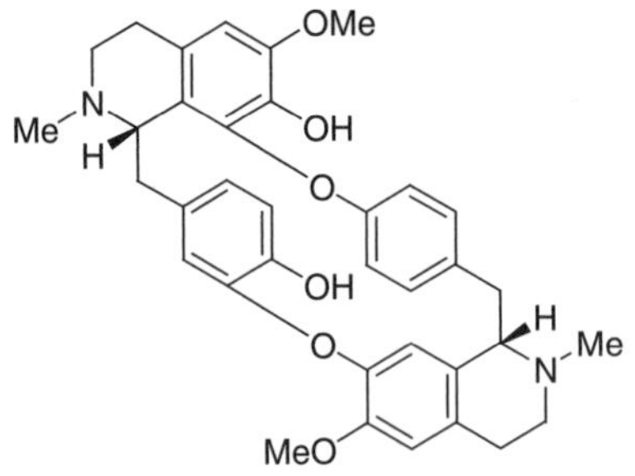

curine (**105**) (R,R)

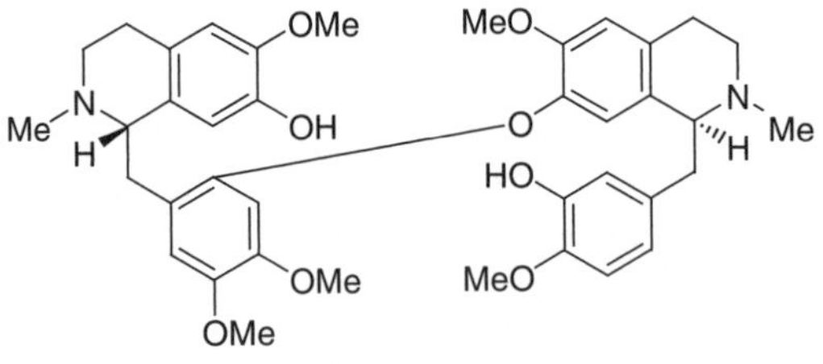

malekulatine (**111**) (S,S)

Two species of the Lauraceae, *Dehaasia incrassata* (Jack) Kosterm. and *Nectandra salicifolia* Nees, have been investigated for their antimalarial activity (9,189,190). The three alkaloids isolated from the leaves and bark of *D. incrassata* were tested against the K1 strain of *P. falciparum*, and one of them, oxyacanthine (**89**), was active with an IC_{50} of 0.50 µM (189). On the other hand, 16 alkaloids were isolated from *N. salicifolia*, but only costaricine (**115**) showed appreciable activity, with IC_{50} values of 85.8 nM against the chloroquine-sensitive D-6 clone and 0.50 μM against the chloroquine-resistant W-2 clone of *P. falciparum* (190). Similarly, two species from Ranunculaceae have been investigated for their antiplasmodial activity. Studies on *Thalictrum faberi* Ulbr. yielded several aporphine-benzylisoquinoline alkaloids and, among those thalifasine (**116**) and thalifaberidine (**117**) showed the strongest antimalarial activity with IC_{50} values of less than 1 μM against the D-6 and W-2 clones of *P. falciparum* (191). Likewise, the roots of *Isopyrum thalictroides* L. (Ranunculaceae) yielded the BBIQ alkaloid penduline (**118**), which is active against chloroquine-resistant strains of *P. falciparum*. However, it is also more toxic than chloroquine (192).

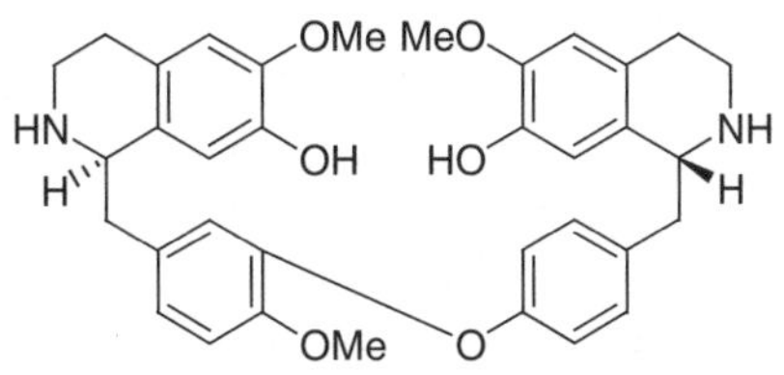

costaricine (**115**) (R,R)

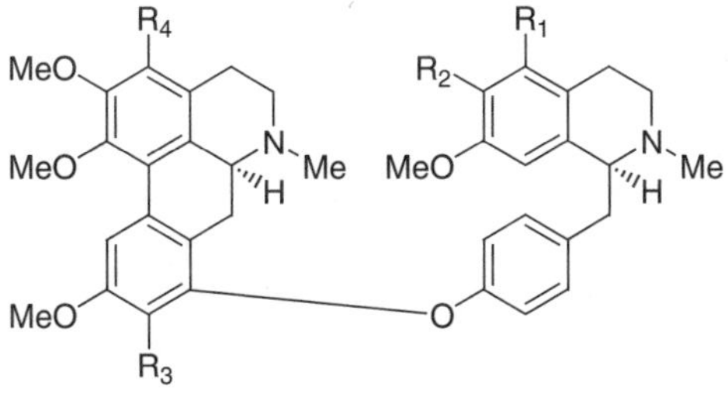

thalifasine (**116**) R_1=R_4=OH, R_2=R_3=OMe (S)
thalifaberidine (**117**) R_1=H, R_2=R_3=OH, R_4=OMe (S)

Optimal structure–antiplasmodial activity requirements of the BBIQ alkaloids are not fully understood and current results do not reveal any clear structure–activity relationships between alkaloid subgroups. Nevertheless, some features have been pointed out. In general, single-bridged BBIQ alkaloids are less active. Likewise, the status of the nitrogen atoms is fundamental to antiplasmodial activity, as evidenced by the quaternization of one or two nitrogen atoms, which leads to a loss of both toxicity and antimalarial activity. The decrease in lipophilicity of BBIQ alkaloids probably contributes to the lower toxicity (176,177). A study of the conformations assumed by compounds of the same subgroup (i.e., modification of conformation with the change of configuration at C-1 and C-1′) should give more information about the structure–activity relationship (176).

Although the mechanism of action of BBIQ alkaloids is not well known, recent studies show the capacity of these compounds to block Ca^{2+} uptake through the L-type Ca^{2+} channels (193–195), an important mechanism for the penetration of the trypomastigote forms of *T. cruzi* into the host cell (196,197). Another hypothesis describes BBIQ alkaloids as possible trypanothione reductase inhibitors. Results obtained with daphnoline (**100**) against this target partially confirm its inhibitor activity against *T. cruzi* in the bloodstream (172) and in acute or chronically *T. cruzi*-infected mice (198). Another possible explanation of the antiprotozoal properties of BBIQ alkaloids could be the DNA intercalation in combination with the inhibition of protein synthesis (142). Nevertheless, with the exception of the single-bridged BBIQ alkaloids, they all possess a large heterocycle of 18 to 20 atoms that is not totally rigid. They could probably act as adaptive intercalators, adopting a strained planar conformation after binding to the DNA.

5. Naphthylisoquinoline Alkaloids

Naphthylisoquinoline (NIQ) alkaloids are axially chiral, natural biaryls isolated from African plants belonging to the Ancistrocladaceae and Dioncophyllaceae families. The Dioncophyllaceae constitute a very small family, with only three species: *Triphyophyllum peltatum* (Hutch & Dalziel.) Airy Shaw (a "part-time" carnivorous liana), *Habropetalum dawei* (Hutch & Dalziel.) Airy Shaw, and *Dioncophyllum thollonii* Baill. Closely related, but not carnivorous, are the Ancistrocladaceae, an even smaller family whose only genus, *Ancistrocladus*, consists of ca. 25 species found in the palaeotropic rain forests of Africa and Asia (199). These natural biaryl alkaloids have been characterized chemically (unique structural framework, axial chirality, unprecedented origin of isoquinoline alkaloids from acetate), and also pharmacologically due to their wide variety of activities related to, e.g., bilharzia, Chagas' disease, sleeping sickness, leishmaniasis, river blindness, elephantiasis and, in particular, malaria (11,199,200). Some of the alkaloids are also active against larvae of the vectors *Anopheles stephensi* and *Aedes aegypti* (199,201).

Among the monomeric NIQ alkaloids tested up to now, dioncophyllines A (**119**), B (**120**), and E (**121**), isolated from the rare West African liana *D. thollonii*, are the most active, with IC_{50} values of 2–3 µM against *T. b. brucei* or *T. b. rhodesiense* bloodstream forms (202,203). Dioncophyllines A and B were previously known from the related plant species *T. peltatum*, which is the most phytochemically investigated species in the Dioncophyllaceae family (204–206). Dioncophylline A is also a strong agent against *T. cruzi* ($IC_{50} = 1.85$ µM), *T. b. rhodesiense* ($IC_{50} < 0.50$ µM), and *L. donovani* ($IC_{50} < 32.05$ µM) (199). Ancistroealaines A (**122**) and B (**123**), two 5,8'-coupled NIQ alkaloids isolated from the stem bark of *Ancistrocladus ealaensis* (207), were tested *in vitro* against *L. donovani*, *T. cruzi*, and *T. b. rhodesiense*. Ancistroealaine A displayed good activity against *T. cruzi* ($IC_{50} = 5.60$ µM) and *T. b. rhodesiense* ($IC_{50} = 8.25$ µM). Ancistroealaine B, on the other hand, showed significant trypanocidal activity, with an IC_{50} of 4.98 µM against *T. b. rhodesiense*. Both alkaloids exhibited cytotoxic effects in mammalian L6 cells (rat skeletal myoblasts) at much higher concentrations ($IC_{50} > 215$ µM for ancistroealaine A and 220 µM for ancistroealaine B).

The first phytochemical investigation of the East African liana *A. tanzaniensis* Cheek & Frim.-Møll., reported the isolation of NIQ alkaloid derivatives with antiprotozoal activity, including two new alkaloids, ancistrotanzanines A (**124**) and B (**125**, atropodiastereomer of ancistroealaine A), and the known alkaloid ancistrotectoriline A (**126**) (208). The most significant activity was found for ancistrotanzanine A and ancistrotanzanine B against *L. donovani*, *T. cruzi*, and *T. b. rhodesiense*, with IC_{50} values ranging from 1.73 to 4.44 µM and 1.67 to 3.81 µM, respectively. Ancistrotectoriline A, also obtained from the Southeast Asian species *A. tectorius* (209), showed weak antitrypanosomal activity against *T. b. rhodeniense* (IC_{50} = 4.98 µM). Other NIQs with similar antiprotozoal activity have since been reported from the same species (210). Ancistrocladidine (**127**) was active against *L. donovani* (IC_{50} = 7.15 µM), weaker than the active ancistrotanzanine B (**125**) by a factor of only 2. Likewise, good antitrypanosomal activity was exhibited by ancistrocladidine and ancistrotanzanine C (**128**) against the pathogen *T. b. rhodesiense* (IC_{50} = 4.93 and 3.19 µM) (210). Unfortunately, all of the natural alkaloids were less active than the standard drugs.

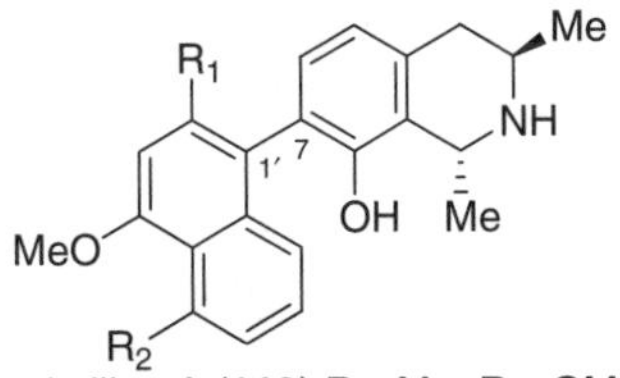

dioncophylline A (**119**) R₁=Me, R₂=OMe (R,R) (*P*)
dioncopeltine A (**138**) R₁=CH₂OH, R₂=OH (R,R)
habropetaline A (**139**) R₁=CH₂OH, R₂=OMe (R,R)
7-epidioncophylline A (**144**) R₁=Me, R₂=OMe (R,R) (*M*)
5′-*O*-demethyldioncophylline A (**145**): R₁=Me, R₂=OH (R,R)

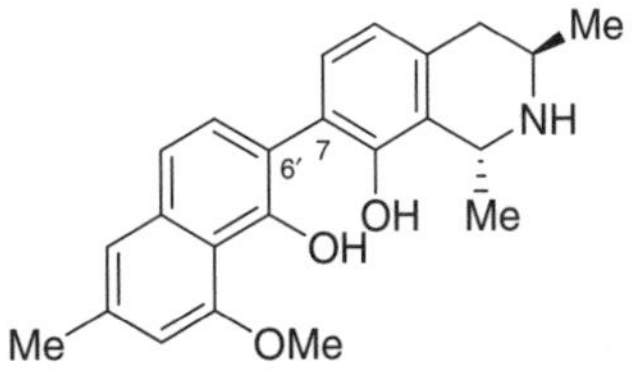

dioncophylline B (**120**) (R,R)

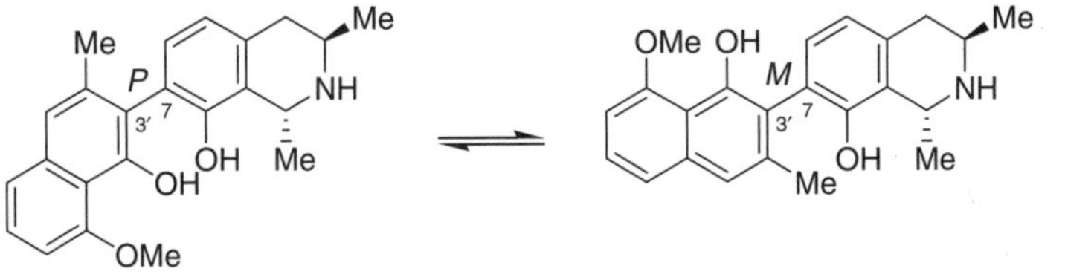

dioncophylline E (**121**) (R,R) (*P* and *M* rotational isomers)

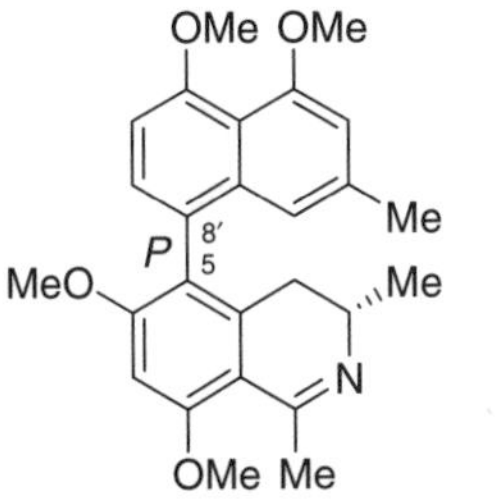

ancistroealaine A (**122**) (S)

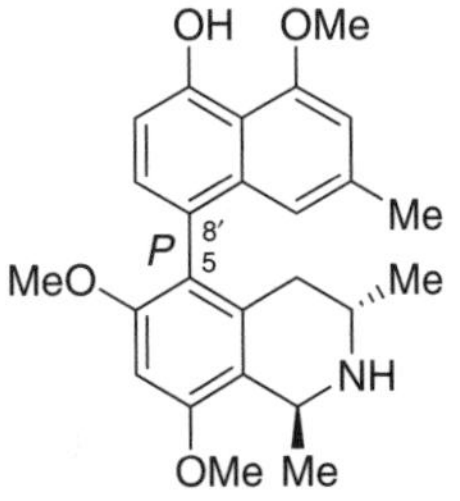

ancistroealaine B (**123**) (S,S)

NIQ alkaloids with similar antiprotozoal activity have been isolated from *A. congolensis* J.Leónard (211), *A. heyneanus* Wall. (212), *A. likoko* J.Leónard (213), and

A. griffithii Planch. (214). The three alkaloids obtained from the first species, named as ancistrocongoline A (**129**), ancistrocongoline B (**130**), and korupensamine A (**131**), were tested for their *in vitro* activity against *T. b. rhodesiense* ($IC_{50} = 7.55$, 6.02, and 4.93 µM, respectively) (211). Three secondary metabolites belonging to the rare 7,3′-coupling type: ancistrocladidine (**127**), ancistroheynine B (**132**), and ancistrotanzanine C (**128**), which has already been previously described from the East African species *A. tanzaniensis* (210), were isolated from the leaves of the Indian liana *A. heyneanus* (212). Antitrypanosomal and antileishmanial activities were exhibited by ancistrocladidine against *T. b. rhodesiense* and *L. donovani* ($IC_{50} = 4.93$ and 7.15 µM). Ancistroheynine B and ancistrotanzanine C, as well as ancistrolikokine D (**133**), which was isolated from *A. likoko*, presented relatively high activity against *T. b. rhodesiense* with IC_{50} values of 7.41, 3.19, and 6.90 µM, respectively (212,213). Likewise, two monomeric NIQ alkaloids, ancistrogriffine A (**134**) and ancistrogriffine C (**135**), and the first dimer of a 7,8′-coupled NIQ, ancistrogriffithine A (**136**), were isolated from the twigs of *A. griffithii* and analyzed for their antiprotozoal activity (214). Ancistrogriffine A also showed good activity against *L. donovani* ($IC_{50} = 7.61$ µM), while all three alkaloids were active against *T. b. rhodesiense* ($IC_{50} = 5.32$, 7.55, and 1.15 µM, respectively). On the contrary, other dimers, including the anti-HIV active michellamines, e.g., michellamine B (**137**), were devoid of any antitrypanosomal activity (11,202). The monomeric dioncopeltine A (**138**), with its additional alcohol function on the methyl group of the naphthalene ring, has a distinctly lower activity against *T. b. brucei* bloodstream trypomastigotes ($IC_{50} = 22$ µM) (11). Habropetaline A (**139**), isolated from the roots of *T. peltatum* and identified previously in the crude extract of the rare and difficult-to-provide, related plant species *H. dawei* (215), was inactive against *T. b. rhodesiense*, *T. cruzi*, and *L. donovani* (216). These findings, although not as good as the results for other NIQs, are an important contribution to the structure–activity relationship investigations.

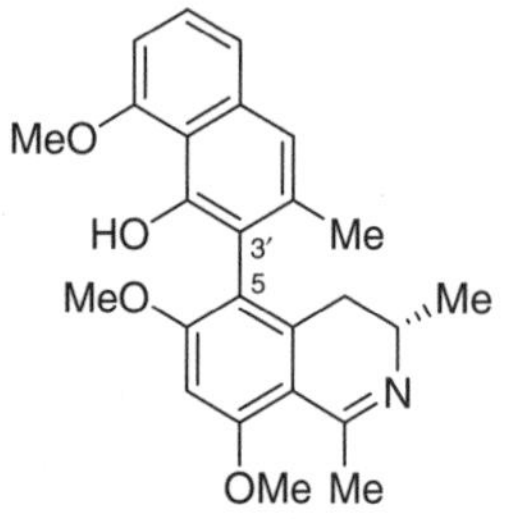

ancistrotanzanine A (**124**) (S)

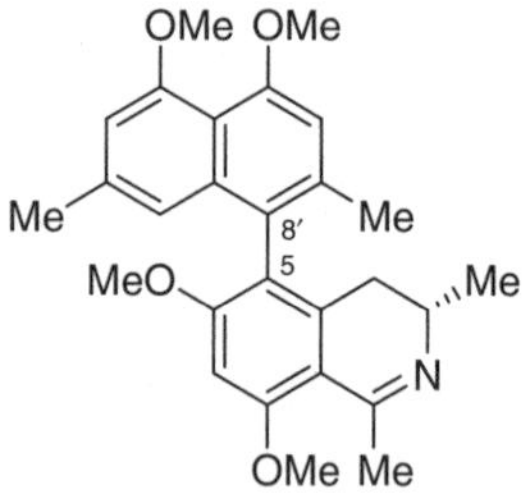

ancistrotanzanine B (**125**) (S)

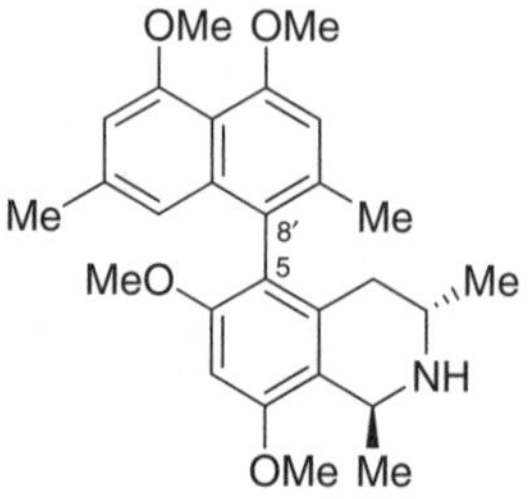

ancistrotectoriline A (**126**) (S,S)

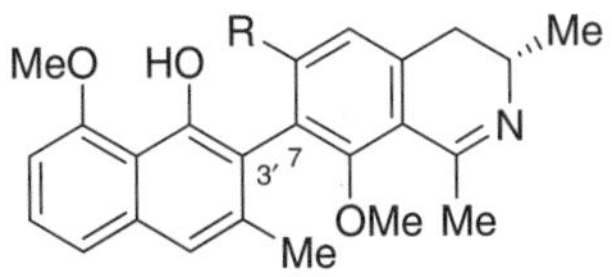

ancistrocladidine (**127**) R=OMe (S)
ancistroheynine B (**132**) R=OH (S)

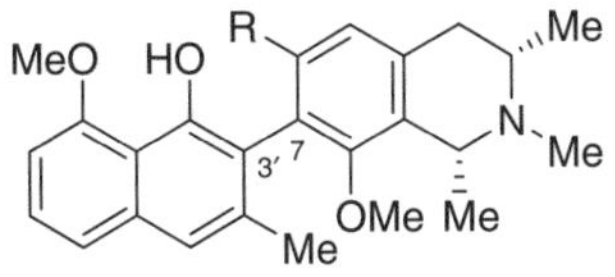

ancistrotanzanine C (**128**) R=OH (R,S)
ancistrotectorine (**146**) R=OMe (R,S)

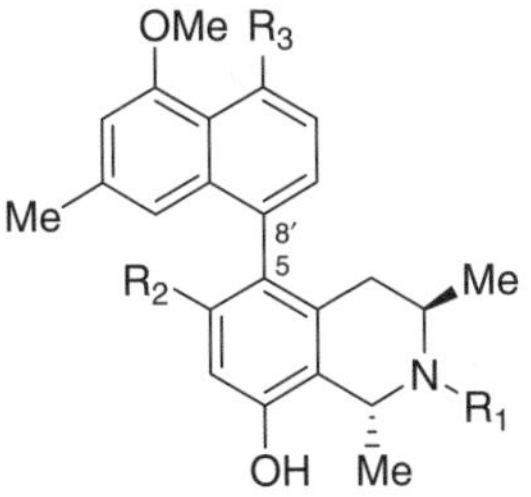

ancistrocongoline A (**129**) R$_1$=Me, R$_2$=R$_3$=OH (R,R)
ancistrocongoline B (**130**) R$_1$=Me, R$_2$=R$_3$=OMe (R,R)
korupensamine A (**131**) R$_1$=H, R$_2$=R$_3$=OH (R,R)

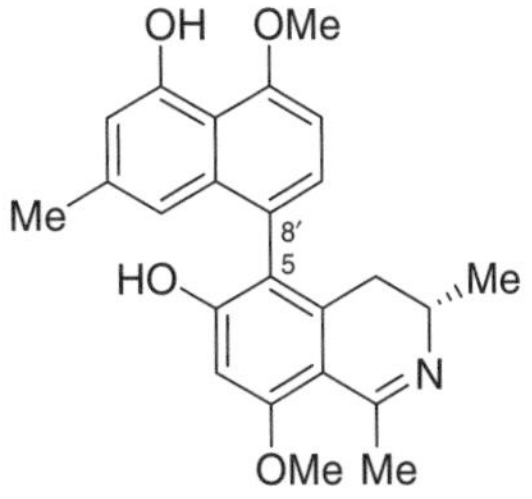

ancistrolikokine D (**133**) (S)

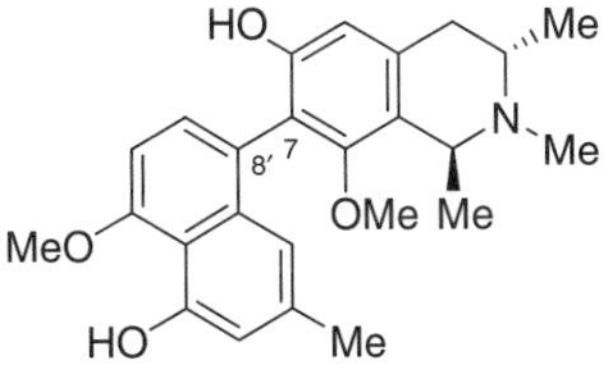

ancistrogriffine A (**134**) (S,S)

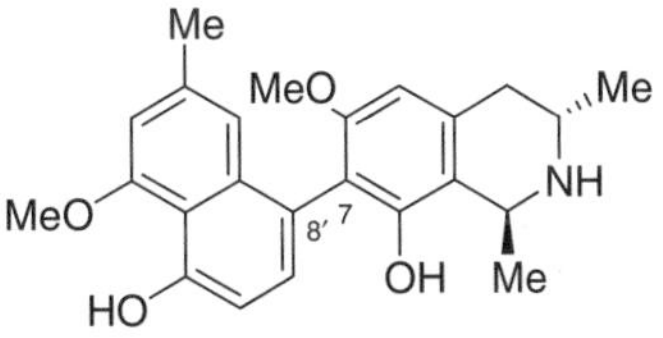

ancistrogriffine C (**135**) (S,S)

Very recently, the NIQ alkaloids ancistrocladinium A (**140**) and ancistrocladinium B (**141**) and a synthetically prepared isoquinolinium salt **142** were found to present potent activity against intracellular amastigotes of *L. major* at concentrations in the low submicromolar range (IC$_{50}$ = 4.90, 1.24, and 2.91 μM, respectively). They are more toxic against J774.1 macrophages and peritoneal macrophages than the reference drug amphotericin B (217). In general, the NIQ alkaloids, with *N,C*-coupled rings and a hetero-biaryl part, exhibit significantly higher leishmanicidal activities. With regard to structural features, these results demonstrate the importance of the quaternary nitrogen atom, which clearly increases antileishmanial activity (217).

Although the results concerning antileishmanial and antitrypanosomal activities of NIQ alkaloids are particularly recent, their antimalarial activities were discovered a few years earlier due to the traditional use of the plants in treating malaria (199). One species of the Dioncophyllaceae, *T. peltatum* (218–220), and several species of Ancistrocladaceae, viz., *A. abbreviatus* Airy Shaw (218,219), *A. barteri* Scott-Elliot (218,219), *A. heyneanus* (221,222), *A. korupensis* D.W.Thomas & Gereau (223–225), *A. likoko* (226), and *A. robertsoniorum* J.Leónard (227) have been investigated for their antiplasmodial activity. In agreement with the results obtained from crude extracts, several NIQ alkaloids are known to be highly active against *P. falciparum*. They were initially described as exhibiting strong growth-inhibiting activities *in vitro* against *P. falciparum* and *P. berghei* (228,229), and *in vivo* against *P. berghei* (230). Another remarkable fact is that NIQ alkaloids act against the blood forms of *Plasmodium* spp., and against the exoerythrocytic forms (199,229,231), a most promising additional perspective for these novel antimalarial agents. For synchronized forms of *P. chabaudi chabaudi*, stage-specific activities have been found (232).

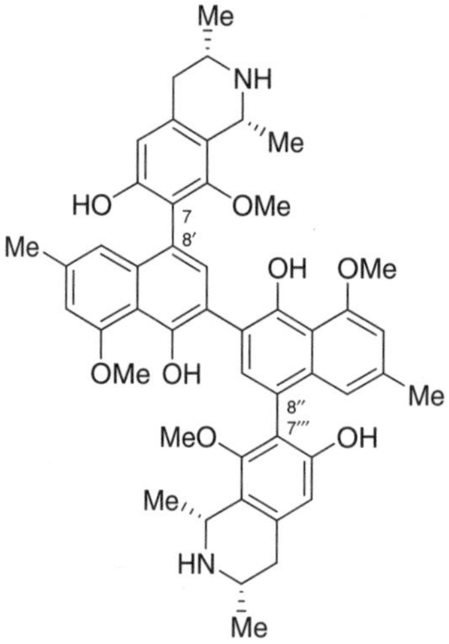

ancistrogriffithine A **136**

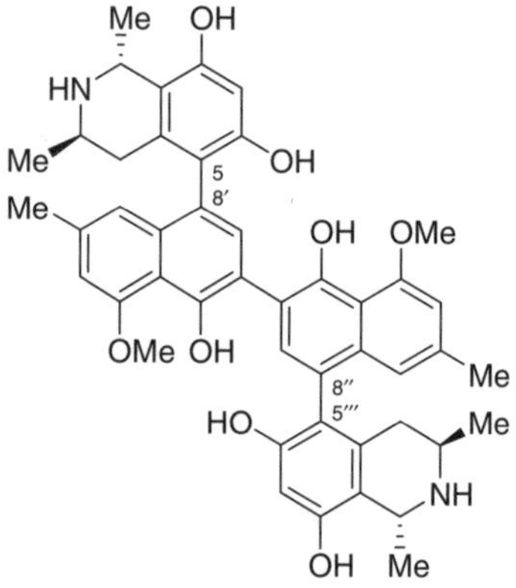

michellamine B **137**

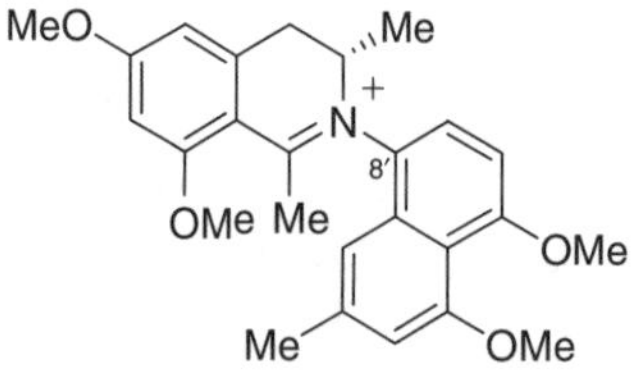

ancistrocladinium A (**140**) (S)

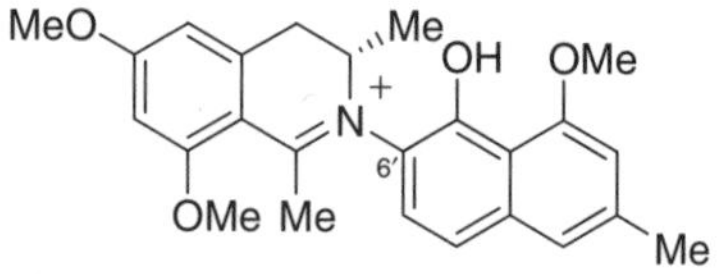

ancistrocladinium B (**141**) (S)

Among the most active alkaloids isolated from *T. peltatum* are dioncophylline B (**120**) ($IC_{50} = 0.616\,\mu M$), dioncopeltine A (**138**) ($IC_{50} = 0.055\,\mu M$), habropetaline A (**139**) ($IC_{50} = 5.8\,nM$), dioncophylline C (**143**) ($IC_{50} = 0.038\,\mu M$), 7-epidioncophylline A (**144**) ($IC_{50} = 0.503\,\mu M$), and 5'-O-demethyldioncophylline A (**145**) ($IC_{50} = 0.935\,\mu M$) with excellent antiplasmodial activities against the chloroquine-sensitive strain of *P. falciparum* (NF54) *in vitro*, and in some cases, *in vivo* (9,206,216,219,230,232). The IC_{50} values obtained for these alkaloids are much lower than those observed for most other plant-derived compounds, and compare well with the IC_{50} values for antiplasmodial drugs currently in use (chloroquine). The good *in vitro* activity exhibited by dioncophylline B, dioncophylline C, and dioncopeltine A led to *in vivo* testing in mice against *P. berghei* or *P. chabaudi chabaudi* (228,230,232). A complete cure was achieved by dioncophylline C in malaria-infected mice after a 4-day oral treatment with 50 mg/kg/day, without noticeable toxic effects. Likewise, dioncophylline E (**121**), an alkaloid isolated from *D. thollonii*, exhibited good activity against *P. falciparum*, with nearly identical IC_{50} values against the chloroquine-sensitive (NF54) and chloroquine-resistant (K1) strains. With values of only 0.06 and 0.058 μM, the activity of the new alkaloid ranges within that of the most active NIQ alkaloids, and is only weaker than the standards artemisinin and chloroquine by a factor of 5–10 (203).

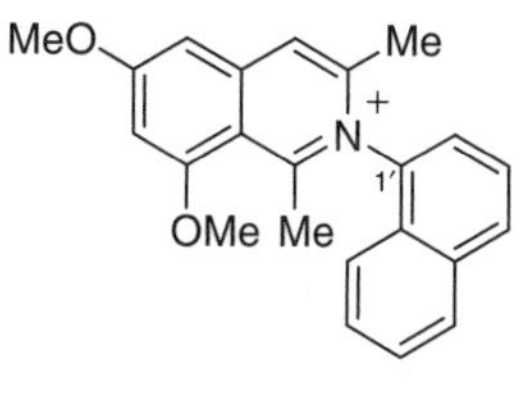

isoquinolinium salt **142**

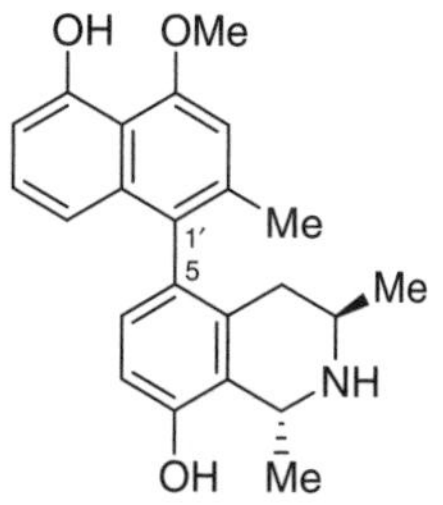

dioncophylline C (**143**) (R,R)

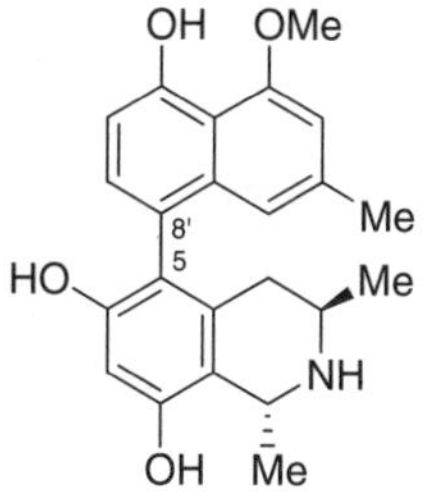

korupensamine B (**147**) (R,R)

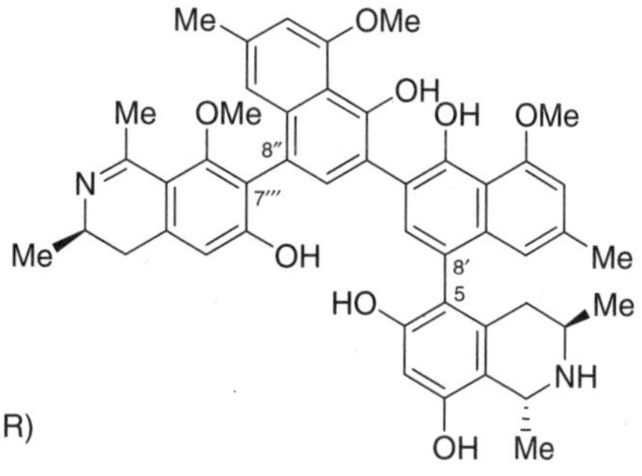

korundamine A **148**

Other NIQs like ancistroealaine B (**123**) (IC_{50} = 1.28 μM), ancistrotanzanine A (**124**) (IC_{50} = 0.74 μM), ancistrotanzanine B (**125**) (IC_{50} = 0.72 μM), ancistrotectoriline A (**126**) (IC_{50} = 1.19 μM), ancistrocladidine (**127**) (IC_{50} = 0.74 μM), ancistrotanzanine C (**128**) (IC_{50} = 0.24 μM), ancistrocongoline A (**129**) (IC_{50} = 0.54 μM), ancistrocongoline B (**130**) (IC_{50} = 0.37 μM), korupensamine A (**131**) (IC_{50} = 0.43 μM), ancistroheynine B (**132**) (IC_{50} = 1.27 μM), ancistrogriffine A (**134**) (IC_{50} = 0.18 μM), ancistrogriffine C (**135**) (IC_{50} = 1.06 μM), ancistrogriffithine A (**136**) (IC_{50} = 0.04 μM), ancistrotectorine (**146**) (IC_{50} = 1.66 μM), and korupensamine B (**147**) (IC_{50} = 0.47 μM) showed significant antimalarial activities with IC_{50} values lower than 2 μM against the chloroquine- and pyrimethamine-resistant K1 strain of *P. falciparum* (207,208,210–214,223). However, in most cases, the IC_{50} values were higher than those of the standards (artemisinin or chloroquine). Only the dimeric NIQ ancistrogriffithine A (136) was found to be more active than the standard chloroquine by a factor of two (214). This result was surprising, since dimeric NIQs rarely show considerable antiplasmodial activities (223). Another exception is korundamine A (148) ($IC_{50} = 1.43\,\mu M$), a dimeric NIQ alkaloid isolated from *A. korupensis* that shows both anti-HIV and antimalarial activity (225). It is noteworthy that some of the studied alkaloids exhibited cytotoxic effects on mammalian cells only at or above $30\,\mu g/mL$, which reflects the selectivity of these natural compounds for the protozoan parasite.

It seems that the compounds from the Dioncophyllaceae are more active than those isolated from the Ancistrocladaceae. Structure–activity considerations indicate that possible criteria for antiplasmodial activity include an *R*-configuration at C-3 associated with the absence of an oxygen substituent at C-6 in the isoquinoline moiety and the absence of *N*-methylation (219,233). Dioncophylline E (121), dioncopeltine A (138), habropetaline A (139), and dioncophylline C (143), maybe the most active NIQs, represent pure Dioncophyllaceae-type alkaloids, all having in common the aforementioned characteristics (234). Nevertheless, not all the alkaloids of the Ancistrocladaceae are of the Ancistrocladaceae-type. In fact, all the West African *Ancistrocladus* spp. investigated so far contain Dioncophyllaceae-type alkaloids, as well as many hybrid-type forms (234,235). Similarly, structure–activity relationship investigations reveal that in NIQ alkaloids the presence of at least one or two free aromatic hydroxy functions is essential for high antiplasmodial activity (228), and this could partly explain the observed activity.

The mechanism of action of NIQ compounds is not very well known. François *et al.* (232) suggest that the antiplasmodial activity of dioncophylline B is due to its inhibition of hemozoin degradation, but this is based only on microscopic observations. Further investigation is required to study the possible mode of action of active alkaloids in the different parasites. Nevertheless, existing results show that NIQ alkaloids are structurally intriguing and pharmacologically promising natural products.

6. Amaryllidaceae Alkaloids

Plants of the Amaryllidaceae family are known to produce structurally unique alkaloids, which have been isolated from plants of all of the genera of this family. The Amaryllidaceae alkaloids represent a large, and still expanding, group of isoquinoline alkaloids, the majority of which are not known to occur in any other family of plants. Since the isolation of the first alkaloid, lycorine, from *Narcissus pseudonarcissus* L. in 1877, substantial progress has been made in examining the Amaryllidaceae plants, although they are still a relatively untapped phytochemical source (236). At present, over 300 alkaloids have been isolated from plants of this family and, although their structures vary considerably, these alkaloids are considered to be biogenetically related (236,237).

The Amaryllidaceae alkaloids present a wide range of interesting physiological effects, including antitumor, antiviral, cytotoxic, acetylcholinesterase inhibitory, immunostimulatory, antiinflammatory, analgesic, and DNA-binding activities, and some have also been used in the treatment of Alzheimer's disease (236). Some of these alkaloids are of particular interest because of their potential antiprotozoal activity. Thus, lycorine (**149**), augustine (**150**), and crinamine (**151**) were found to be the principal antimalarial constituents in *Crinum amabile* Donn bulbs (238). Augustine, a 5,10b-ethano-phenanthridine, appeared to be the most active alkaloid demonstrating significant antimalarial activity in both chloroquine-sensitive and chloroquine-resistant strains of *P. falciparum* (IC_{50} = 0.46 and 0.60 μM, respectively). Similarly, the pyrrolophenanthridine-type alkaloid lycorine was found to be very active against different strains of *P. falciparum* (IC_{50} of around 1.04–2.44 μM) (238,239). However, compared with the antimalarial control compounds, its selectivity indices were low (238). In other studies of the antimalarial properties of Amaryllidaceae alkaloids (240), hemanthamine (**152**) and hemanthidine (**153**) exhibited activity against the chloroquine-sensitive strain T9/96 of *P. falciparum* (IC_{50} = 2.33 and 1.10 μM, respectively). In addition to hemanthamine and hemanthidine, lycorine, 3-epihydroxybulbispermine (**154**), galanthine (**155**), and pancracine (**156**) were also highly active against the chloroquine-resistant strain K1 of *P. falciparum*, with IC_{50} values lower than 1 μg/mL. Some of these alkaloids were more active than the standards used (240–242). Likewise, hemanthamine and hemanthidine isolated from the bulbs of *Cyrtanthus elatus* (Jacq.) Traub, and pancracine isolated from *N. angustifolius* Curt. subspecies *transcarpathicus*, presented antimalarial activity against the NF54 strain of *P. falciparum* (IC_{50} = 2.22, 2.20, and 2.43 μM, respectively) (242,243).

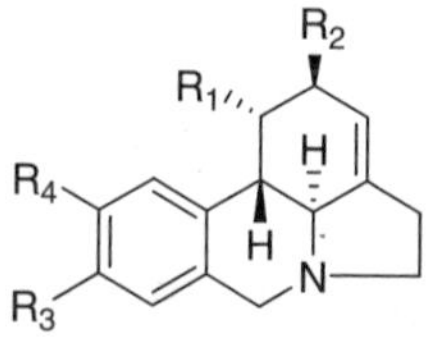

lycorine (**149**) $R_1=R_2=OH$, $R_3+R_4=OCH_2O$
galanthine (**155**) $R_1=OH$, $R_2=R_3=R_4=OMe$
1,2-*O*-diacetyllycorine (**159**) $R_1=R_2=OAc$, $R_3+R_4=OCH_2O$
pseudolycorine (**161**) $R_1=R_2=R_4=OH$, $R_3=OMe$

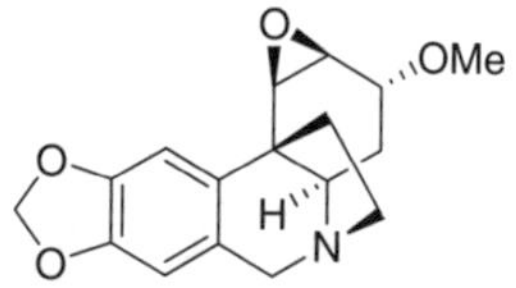

augustine **150**

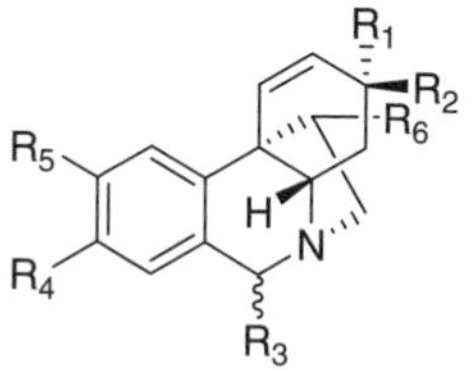

crinamine (**151**) $R_1=OMe$, $R_2=R_3=H$, $R_4+R_5=OCH_2O$, $R_6=OH$
hemanthamine (**152**) $R_1=R_3=H$, $R_2=OMe$, $R_4+R_5=OCH_2O$, $R_6=OH$
hemanthidine (**153**) $R_1=H$, $R_2=OMe$, $R_3=R_6=OH$, $R_4+R_5=OCH_2O$
3-epihydroxybulbispermine (**154**) $R_1=R_3=H$, $R_2=R_6=OH$, $R_4+R_5=OCH_2O$
oxomaritidine (**158**) $R_1+R_2=O$, $R_3=R_6=H$, $R_4=R_5=OMe$

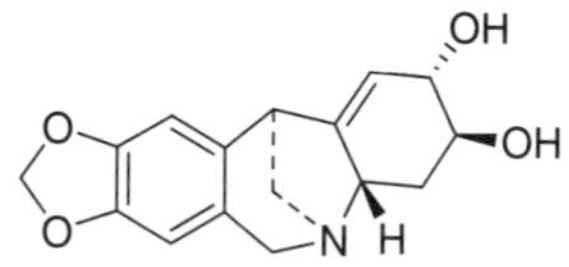

pancracine **156**

Hemanthidine, also isolated from *Zephyranthes citrina* Baker, and 3-*O*-acetylsanguinine (157) isolated from *Crinum kirkii* Baker bulbs, showed *in vitro* biological activity against *T. b. rhodesiense* (strain STIB-900). with similar IC_{50} values of 3.47 and 3.48 μM, respectively, and against *T. cruzi* (Tulahuen C4 strain) with IC_{50} values of 4.35 and 7.29 μM, respectively. On the other hand, galanthine, pancracine, oxomaritidine (158) and 1,2-*O*-diacetyllycorine (159) showed activity only against *T. b. rhodesiense* with IC_{50} values of 9.77, 2.44, 9.81, and 2.69 μM, respectively (241–244). It is interesting to note that pancracine showed no cytotoxicity for L6 cells (rat skeletal myoblasts), which confirms the selective activity of this alkaloid for *T. b. rhodesiense* and *P. falciparum* (242).

3-*O*-acetylsanguinine 157   ungeremine 160

These results prompted us to investigate the alkaloids from the bulbs of *Phaedranassa dubia* (H.B. & K.) J.F.Macbr., and to assay the *in vitro* antiprotozoal activity against *T. b. rhodesiense*, *T. cruzi*, *L. donovani*, and *P. falciparum*. Eight alkaloids were obtained, and among them, ungeremine (160) showed the highest activity against three of the four protozoan parasites tested. It was active against *T. b. rhodesiense* ($IC_{50} = 3.41$ μM), *T. cruzi* ($IC_{50} = 2.99$ μM), and *P. falciparum* ($IC_{50} = 0.32$ μM), but not against *L. donovani*. In spite of showing a degree of cytoxicity against L6 cells (rat skeletal myoblasts, $IC_{50} = 60.85$ μM), the SI of ungeremine (IC_{50} L6/IC_{50} K1) for *P. falciparum* was around 190, which confirms its selective activity for this protozoa. Pseudolycorine (161) showed some *in vitro* biological activity against *P. falciparum* ($IC_{50} = 0.83$ μM), as did hemanthamine against *T. b. rhodesiense* ($IC_{50} = 1.62$ μM) and *P. falciparum* ($IC_{50} = 2.29$ μM), the latter was in agreement with previous reports (240,243). For the last two alkaloids, no cytotoxicity was found for L6 cells, which confirms their selective antiprotozoal activity (data not published).

Structure–antiprotozoal activity relationships have not been studied in Amaryllidaceae alkaloids. Some results suggest that the methylenedioxy group and a tertiary non-methylated nitrogen contribute to a higher activity in these alkaloids (240). Nevertheless, we have observed strong activity in alkaloids with quaternary nitrogen. We have also found that the mechanism of antiplasmodial action of Amaryllidaceae alkaloids does not depend on interactions with heme, at least in the case of alkaloids like ungeremine, pseudolycorine, and hemanthamine (data not published). Additional investigation is required to specify the necessary structural requirements for antiparasitic activity, and, in particular, the possible mode of action.

B. Indole Alkaloids

1. β-Carboline and Structurally Related Alkaloids

β-Carboline alkaloids, also known as harmala alkaloids since they were first isolated from *Peganum harmala* L. (Zygophyllaceae), constitute a large group of natural and synthetic indole alkaloids with varying degrees of aromaticity and a tricyclic pyrido[3,4-*b*]indole basic structure, including two heterocyclic nitrogen atoms. In addition to their natural occurrence in plants, β-carbolines have been found in a number of other sources, including foodstuffs, marine organisms, insects, and mammals, as well as human tissues and body fluids (245). They are compounds of great interest because of the variety of actions they evoke in biological systems, such as intercalating into DNA, inhibiting CDK, topoisomerase, and monoamine oxidase, and interacting with benzodiazepine and 5-hydroxy serotonin receptors. Furthermore, β-carbolines also possess a broad spectrum of pharmacological properties including sedative, spasmolytic, anxiolytic, hypnotic, antiplatelet, vasorelaxant, anticonvulsant, antitumor, antiviral, antimicrobial, as well as antiparasitic activities (245–250).

One of the simple β-carboline alkaloids reported to possess antiprotozoal activity is the dihydro-β-carboline harmaline (**162**), the main constituent of a number of plants used in traditional medicine to cure leishmaniasis, including *P. harmala* and *Passiflora incarnata* L. (6,251). Harmaline has shown antiparasitic action against the human pathogen *L. mexicana amazonensis* both *in vitro* and *in vivo* (251). Recently, harmaline and two aromatic β-carbolines, harmane (**163**) and harmine (**164**), were investigated for their *in vitro* antileishmanial activity toward parasites of the species *L. infantum* and *L. donovani* (252,253). Harmaline weakly inhibited the growth of *L. infantum* promastigotes ($IC_{50} = 116.8\,\mu M$), but it showed an interesting amastigote-specific activity with an IC_{50} of $1.16\,\mu M$ and a SI greater than 170. Harmane and harmine displayed weak antileishmanial activity toward both the promastigote and amastigote forms of *L. infantum*. Harmine was the most efficient alkaloid against the promastigote stage of the parasite ($IC_{50} = 3.7\,\mu M$), but both harmane and harmine strongly inhibited the growth of intracellular amastigotes ($IC_{50} = 0.27$ and $0.23\,\mu M$, respectively) with moderate selectivity (SI = 81 and 73) (252). Results observed in some studies demonstrated that harmaline, harmane, harmine, and other simple β-carboline alkaloids, like harmalol (**165**) and harmol (**166**), could also exert antiproliferative effects on parasites of the genus *Trypanosoma*. The most potent were harmane, which presented trypanocidal activity against *T. b. rhodesiense*, strain STIB 900 ($IC_{50} = 12.63\,\mu M$), and harmine against epimastigotes of two different strains of *T. cruzi*: Tulahuen and LQ ($IC_{50} = 19.69$ and $17.99\,\mu M$, respectively) (254,255). On the other hand, harmaline, harmalol, and harmol inhibited the *in vitro* growth of *T. cruzi* epimastigotes by 50–90% at around $0.23\,mM$ (5,256).

Bioassay-guided fractionation of the bark extract of *Annona foetida* Mart. (Annonaceae) afforded two antileishmanial pyrimidine-β-carboline alkaloids, annomontine (**167**) and *N*-hydroxyannomontine (**168**). Annomontine was six times more active against the promastigote forms of *L. braziliensis*

(IC$_{50}$ = 34.8 µM) than N-hydroxyannomontine (IC$_{50}$ = 252.7 µM). Both alkaloids were inactive against *L. guyanensis* (257).

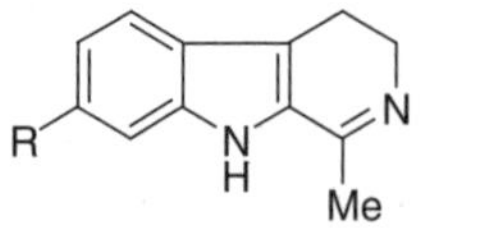

harmaline (**162**) R=OMe
harmalol (**165**) R=OH

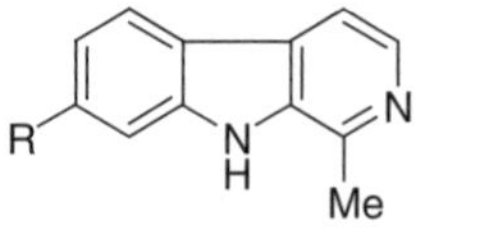

harmane (**163**) R=H
harmine (**164**) R=OMe
harmol (**166**) R=OH

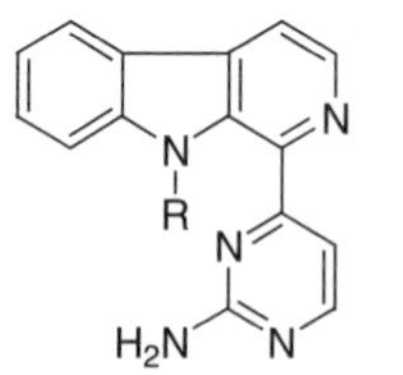

annomontine (**167**) R=H
N-hydroxyannomontine (**168**) R=OH

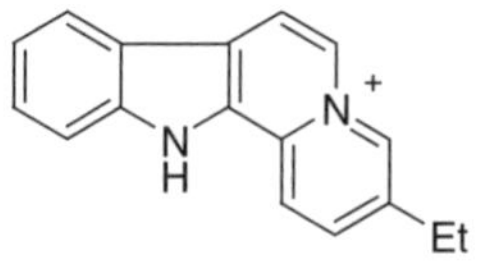

flavopereirine **169**

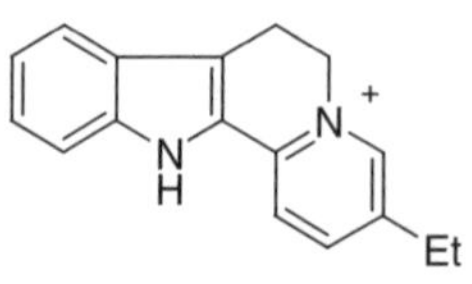

6,7-dihydroflavopereirine **170**

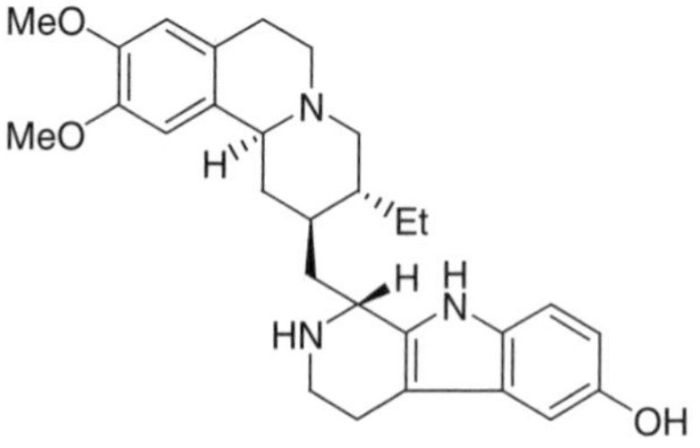

tubulosine **171**

There are fewer reports of β-carboline alkaloids exhibiting antimalarial activity than antileishmanial and/or trypanocidal effects. A series of indole alkaloids was obtained from the bark of *Geissospermum sericeum* Miers (Apocynaceae), a plant that belongs to a small genus of Amazonian trees native to northern South America, several of which are recognized locally as having antimalarial properties. Among them, the best antiplasmodial activity was observed for the β-carboline alkaloid flavopereirine (**169**) against the K1 and T9/96 strains of *P. falciparum* (IC$_{50}$ = 11.53 and 1.83 µM, respectively) (258,259). Flavopereirine also showed moderate cytotoxicity (IC$_{50}$ = 10.7 µM) against the human cell line KB. This cytotoxicity had been noted previously because of the ability of the alkaloid to selectively inhibit the synthesis of cancer cell DNA (260). Wright *et al.* obtained an IC$_{50}$ value of 3.02 µM for the related derivative 6,7-dihydroflavopereirine (**170**) against chloroquine-resistant K1 *P. falciparum* (261). In addition, *Pogonopus tubulosus* K.Schum. (Rubiaceae), a Bolivian plant used by traditional healers in the treatment of malaria, was found to contain three active alkaloids, tubulosine (171), psychotrine, and cephaeline. Tubulosine, a β-carboline-benzoquinolizidine alkaloid derivative, was the most active compound with an IC_{50} value of 0.012 µM against a chloroquine-sensitive strain of *P. falciparum*, and an IC_{50} value of 0.023 µM against a resistant strain. Tubulosine was also tested *in vivo* against *P. vinckei petteri* and *P. berghei* in mice, and in each case, good results were obtained at concentrations lower than the lethal dose (262). However, this alkaloid also displayed some toxicity against KB cells. The results showed some support for the traditional use of the bark of this plant in the treatment of malaria (9).

The mechanisms of action of β-carboline alkaloids as antiprotozoal agents are not known. However, β-carbolines have shown a variety of actions in biological systems related to their interaction with DNA, so it has been postulated that some are able to intercalate DNA (133,252,263). Regarding structure–activity relationships, the results obtained in some studies suggest that unsaturation of the pyridine ring increases the antiprotozoal activity. 1,2,3,4-Tetrahydro-β-carbolines have quite low activities, while 1,2-dihydro-β-carbolines are better growth inhibitors. A fully unsaturated pyridoindole ring system seems to be a necessary condition for optimal activity in these compounds. This could be associated with planarity of the molecule, its redox behavior or its electron density distribution, to mention three possible structural desiderata (254,256). Nevertheless, other mechanisms could be involved, since compounds such as harmaline (162), a dihydro-β-carboline, present greater antiprotozoal activity than fully aromatic compounds. It has been reported that these alkaloids form intermolecular associations with the prosthetic groups of the flavoenzymes FAD and FRN, and that they are also able to interfere with topoisomerase II, or the metabolism of aromatic amino acids, or inhibit the respiratory chain in the parasite (6,254,264,265). These processes would play an important role in the mechanism of action of β-carbolines. In spite of this, due to their possible mutagenic potential, and their activity as inhibitors of monoamino oxidases A and B, which could produce psychopathic effects, β-carbolines cannot be used as therapeutic agents (266,267).

2. Indolemonoterpene Alkaloids

The indolemonoterpene alkaloids have been extensively investigated for a wide variety of pharmacological effects, such as contraceptive, antitumor, antiinflammatory, anti-HIV, bactericide, leishmanicidal, and antimalarial activities, as well as a stimulatory action on the CNS (12,268–273). Until now, more than 2000 different alkaloids of this class have been isolated. Due to their effect on the CNS and their potential use as antiaddiction agents (274), together with the anticancer Vinca alkaloids, the Apocynaceae alkaloids coronaridine (172) and ibogaine (173) have been the most studied. Coronaridine was purified by bioassay-guided fractionation from a stem ethanolic extract of *Peschiera australis* Miers (syn. *Tabernaemontana australis*, Apocynaceae), which inhibited the growth of *L. amazonensis* in axenic cultures and infected murine macrophages. The pure alkaloid exhibited potent antileishmanial activity against *L. amazonensis* promastigotes (IC_{97} = 35.25 µM) (275). Coronaridine, with an isoquinuclidine ring fused to an indole moiety, is an iboga-type indole alkaloid found in species used traditionally in folk medicine for the treatment of leishmaniasis, like *Peschiera laeta* Miers and *P. vanheurckii* (Müll.Arg.) L.Allorge (271,276–278). Another monoterpene indole alkaloid from the Apocynaceae, but not widely studied, is the aspidofractinine-type alkaloid, pleiocarpine (**174**) (IC$_{50}$ < 63 µM), one of three leishmanicidal alkaloids from *Kopsia griffithii* King & Gamble was found to show activity against the promastigote forms of *L. donovani* (253).

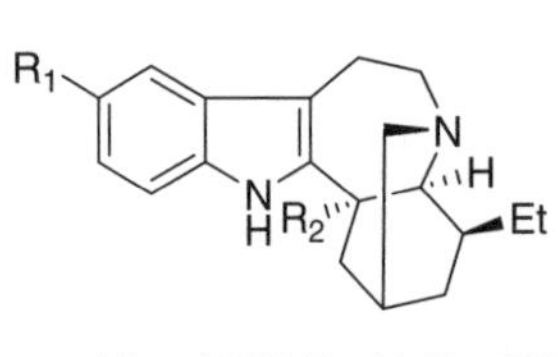

coronaridine (**172**) R$_1$=H, R$_2$=COOMe
ibogaine (**173**) R$_1$=OMe, R$_2$=H

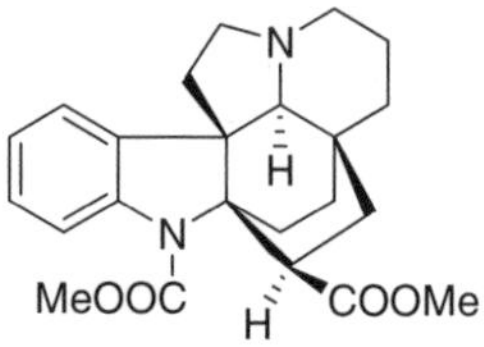

pleiocarpine **174**

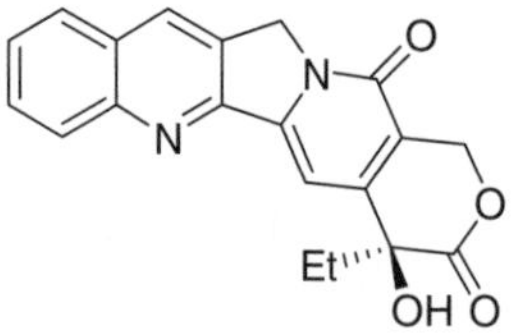

camptothecin **175**

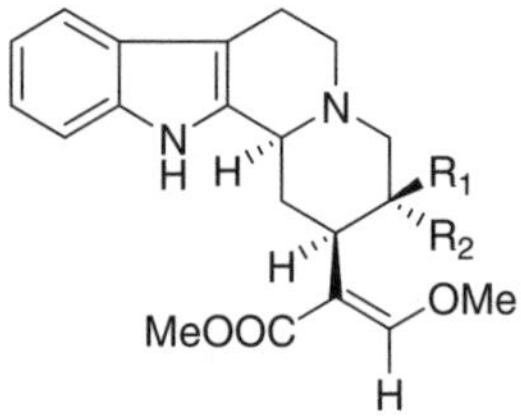

dihydrocorynantheine (**176**) R$_1$=H, R$_2$=Et
corynantheine (**177**) R$_1$=H, R$_2$=CHCH$_2$
corynantheidine (**178**) R$_1$=Et, R$_2$=H

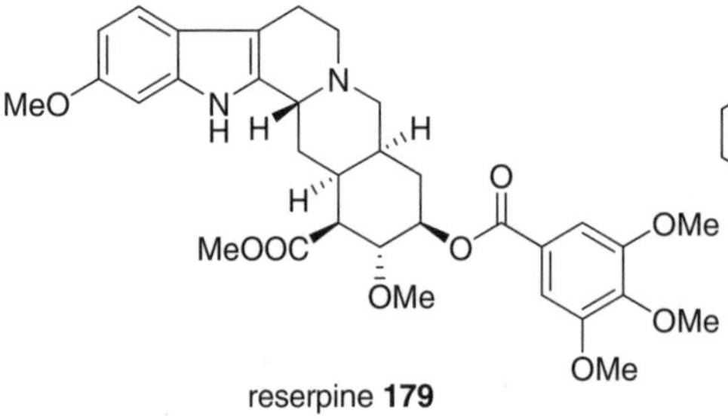

reserpine **179**

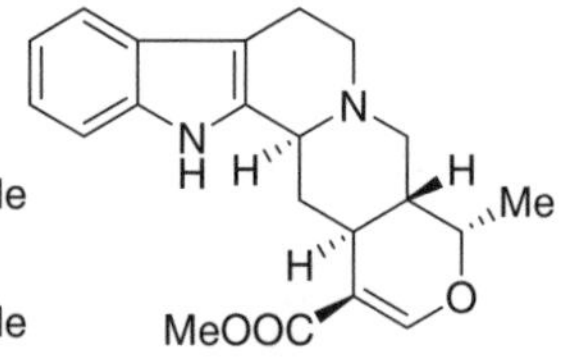

ajmalicine **180**

Camptothecin (**175**), a modified indolemonoterpene alkaloid produced by *Camptotheca acuminata* Decne. (Cornaceae), *Nothapodytes foetida* (Wight) Sleumer, *Pyrenacantha klaineana* Pierre et Exell & Mendonca, *Merrilliodendron megacarpum* (Hemsl.) Sleumer (Icacinaceae), *Ophiorrhiza pumila* Champ. ex Benth. (Rubiaceae), *Ervatamia heyneana* Cooke (Apocynaceae), and *Mostuea brunonis* Didr. (Logania-ceae), is a known antitumor agent with an unusual heterocyclic structure (279,280). Its structure was elucidated in the mid-1960s (281), and the first total synthesis of optically active 20(S)-camptothecin was reported in 1975 (282). Camptothecin (175) is lethal to *T. brucei, T. cruzi*, and *L. donovani*, with EC$_{50}$ values of 1.5, 1.6, and 3.2 µM, respectively (279). In erythrocytic *P. falciparum*, camptothecin was cytotoxic with an EC$_{50}$ value of 32 µM for NF54 (chloroquine-sensitive strain) and 36 µM for FCR3 (chloroquine-resistant strain) (283). The rather broad spectrum of antiparasitic activity of camptothecin is probably due to its inhibitory activity against topoisomerase I. Camptothecin is an established natural antitumor drug and a well-characterized inhibitor of eukaryotic DNA topoisomerase I (280,284). When trypanosomes or leishmania were treated with this alkaloid, both nuclear and mitochondrial DNA were cleaved and covalently linked to protein. This is consistent with the existence of drug-sensitive topoisomerase I activity in both compartments. Camptothecin is an important lead for much needed new chemotherapy, as well as a valuable tool for studying topoisomerase I activity (279).

Five indolemonoterpene alkaloids of the yohimbine- and corynantheine-type, obtained from the bark of *Corynanthe pachyceras* K. Schum. (Rubiaceae), presented marked activity against promastigotes of *L. major*. Among these alkaloids are dihydrocorynantheine (176), corynantheine (177), and corynantheidine (178), which present IC$_{50}$ values of below 3 µM. These metabolites do not show significant cytotoxic activity against the drug-sensitive KB-3-1 and multidrug-resistant KB-V1 cell lines, indicating an important selectivity in their antiprotozoal activity (285). Due to the absence of previous reports about the leishmanicidal activity of yohimbine- and corynantheine-type indole alkaloids, several structurally related alkaloids were added to this last study, including reserpine (179), ajmalicine (180), and ajmaline (181). The leishmanicidal activity of ajmalicine corresponded to an IC$_{50}$ value of 0.57 µM, and interestingly, ajmaline was inactive against *L. major* promastigotes, suggesting that the active alkaloids contain a relatively planar tetracyclic structure. It is proposed that the mechanism of action of the *C. pachyceras* alkaloids could be based on the inhibition of the respiratory chain of the parasite (285).

ajmaline 181   strychnine 182   akagerine 183

Several research groups have studied the antimalarial activity of indolemonoterpene alkaloids to find new antimalarial compounds of plant origin. In this field, indole alkaloids with different skeletons have been tested (*Aspidosperma* and *Picralima* from the Apocynaceae family and *Strychnos* from Loganiaceae), but no clear relationships between structure and activity have been deduced. Interestingly, indolemonoterpene alkaloids such as strychnine (182) from

Strychnos have been tested against *Plasmodium*, but are devoid of any antiplasmodial activity, as previously observed for other *Strychnos* monomers (286–289) (see in following section). Only akagerine (**183**), a known corynanthean-type indole alkaloid with an N–C17 linkage (290), isolated from *S. usambarensis* Gilg ex Engl., displayed antiprotozoal activity in mice (286,291).

During a phytochemical investigation of two *Aspidosperma* spp., *A. pyrifolium* Mart. and *A. megalocarpon* Müll.Arg., monoterpene indole alkaloids possessing an aspidospermane skeleton were isolated (292,293). In this series of alkaloids, 11 were chosen for their close structural similarity, and their *in vitro* antiplasmodial activity and toxicity were examined. The most active alkaloid, aspidospermine (**184**), 10-methoxyaspidospermidine (**185**), and vallesine (**186**), presented IC_{50} values between 3.2 and 6.6 µM after incubation for 72 h against chloroquine-sensitive and chloroquine-resistant strains of *P. falciparum* (294). The SI of the alkaloids 10-methoxyaspidospermidine and vallesine were interesting, with values of 22.7 and 15.6, respectively. On the other hand, extracts from *Picralima nitida* Th. & H.Dur., a plant extensively used in many folk remedies, e.g., against malaria, have been studied experimentally against *Plasmodium* both *in vitro* and *in vivo* (295–298), with encouraging results. A number of alkaloid constituents have been isolated and identified (299), but only one active alkaloid, akuammine (**187**), has been reported from the seed extract (300).

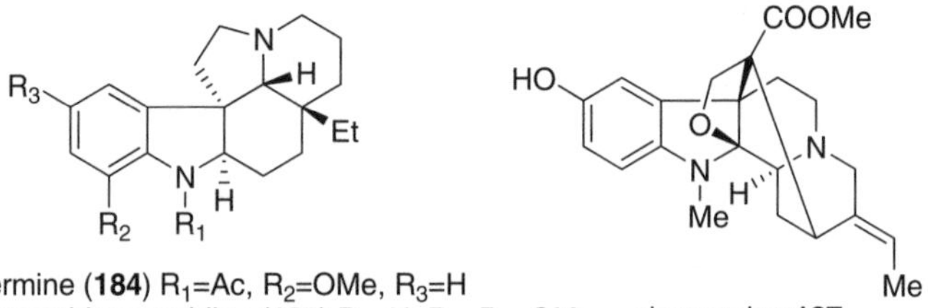

aspidospermine (**184**) R_1=Ac, R_2=OMe, R_3=H
10-methoxyaspidospermidine (**185**) R_1=H, R_2=R_3=OMe   akuammine **187**
vallesine (**186**) R_1=CHO, R_2=OMe, R_3=H

3. Bisindole Alkaloids

Members of the Apocynaceae family have been used for centuries in folk medicine, and several of their alkaloids have been isolated and are now in clinical use as separate drugs, such as vinblastine, vincristine, and reserpine (**179**) (275). Ethnobotanical sources mention that the most common medicinal uses of this family of plants involve its antimicrobial action against infectious diseases such as syphilis, leprosy, and gonorrhea, as well as its antiparasitic action against worms, dysentery, diarrhea, cutaneous leishmaniasis, and malaria (275,301). These effective uses described in traditional medicine are probably due to the presence of indole alkaloids, which are the main secondary metabolites in Apocynaceae (275). Included in the family is the genus *Kopsia* with some 30 species found mostly in tropical Asia (302). Preliminary screening of extracts from *K. griffithii* showed strong antileishmanial activity, which had been previously traced to the basic fraction from the ethanol extract of the leaves (303). The active principles responsible for this activity were subsequently identified (253). Although *K. griffithii* is notable for furnishing a remarkable array of alkaloidal types, only three alkaloids were found to show activity against the promastigote forms of *L. donovani*: the simple β-carboline, harmane (**163**), the aspidofractinine-type alkaloid pleiocarpine (**174**), which was previously discussed, and the quasidimer buchtienine (**188**) (IC$_{50}$ between 0.79–3.16 μM) (253). This result, which identifies buchtienine as the principal alkaloid responsible for the leishmanicidal activity, is in agreement with the isolation of buchtienine along with other indole alkaloids from the Bolivian plant *Peschiera buchtienii* (H.Winkler) Markgr. (syn. *Tabernaemontana buchtienii*), which is also used locally for the treatment of leishmaniasis (278). Natives of the tropical Bolivian Chapare region also use *P. vanheurckii* to treat cutaneous leishmaniasis. Preliminary antileishmania screening against *L. amazonensis* and *L. braziliensis* showed that extracts from the leaves and stem bark of *P. van heurkii* exhibited significant activity against parasites in relation to the presence of conodurine-type bisindole alkaloids (277). The strongest leishmanicidal activity was observed with the bisindole alkaloids conodurine (**189**), *N*-demethylconodurine (= gabunine) (**190**), and conoduramine (**191**). Weak toxicity toward macrophage host cells, and strong activity against the intracellular amastigote form of *Leishmania*, were observed for conodurine and *N*-demethylconodurine, which confirms that the leishmanicidal activity exhibited by these alkaloids is not due to a general antiproliferative effect. *In vivo*, conodurine was less active than Glucantime®, while *N*-demethylconodurine was devoid of activity at 100 mg/kg (277). On the other hand, a related bisindole alkaloid, voacamine (**192**), was the most active alkaloid isolated from *Peschiera fuchsiaefolia* (A.DC.) Miers (304). It showed good antiplasmodial activity against both the chloroquine-sensitive D-6 strain (IC$_{50}$ = 0.34 μM) and the chloroquine-resistant W-2 strain of *P. falciparum* (IC$_{50}$ = 0.41 μM). Apparently, the crude alkaloid extract from the root bark of the same plant is more active than voacamine (IC$_{50}$ = 0.18 μg/mL for D-6 and 0.28 μg/mL for W-2 strain), and is particularly rich in bisindole alkaloids (0.22% of the plant material) (304).

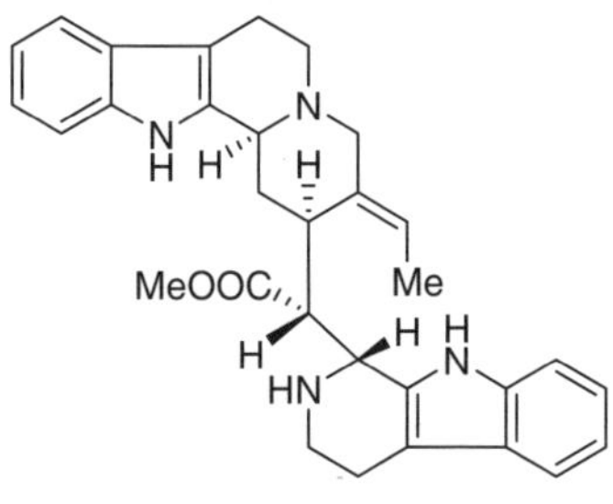

buchtienine **188**

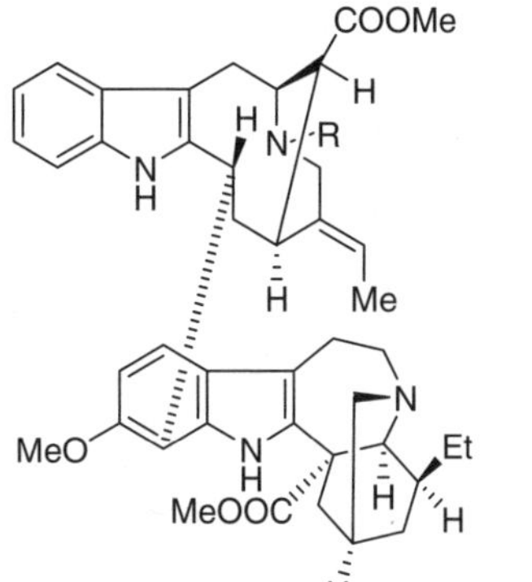

conodurine (**189**) R=Me
N-demethylconodurine (**190**) R=H

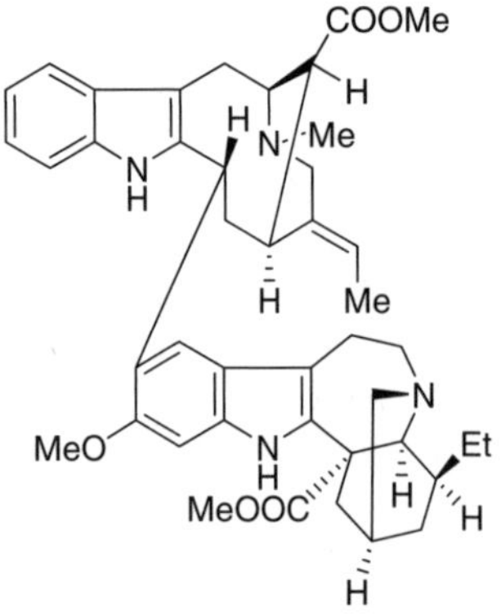

conoduramine **191** 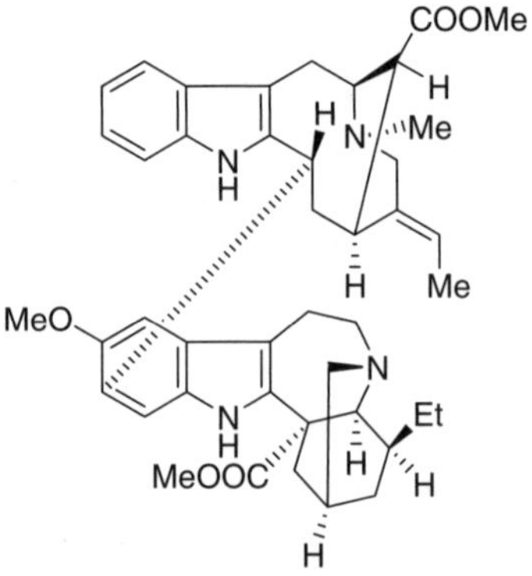   voacamine **192**

Another genus of Apocynaceae investigated for its antiprotozoal activity is *Alstonia*, whose 43 species are distributed throughout Africa, Central America, Asia, and the Pacific region. A number of species are known to be used by traditional healers in the treatment of malaria though there are conflicting reports in the literature as to their activity (305). More than 130 alkaloids have been isolated from *Alstonia* spp., only a few of which have been assessed for antimalarial activity. From *A. angustifolia* Wall., a plant used in Southeast Asia to treat malaria and dysentery, nine alkaloids were isolated and tested, and the one most active against *P. falciparum* was villalstonine (**193**) (306). Likewise, 13 indole alkaloids were isolated from the active extract of *A. macrophylla* Wall., and again villalstonine ($IC_{50} = 0.27\,\mu M$) was the most active. This alkaloid, as well as macrocarpamine (**194**), with an IC_{50} of $0.36\,\mu M$, presented high antiplasmodial activity against the multidrug-resistant K1 strain of *P. falciparum* cultured in human erythrocytes (307).

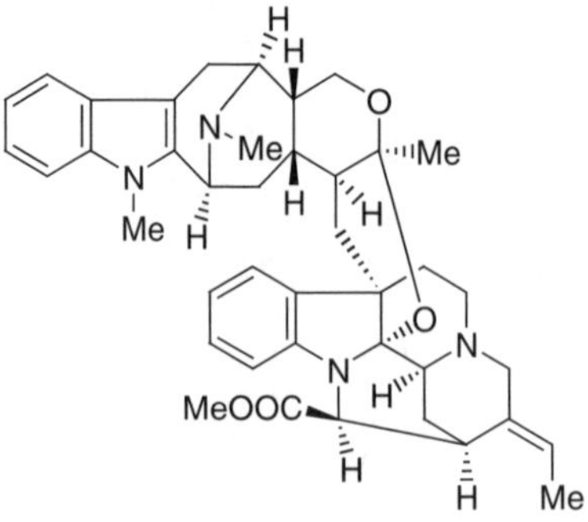

villalstonine **193**

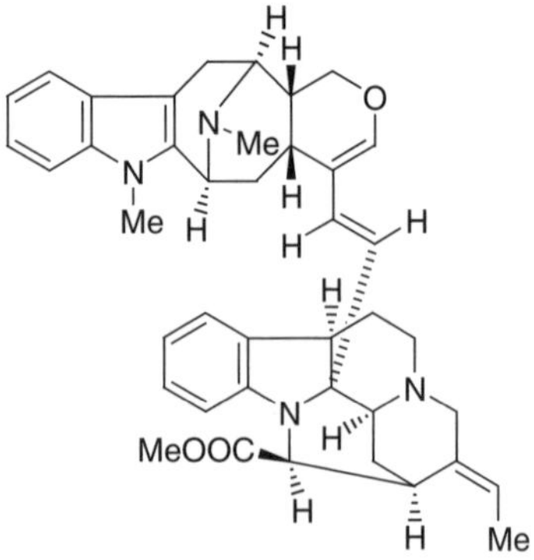

macrocarpamine **194**

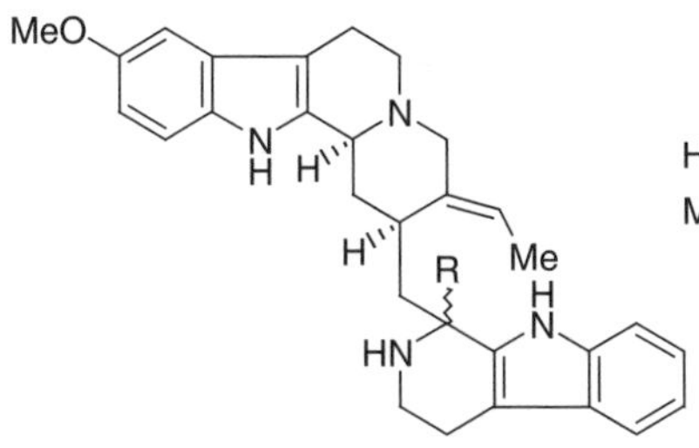

ramiflorine A (**195**) R=Hα
ramiflorine B (**196**) R=Hβ

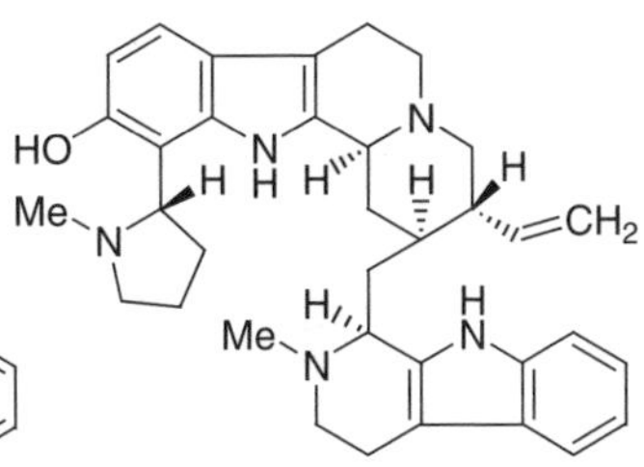

isostrychnopentamine **197**

The crude basic extract from the stem bark of *Aspidosperma ramiflorum* Müll.Arg. showed good antileishmanial activity, which was attributed to the presence of indole alkaloids (308). The fractionation, purification procedures, and the activity of the isolated alkaloids of this species against promastigote forms of the *L. amazonensis* parasite were described recently (309). The bisindole alkaloids ramiflorines A (**195**) and B (**196**) showed good antipromastigote activity, with calculated IC_{50} values of 34.69 and 10.51 μM, respectively. In comparison with pentamidine, these alkaloids were very active. Although the mode of action of these alkaloids is not known, they are similar in structure to several other bisindole alkaloids, which indicates that their mode of action may also be similar; that is, they would act as inhibitors of protein synthesis, DNA intercalating agents, or as topoisomerase inhibitors (310).

Alkaloids with a dihydrotchibangensine structure, e.g., buchtienine (**188**), are also common in the genus *Strychnos*. Belonging to the Loganiaceae family, the pantropical *Strychnos* genus comprises about 200 species and can be subdivided into three geographically separated groups found in Central and South America (at least 73 species), Africa (75 species), and Asia (about 44 species). Plants in this genus have been used in folk medicine, and in arrow and dart poisons, in many parts of the world, and are very well known for providing one of the most famous of all poisons. Strychnine (**182**), as it was logically named, is one of the numerous indolemonoterpene alkaloids possessing the strychnan skeleton produced by the genus *Strychnos* (291,311). However, the results obtained in previous studies largely confirm that the monomeric indole alkaloid strychnine and its derivative monomers are devoid of any antiplasmodial activity (286–289), in comparison to the *Strychnos* bisindole derivatives.

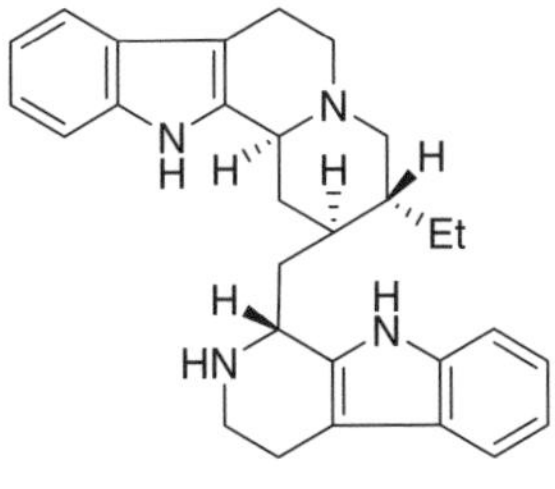

ochrolifuanine A **198**

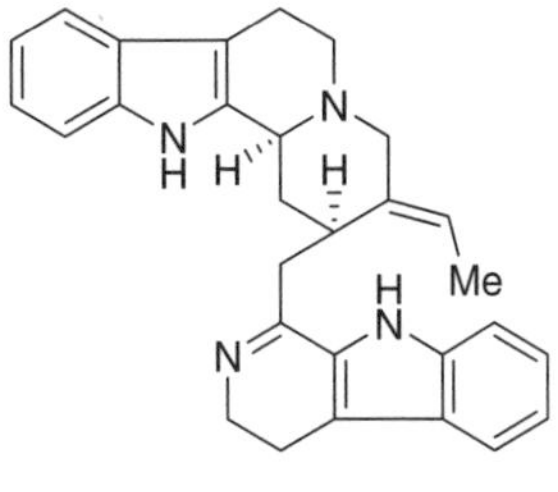

dihydrousambarensine **199**

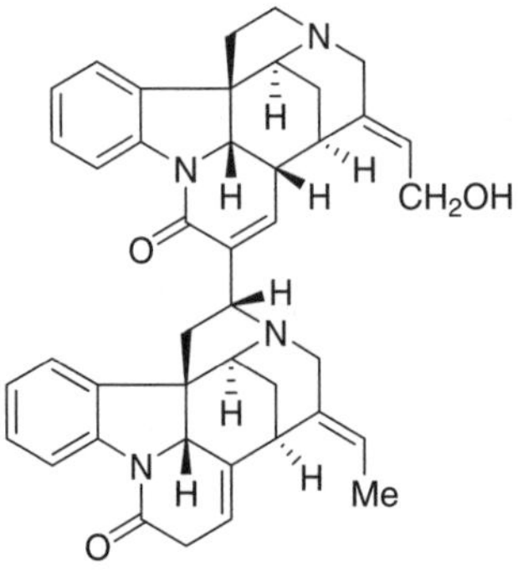

18-hydroxyisosungucine **200** 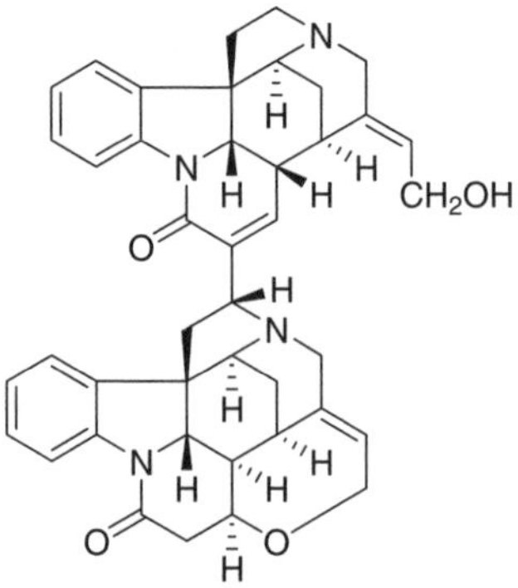   strychnogucine B **201**

The *in vitro* antiplasmodial activities of 69 alkaloids from various *Strychnos* species were evaluated against chloroquine-resistant and chloroquine-sensitive lines of *P. falciparum* (9). Twelve bisindole alkaloids showed IC$_{50}$ values of <2 μM against all of the *Plasmodium* lines tested. Among these, it was possible to distinguish four principal structural classes: usambarine-type (i.e., comprised of two tryptamine units with a single iridoid unit), matopensine-type, longicaudatine-type, and sungucine-type alkaloids. Two alkaloids, isostrychnopentamine (**197**), an asymmetric bisindole monoterpene alkaloid obtained from the leaves of *S. usambarensis*, whose pyrrolidine ring is joined to the widely distributed β-carbolinylindolo[2,3-*a*]quinolizinyl-methane system (312), and ochrolifuanine A (**198**) exhibited very potent activity of <500 nM against all lines tested. Dihydrousambarensine (**199**) was selectively highly active (IC$_{50}$ = 0.032 μM) against the chloroquine-resistant strain W-2. These three alkaloids are usambarine-type bisindole alkaloids (313), but as they possess highly varied structures, no structure–activity relationships could be deduced from these data. However, some structural characteristics related to antiplasmodial activity can be kept in mind: all of the active alkaloids were tertiary dimers, and a certain degree of basicity seems necessary for the antiplasmodial activity of this family of alkaloids. As has been hypothesized for chloroquine, these observations are consistent with the ability of basic alkaloids to accumulate at higher levels in the acidic food vacuole of the parasite, where digestion of hemoglobin takes place.

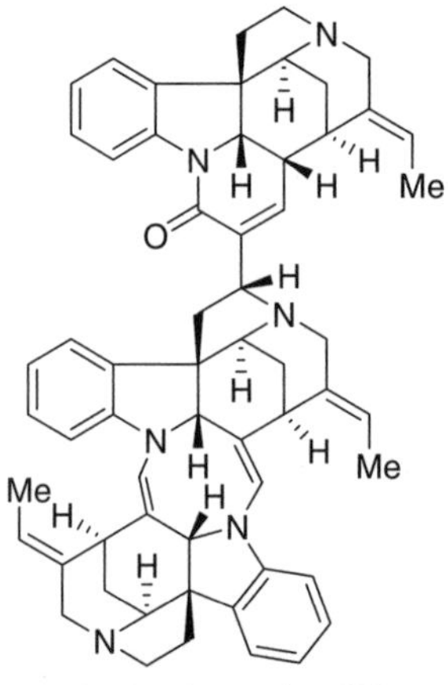

strychnohexamine **202**

The liana *Strychnos icaja* Baill., besides being used as an ordeal poison, is also occasionally used in African traditional medicine to treat malaria. The roots of this species are particularly rich in antiplasmodial bisindole alkaloids. Some bisindole alkaloid derivatives of sungucine, such as 18-hydroxyisosungucine (**200**) and strychnogucine B (**201**), and the trisindolomonoterpene alkaloid strychnohexamine (**202**), all possessing a strychnine substructure, as well as potent and selective antiplasmodial properties, have been isolated from *S. icaja* (289,314,315). 18-Hydroxyisosungucine and strychnogucine B were highly active against four *P. falciparum* strains (IC$_{50}$ values of 0.14–1.26 and 0.08–0.62 μM, respectively). Strychnogucine B was more active against a chloroquine-resistant strain than against a chloroquine-sensitive one (best IC$_{50}$ was 85 nM against the W-2 strain). In addition, this alkaloid showed selective antiplasmodial activity with 25–180 times greater toxicity toward *P. falciparum*, relative to cultured human cancer cells (KB) or human fibroblasts (WI38) (315). Likewise, strychnohexamine presented interesting antiplasmodial activity with an IC$_{50}$ value of nearly 1 μM against the FCA chloroquine-sensitive strain of *P. falciparum* (289). These results confirmed the reported antiplasmodial activities of *Strychnos* bisindole alkaloids. Since they are derivatives of strychnine, it was important, in view of their potential use as antimalarial drugs, to assess their possible strychnine-like convulsant properties (316). The convulsant effects of most monomeric derivatives were described in mice many years ago (317), but some bisindolic and trisindolic alkaloids from *S. icaja* are devoid of such strychnine-related properties, at least *in vitro*. Further studies will involve *in vivo* assays aimed at confirming the absence of pro-convulsant or toxic effects of these alkaloids, together with examining their *in vivo* antimalarial potency (316).

4. Indoloquinoline Alkaloids

Cryptolepine (**203**) is an indoloquinoline alkaloid first isolated from the roots of *Cryptolepis triangularis* N.E.Br. (Asclepiadaceae) collected in the Congo. Extracts of the roots of the related climbing liana *C. sanguinolenta*, in which cryptolepine is the main alkaloid, have been used in traditional medicine in West and Central Africa, and clinically in Ghana, for the treatment of malaria (318,319). Cryptolepine, which is an indolo[3,2-*b*]quinoline (or a benzo-δ-carboline), showed potent *in vitro* activity (IC$_{50}$ = 0.13 μM) against the multidrug-resistant K1 strain of *P. falciparum* (261,320). Cimanga *et al.* reported an *in vitro* IC$_{50}$ of 0.14 μM for cryptolepine against the same *P. falciparum* strain (K1), and also *in vivo* antiplasmodial activity (significant reduction of parasitemia) against *P. berghei* in subcutaneously infected mice when cryptolepine or its hydrochloride were orally administered (dissolved in 2.5% Tween 80) daily for 4 days at a dose of 50 mg/kg body weight (321). Grellier *et al.* also confirmed the antiplasmodial activity of cryptolepine *in vivo* in a similar 4-day suppressive test in mice infected with *P. vinckei* petteri or *P. berghei* (intraperitoneal treatment with 1.25–10 mg/kg body weight) (322). However, this alkaloid also has cytotoxic properties probably due to its ability to intercalate into DNA, as well as inhibit topoisomerase II and DNA synthesis (310). Although initially neocryptolepine (**204**), a minor alkaloid from *C. sanguinolenta* (Lindl.) Schltr., was reported to show activity comparable to cryptolepine (IC$_{50}$ of 0.22 μM, K1 strain) (321), more recent investigations have shown that it is about seven times less active against the chloroquine-resistant

P. falciparum Ghana-strain (323). Another minor alkaloid from the same plant with an indolo[3,2-*c*]quinoline (or a benzo-*γ*-carboline) structure, which was named isocryptolepine (**205**), also showed *in vitro* antiplasmodial properties against various chloroquine-sensitive or -resistant strains of *P. falciparum* (IC$_{50}$ values of about 0.8 µM) (322).

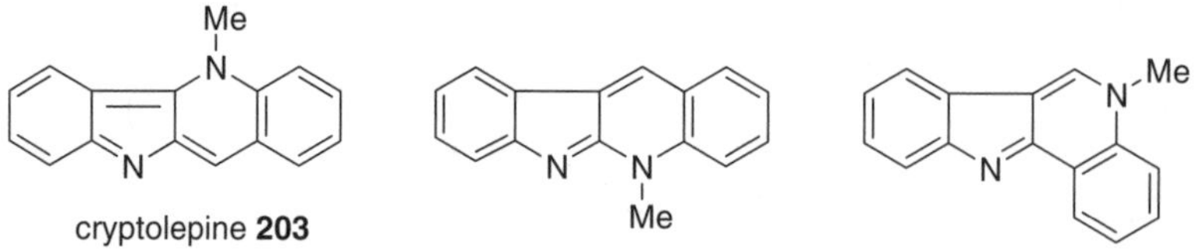

cryptolepine **203**   neocryptolepine **204**   isocryptolepine **205**

As in the case of the *β*-carboline alkaloids, DNA and topoisomerase II are two potential targets for other indole alkaloids, such as indolemonoterpenes and indoloquinolines. These alkaloids behave like typical DNA intercalating agents, but their affinities for double-stranded DNA can vary significantly depending on their structural complexity (324). Nevertheless, the activity of cryptolepine (**203**) is due, at least in part, to a chloroquine-like action that does not depend on intercalation into DNA (325). It has recently been postulated that the antiplasmodial activity of the indoloquinoline alkaoids is due to a combination of at least two mechanisms of action. Inhibition of the heme detoxification process is a selective mechanism, whereas DNA intercalation, a non-selective mechanism, is responsible for the cytotoxicity, and probably also for the activity against the other parasites tested (319), i.e., *T. b. brucei* bloodstream forms (11).

C. Steroidal and Diterpenoid Alkaloids

Several different steroidal glycoalkaloids of the solanidane or spirosolane type from plants belonging to the genus *Solanum* (Solanaceae) have been tested against epimastigote and trypomastigote forms of *T. cruzi*. Among them, the spirosolanes α-solasonine (**206**) and α-solamargine (**207**), as well as the solanidanes, α-chaconine (**208**) and α-solanine (**209**), four glycoalkaloids composed of a lipid core and a trisaccharide carbohydrate moiety, almost completely inhibited the growth of the epimastigote form of *T. cruzi* at 5.7 µM. They were even more effective than ketoconazole at the same concentration (326). Although without providing supporting data, this study suggests that stage-specific membrane constituents are responsible for susceptibility to these alkaloids. Likewise, it was established that the role of the sugar moiety is very important in antitrypanosomal activity. In fact, recent studies confirm that the carbohydrate moiety clearly plays a significant role in the cytolytic properties of α-solasonine and α-solamargine against *T. cruzi*, by means of specific carbohydrate interactions that lead to the formation and intercalation of sterol complexes into the parasite plasma membrane (327).

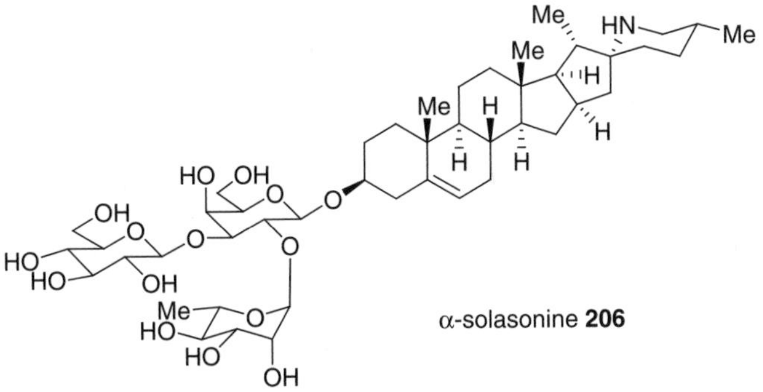

α-solasonine 206

Sarachine (**210**), another steroidal alkaloid isolated from leaves of the Bolivian plant *Saracha punctata* Ruiz & Pav. (Solanaceae), completely inhibits the growth of the promastigote forms of *L. braziliensis*, *L. donovani*, and *L. amazonensis* and also the growth of epismastigote forms of *T. cruzi* (50% inhibition) at a concentration of 25 μM. However, at the same concentration it shows strong toxicity against mouse peritoneal macrophages (328). Sarachine also showed *in vitro* antiplasmodial activity with an IC_{50} of 25 nM against a chloroquine-sensitive strain, and an IC_{50} of 176 nM against INDO-resistant strains of *P. falciparum*. Despite its high cellular toxicity toward macrophages, sarachine exhibited more selective toxicity against *Plasmodium* than against KB carcinoma cells ($IC_{50} = 50$ μM), thus giving cytotoxicity-activity ratios of 2000 and 286, respectively, for the two strains. The alkaloid was also active *in vivo* against *P. vinckei*, with 83% inhibition of the parasitemia at 100 mg/kg/2 days (328). All the above-mentioned studies showed interesting preliminary biological activities for this alkaloid.

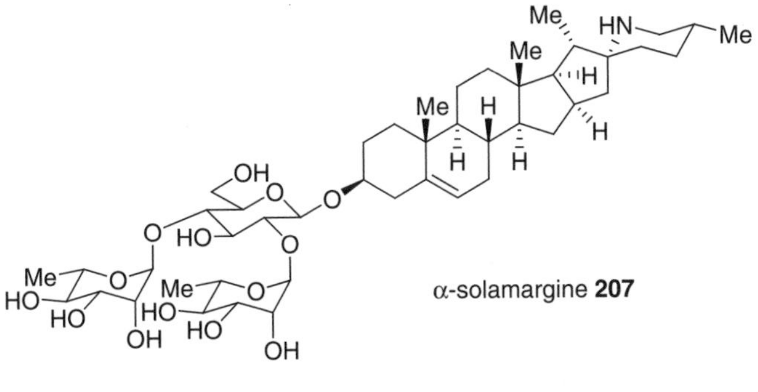

α-solamargine 207

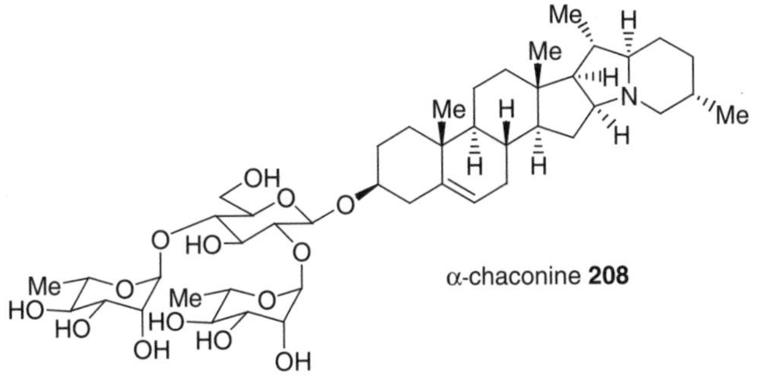

α-chaconine 208

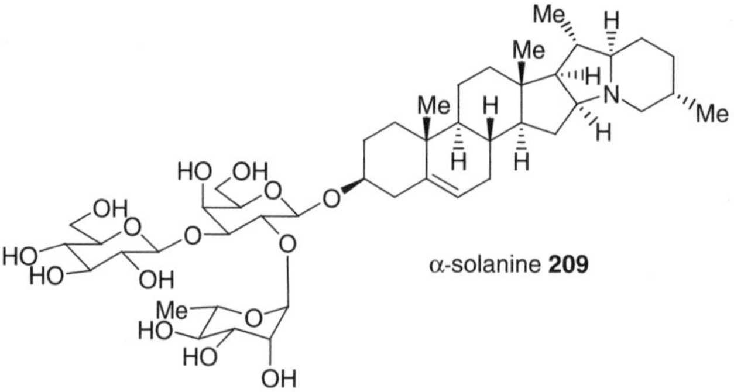

α-solanine **209**

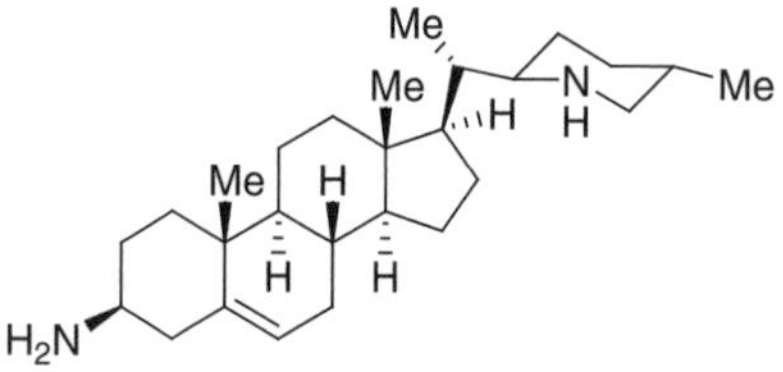

sarachine **210**

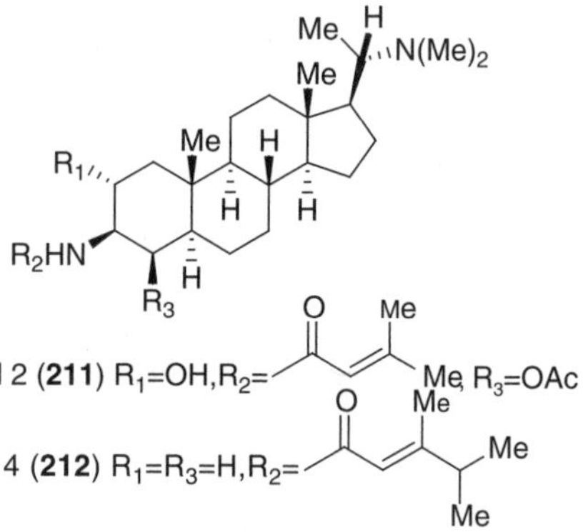

compound 2 (**211**) R$_1$=OH,R$_2$=
compound 14 (**212**) R$_1$=R$_3$=H,R$_2$=

Seventeen natural pregnane-type steroidal alkaloids isolated from *Sarcococca hookeriana* Baill. (Buxaceae) were recently evaluated for their antileishmanial activity. Some of them, such as compound 2 (**211**) and compound 14 (**212**), displayed very similar activity to that of the standard drug amphotericin B against *L. major* promastigotes (IC$_{50}$ = 0.40 and 1.01 µM, respectively) (329). In further exploration to develop antileishmanial drugs, eight steroidal alkaloids obtained from the leaves of *Holarrhena curtisii* King & Gamble (Apocynaceae) exhibited leishmanicidal activity against promastigotes of *L. donovani*. All eight steroidal alkaloids of the pregnane-type were effective in the concentration ranges investigated, in particular the aminoglycosteroids holacurtine (**213**) and *N*-demethylholacurtine (**214**), and the aminosteroids holamine (**215**) and 15*R*-hydroxyholamine (**216**) (IC$_{50}$ values below 20 µM) (330). Additional bioassays, however, revealed that these alkaloids also showed cytotoxic activity against the HL-60 cell line. Another species of the family Apocynaceae, *Funtumia elastica* Stapf, which is commonly used in traditional medicine in West Africa to treat infectious diseases including malaria, was recently analyzed through bioassay-guided fractionation, and four steroidal alkaloids, holarrhetine (**217**), conessine (**218**), holarrhesine (**219**), and isoconessimine (**220**), were isolated (331).

They were identified as the active constituents against the chloroquine-resistant strain FcB1 of *P. falciparum* (IC$_{50}$ values of 1.13, 1.04, 0.97, and 3.39 µM, respectively); conessine showed the highest SI.

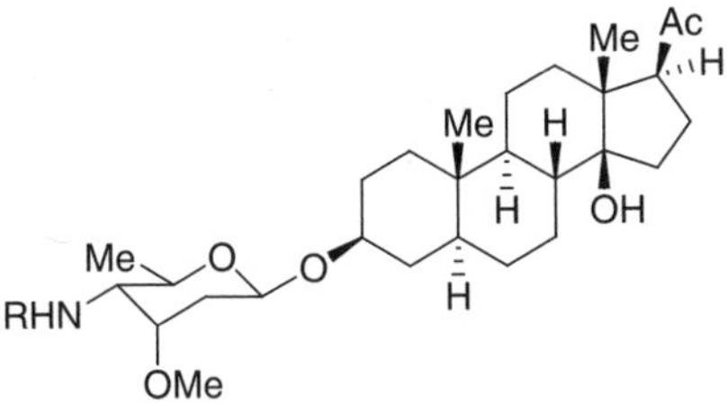

holacurtine (**213**) R=Me
N-demethylholacurtine (**214**) R=H

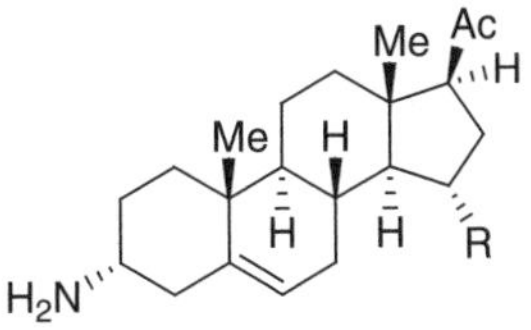

holamine (**215**) R=H
15 -hydroxyholamine (**216**) R=OH

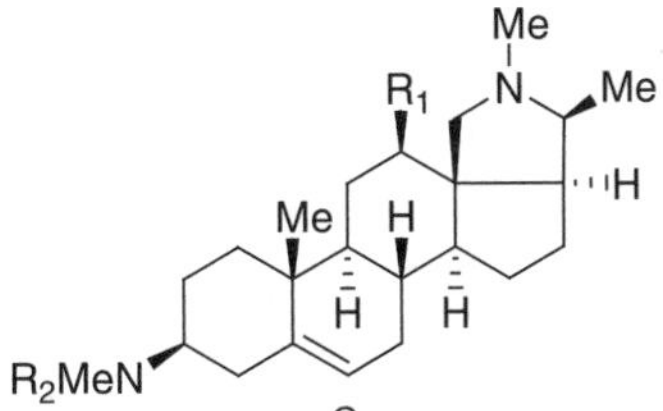

holarrhetine (**217**) R$_1$=
conessine (**218**) R$_1$=H,R$_2$=Me

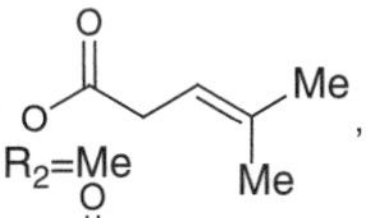

holarrhesine (**219**) R$_1$=
isoconessimine (**220**) R$_1$=R$_2$=H

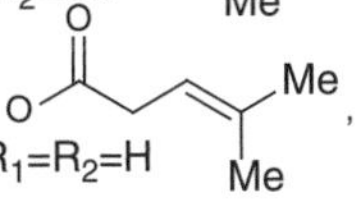

On the other hand, there are few reports of antiprotozoal activity in C$_{19}$-norditerpene (NDAs) and C$_{20}$-diterpene alkaloids (DAs). The only study of diterpenoid alkaloids suggests that DAs are a class of alkaloids with potential for further development in antiprotozoal therapy (332). From a total of 43 diterpene alkaloids (both NDAs and DAs from several chemical groups), only three atisine-type DAs markedly inhibited the growth of the *L. infantum* promastigotes. Isoazitine (**221**) exhibited the highest toxicity against the extracellular *L. infantum* parasites (IC$_{50}$ of 22.58 µM) with an IC$_{50}$ lower than that obtained by the antileishmanial reference drug, pentamidine isethionate. Azitine (**222**) and 15,22-*O*-diacetyl-19-oxodihydro-atisine (**223**) also showed pronounced effects against promastigotes (IC$_{50}$ of 30.92 and 27.16 µM, respectively). In general, this leishmanicidal activity was associated with a lack of toxicity to murine macrophages (332). It seems that these alkaloids act fundamentally at the level of the cytoplasmic membrane of the parasites, as well as in some organelle membranes. Additional studies are required to confirm the mechanism of action.

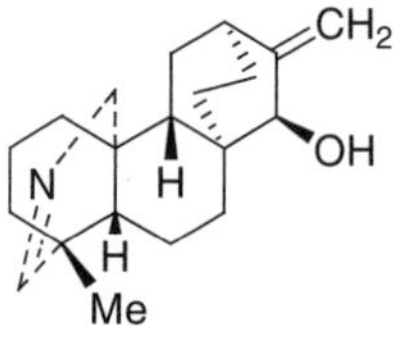

isoazitine **221** 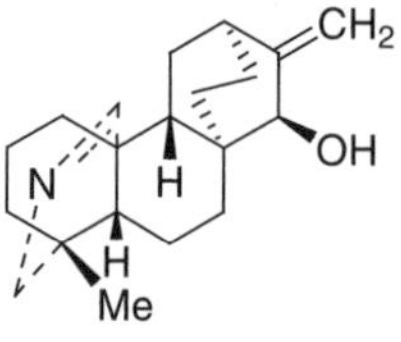   azitine **222** 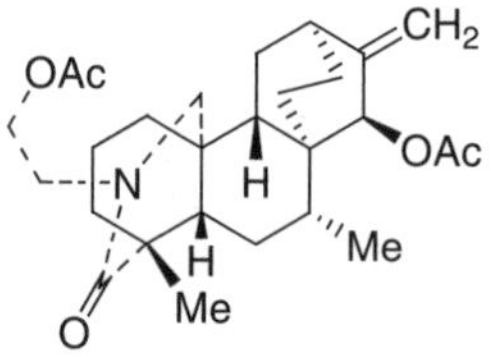   15, 22-*O*-diacetyl-19-oxodihydroatisine **223**

D. Alkaloid Marine Natural Products

In addition to a wide variety of bioactivities, the structural diversity of marine natural products also provides a rich template for the design and development of new pharmaceutical agents (333,334). Among the alkaloids of marine origin, the manzamines are a group of β-carboline alkaloids first reported in 1986 from the Okinawan sponge of the genus *Haliclona* (335). These complex metabolites usually contain several fused nitrogen-polycyclic rings, which are attached to a β-carboline moiety (336,337). Since the first report of manzamine A, an additional 30 manzamine-type alkaloids have been reported from nine different sponge genera (338,339). It has been demonstrated that manzamines exhibit a variety of bioactivities including antitumor (340), insecticidal, antibacterial, antimalarial, and leishmanicidal (341,342). Manzamine A (**224**) and 8-hydroxymanzamine A (**225**) exhibited potent *in vitro* bioactivity against D-6 chloroquine-sensitive ($IC_{50} = 0.0082$ and $0.011 \mu M$, respectively) and W-2 chloroquine-resistant ($IC_{50} = 0.0014 \mu M$) strains of *P. falciparum*. Other manzamine alkaloids such as 6-hydroxymanzamine E (**226**) and manzamine Y (**227**) also presented antimalarial activity with IC_{50} values of 1.34–1.50 and 0.74–1.50 μM, respectively, against the same species of *P. falciparum* (342). In addition to their antiplasmodial activity, the manzamines have been evaluated for their activity against *L. donovani* promastigotes. The alkaloids manzamine A, manzamine Y, 6-hydroxymanzamine E, manzamine E (**228**), and manzamine F (**229**) exhibited *in vitro* bioactivity with IC_{50} values below 9.11 μM (342).

Other studies have further demonstrated the potential of manzamine alkaloids as novel antimalarial agents that inhibit malaria parasites *in vivo* (335,343). Manzamine A and its hydroxy derivative, 8-hydroxymanzamine A (**225**), significantly prolonged survival in *P. berghei*-infected mice when administered in a single dose of 50 or 100 μmoles/kg (343). On the other hand, the alkaloids *ent*-8-hydroxymanzamine A (**230**), *ent*-manzamine F (**231**), along with the unprecedented manzamine dimer, *neo*-kauluamine (**232**) were assayed *in vivo* against *P. berghei* with a single intraperitoneal (i.p.) dose of 100 μmoles/kg. Parasitemia was efficiently reduced by *ent*-8-hydroxymanzamine

A and *neo*-kauluamine, prolonging the survival of *P. berghei*-infected mice (by 9–12 days), as compared with untreated controls (2–3 days), and mice treated with artemisinin (2 days) and chloroquine (6 days). The increase in survival of mice treated with manzamines was partly attributed to an observed immunostimulatory effect (343).

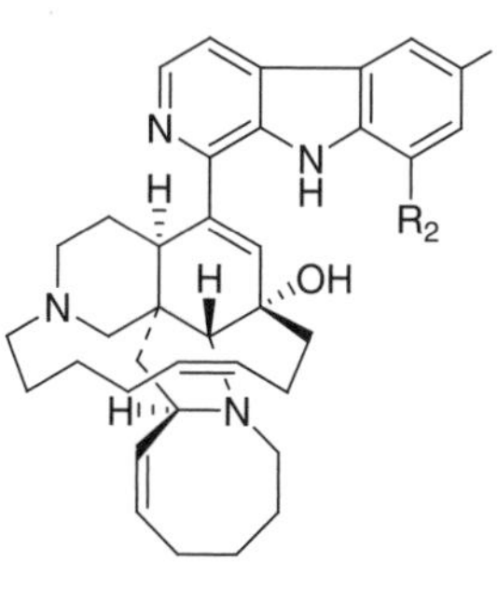

manzamine A (**224**) R$_1$=R$_2$=H
8-hydroxymanzamine A (**225**) R$_1$=H, R$_2$=OH
manzamine Y (**227**) R$_1$=OH, R$_2$=H

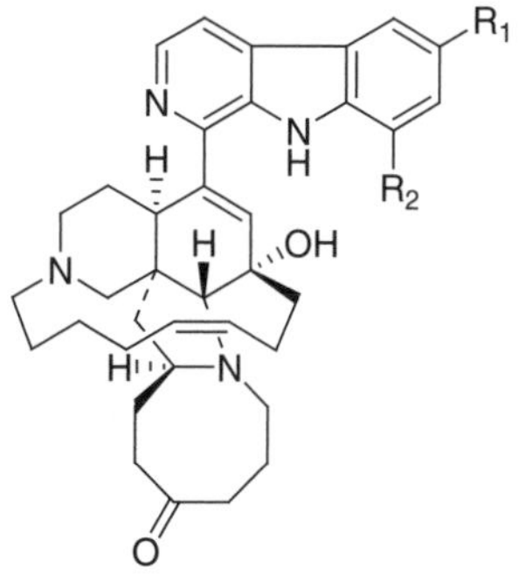

6-hydroxymanzamine E (**226**) R$_1$=OH, R$_2$=H
manzamine E (**228**) R$_1$=R$_2$=H
manzamine F (**229**) R$_1$=H, R$_2$=OH

Considering their greater activity compared to the currently used drugs, artemisinin and chloroquine, their immunostimulatory properties, and the absence of significant *in vivo* toxicity (335,343), the manzamines are clearly valuable candidates for further investigation and development as promising leads against malaria. Furthermore, oral and intravenous pharmacokinetic studies of manzamine A in rats show the alkaloid has low metabolic clearance, a reasonably long pharmacokinetic half-life, and good absolute oral bioavailability of 20.6% (344). Understanding the structure–activity relationships and molecular mechanism of action of this unique class of alkaloids is certain to lead to more effective and safer manzamine-related antimalarial drugs (345).

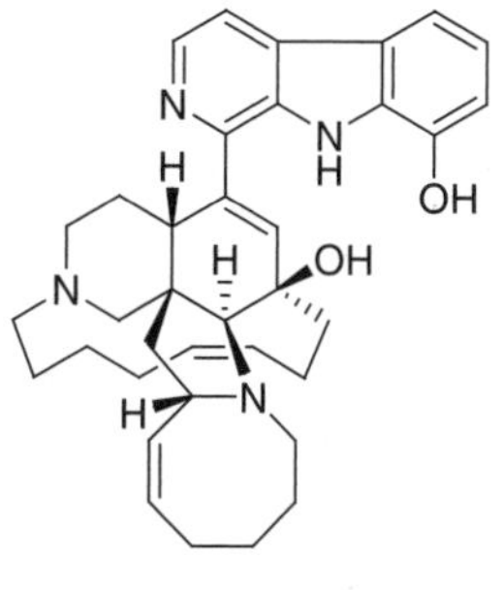

ent-8-hydroxymanzamine A **230**

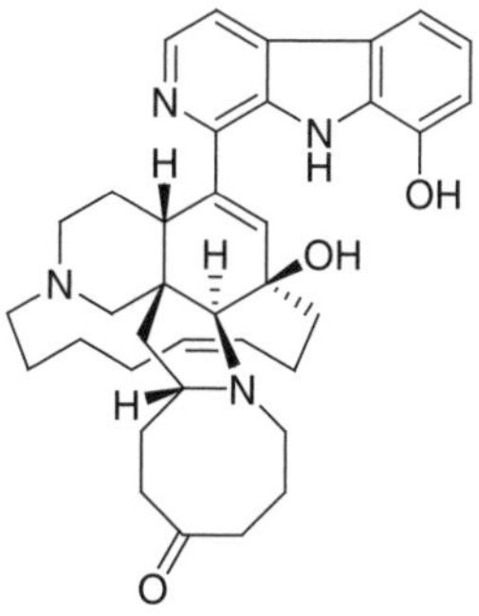

ent-manzamine F **231**

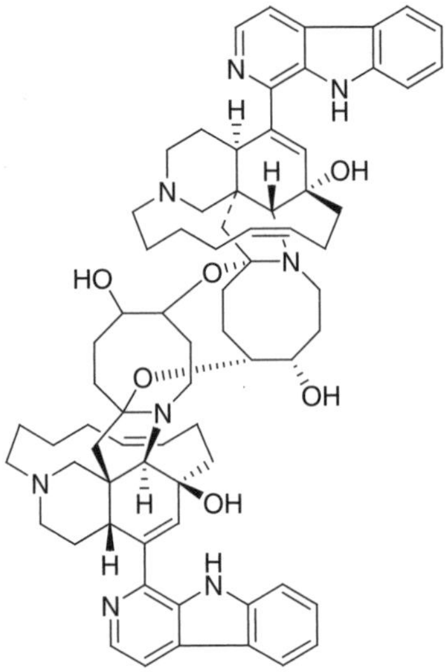

neo-kauluamine **232**

Other marine indole alkaloids have also drawn attention due to the significant activity they have shown in cancer or cytotoxicity assays (346), and this class of secondary metabolites may also have tremendous unexplored potential in the treatment of antiprotozoal diseases. Among the investigated indole alkaloids, homofascaplysin (**233**) and fascaplysin (**234**), two quaternary β-carboline alkaloids isolated from the sponge *Hyrtios erecta*, have proven to be potent *in vitro* inhibitors of NF54 chloroquine-susceptible (0.07 and 0.12 μM) and K1 chloroquine-resistant (0.04 and 0.18 μM) strains of *P. falciparum* (347). Further biological activity of fascaplysin was found against *T. b. rhodesiense* (IC_{50} = 0.60 μM), but it was cytotoxic to L6 cells, which confirms that the antitrypanosomal activity exhibited by this alkaloid is probably due to a general antiproliferative effect (347).

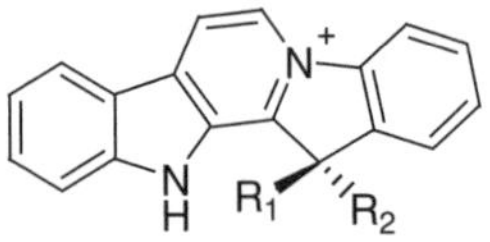

homofascaplysin (**233**) R_1 = CH$_2$COMe, R_2 = OH
fascaplysin (**234**) R_1 + R_2= O

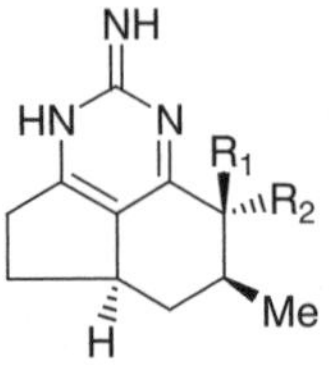

1,8a;8b,3a-didehydro-8 -hydroxyptilocaulin (**235**) R_1 = OH, R_2 = n-Bu
1,8a;8b,3a-didehydro-8 -hydroxyptilocaulin (**236**) R_1 = n-Bu, R_2 = OH
mirabilinB (**237**) R_1 = H, R_2 = n-Bu

Guanidine alkaloids are a unique class of sponge-derived metabolites exhibiting a broad range of biological activities (348). Members of this class have been evaluated against different protozoan parasites, for example, three ptilocaulin-type tricyclic guanidine alkaloids: 1,8a;8b,3a-didehydro-8β-hydroxyptilocaulin (**235**), 1,8a;8b,3a-didehydro-8α-hydroxyptilocaulin (**236**), and mirabilin B (**237**), which were identified from the marine sponge *Monanchora unguifera* (349). The mixture of **235** and **236** was active against the parasite *P. falciparum* with an IC$_{50}$ of 3.8 μg/mL, while mirabilin B exhibited antiprotozoal activity against *L. donovani* with an IC$_{50}$ of 17 μg/mL (65 μM). Another sponge-derived metabolite, renieramycin A (**238**), is a new alkaloid from the Japanese sponge *Neopetrosia* sp., which dose-dependently inhibited recombinant *L. amazonensis* proliferation (IC$_{50}$ = 0.34 μM) while showing cytotoxicity at a 10 times higher concentration (350). Jasplakinolide (**239**), a cyclic peptide alkaloid isolated from the marine sponge *Jaspis* sp., markedly decreased parasitemia of *P. falciparum in vitro* due to merozoite interference with the erythrocyte invasion (351). The decrease became evident at day 2 at concentrations of 0.3 μM and above, and the parasites finally disappeared at day 4.

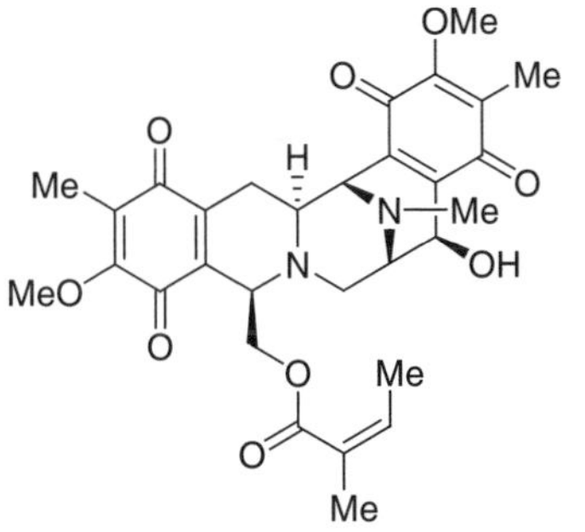

renieramycin A (**238**)

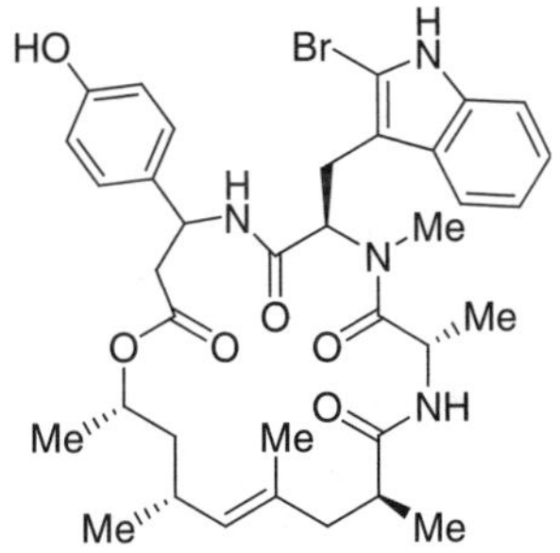

jasplakinolide **239**

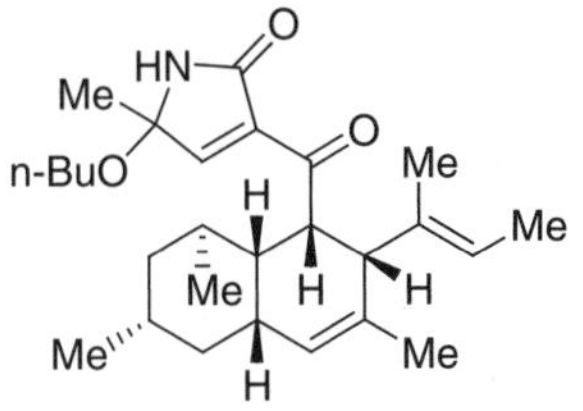

ascosalipyrrolidinone A **240**

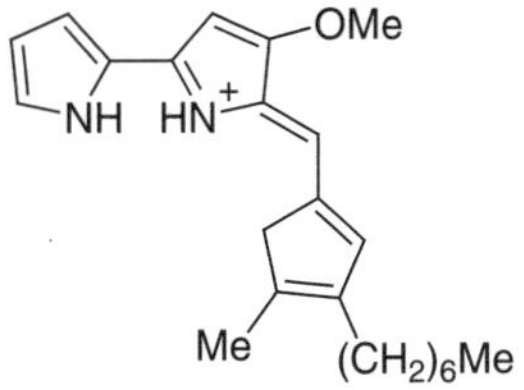

prodigiosin **241**

This present review also includes some other nitrogen-containing compounds of marine origin with interesting bioactivity. As a result of studies with fungal strains associated with marine algae, the novel nitrogen-containing compound ascosalipyrrolidinone A (**240**) was isolated from the obligate marine fungus *Ascochyta salicorniae*. It is noteworthy that ascosalipyrrolidinone A was active against *T. cruzi* with a MIC of 2.57 μM, whereas the control (benznidazole) had a MIC of 115.3 μM. It also displayed antiplasmodial activity toward both the chloroquine-resistant strain K1 and chloroquine-susceptible strain NF54 of *P. falciparum* (1.72 and 0.88 μM, respectively) (352). Further development of this compound is limited by its cytotoxic effects on rat skeletal muscle myoblast cells. On the other hand, *in vitro* and *in vivo* antimalarial studies were conducted with the tripyrrole bacterial pigment heptyl prodigiosin (**241**), purified from the culture of a marine tunicate proteobacteria in the Philippines (353,354). The investigators reported that the *in vitro* activity of heptyl prodigiosin against *P. falciparum* 3D7 was similar to chloroquine ($IC_{50} = 0.07$ vs. 0.015 μM, respectively). Interestingly, a single subcutaneous administration of 5–20 mg/kg heptyl prodigiosin significantly extended the survival of *P. berghei* ANKA strain-infected mice, suggesting that the molecule might be used as a lead compound (353,354).

E. Other Alkaloids

Quinazolinones and their derivatives are now known to have a wide range of useful biological properties, e.g., protein tyrosine kinase inhibitory, cholecystokinin inhibitory, antimicrobial, anticonvulsant, sedative, hypotensive, antidepressant, antiinflammatory, and antiallergy activities (355–357), but the interest of medicinal chemistry in quinazolinone derivatives was stimulated in the early 1950s with the elucidation of a quinazolinone alkaloid, febrifugine (**242**), from the Asian plant *Dichroa febrifuga* Lour., an ingredient of a traditional Chinese herbal remedy effective against malaria (358). This was one of the earliest studies of novel antimalarials from a plant source, and also reported the inactive interconvertible isomeric alkaloid isofebrifugine (**243**). Clinical trials showed that febrifugine was such a powerful emetic that it could not be used successfully as an antimalarial drug. Recently, quinazolinones fused with a simple β-carboline have also been reported to have antiplasmodial activity (359). Bioassay-guided fractionation of *Araliopsis tabouensis* Aubrév. & Pellegr. (Rutaceae) resulted in the isolation of the two alkaloids, 2-methoxy-13-methylrutaecarpine (**244**) and 5,8,13,14-tetrahydro-2-methoxy-14-methyl-5-oxo-7H-indolo[2′,3′:3,4]pyrido[2,1-b]quinazolin-6-ium chloride (**245**), which exhibited significant antimalarial activity against the chloroquine-sensitive D-6 strain of *P. falciparum* ($IC_{50} = 5.43$ and 9.92 μM, respectively) (359). In addition, tryptanthrin (**246**), 4-azatryptanthrin (**247**), and a series of substituted derivatives were studied for their *in vitro* activity against bloodstream forms of *T. b. brucei* (360). The unsubstituted derivatives were weakly active ($IC_{50} = 23$ and 40 μM for tryptanthrin and 4-azatryptanthrin, respectively). The antitrypanosomal activity was markedly improved by the presence of an electron-withdrawing group (halogen or nitro) at position 8 of the (aza)tryptanthrin ring system: the most active analog of the series, 4-aza-8-bromotryptanthrin (**248**), was up to 100 times more active than the unsubstituted

4-azatryptanthrin (IC$_{50}$ = 0.4 µM) (360).

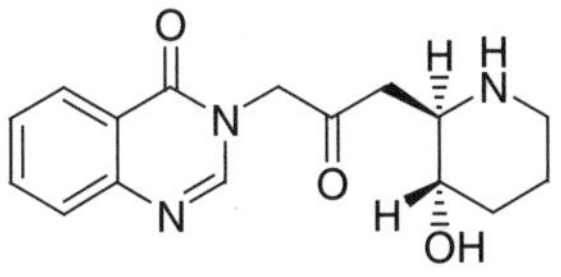

febrifugine **242**

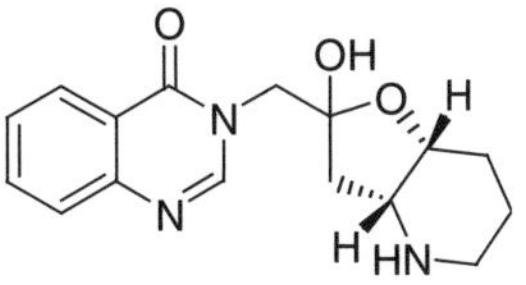

isofebrifugine **243**

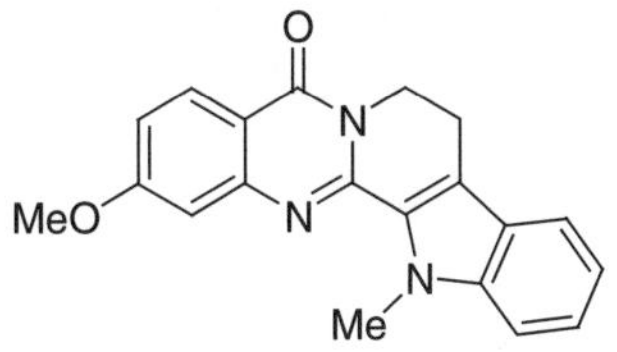

2-methoxy-13-methylrutaecarpine **244**

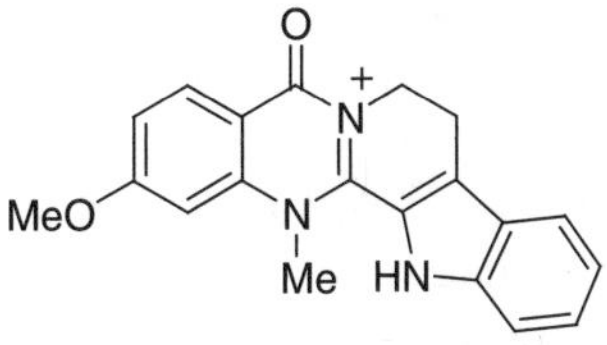

5,8,13,14-tetrahydro-2-methoxy-14-methyl-
5-oxo-7H-indolo[2',3':3,4]pyrido[2,1-b]
quinazolin-6-iumchloride **245**

.The well-known component of *Piper* sp., the alkaloid piperine (**249**), is the main secondary metabolite in *Piper nigrum* L. Various biological activities have been attributed to piperine, including insecticidal (361) and nematocidal activity (362), inhibition of liver metabolism (363), and antiprotozoal properties (364,365). Kapil reported a study of piperine activity against promastigote forms of *L. donovani* (364). Encouraging data were later obtained when testing *L. donovani*-infected hamsters *in vivo* with piperine intercalated into mannose-coated liposomes (366). More recently, lipid nanosphere formulations of piperine have also been assessed in BALB/c mice infected with *L. donovani* AG83 for 60 days. It was observed that a single dose of 5 mg/kg of these piperine formulations significantly reduced the liver and splenic parasite burden (367). Finally, the trypanocidal effects of piperine and some synthetic derivatives against epimastigote and amastigote forms of *T. cruzi* (365) have also been reported. Piperine proved to be a potent *T. cruzi* inhibitor, being more toxic to intracellular amastigotes (IC$_{50}$ = 4.91 µM) than epimastigotes (IC$_{50}$ = 7.36 µM).

tryptanthrin (**246**) X = CH$_2$,R = H
4-azatryptanthrin (**247**) X = N,R = H
4-aza-8-bromotryptanthrin (**248**) X = N,R = Br

piperine **249**

Emetine (**250**), a benzoquinolizidine alkaloid from *Cephaelis ipecacuanha* (Brot.) Rich. (Rubiaceae) used in the treatment of amebiasis, exhibited substantial trypanocidal activity (IC_{50} = 0.039 μM and 0.43 μM for *T. b. brucei* and *T. congolense*, respectively), which was also accompanied by cytotoxic effects in HL-60 cells (142). The *in vitro* antiplasmodial and antileishmanial activities of other emetine-like alkaloids such as cephaeline (**251**), psychotrine (**252**), klugine (**253**), and isocephae-line (**254**), four benzoquinolizidine derivatives isolated from the Bolivian medicinal plants *Pogonopus tubulosus* K.Schum. and *Psychotria klugii* Standl. (Rubiaceae), have also been reported. Cephaeline, with an IC_{50} of 0.058 μM against the chloroquine-sensitive strain 2087, and an IC_{50} of 0.023 μM against the chloroquine-resistant strain INDO of *P. falciparum*, was as potent as chloroquine (262). Likewise, cephaeline and klugine were equally potent against the chloroquine-resistant *P. falciparum* strain W-2 (IC_{50} 0.099 and 0.059 μM) and the chloroquine-sensitive strain D-6 (IC_{50} 0.081 μM) (368). However, psychotrine was less active, its IC_{50} values against the strains 2087 and INDO of *P. falciparum* being below 0.84 μM (262). The relative *in vitro* inactivity of psychotrine in comparison with cephaeline can be explained by its double bond in ring C, which enhances the coplanar conformation and electron environment (262). This suggests that minor changes in alkaloid structures may significantly affect their antiparasitic activities. Cephaeline was reported to inhibit *in vivo* parasitemia against *P. berghei* at a dose of 6 mg/kg/day (262). When tested for *in vitro* cytotoxic activity against the human cancer cell lines SK-MEL, KB, BT-549, and SK-OV-3 human cancer cell lines, cephaeline was more potent than doxorubicin, while klugine and isocephaeline were devoid of cytotoxic activity against these cell lines (368). The antileishmanial activity evaluation revealed that the alkaloids cephaeline, klugine, and isocephaeline showed strong activity against *L. donovani* promastigotes. Among these, cephaeline was the most potent (IC_{50} 0.06 μM), being >20- and >5-fold more active than the antileishmanial agents pentamidine and amphotericin B, respectively. On the other hand, emetine displayed potent activity against *L. donovani* (IC_{50} 0.062 μM), but was more toxic than cephaeline against VERO cells. In addition, klugine and isocephaeline also demonstrated strong activities against *L. donovani*, with IC_{50} values of 0.85 and 0.96 μM, respectively, and were devoid of significant toxicity against VERO cells (368). DNA intercalation, in combination with the inhibition of protein synthesis, could be responsible for the observed antiprotozoal and cytotoxic effects of these different alkaloids (142).

emetine (**250**) R_1=R_2=OMe,R_3=H
cephaeline (**251**) R_1=OMe,R_2=OH,R_3=H
klugine (**253**) R_1=R_2=OH,R_3= H
isocephaeline (**254**) R_1=OMe,R_2=OH,R_3=H

psychotrine **252**

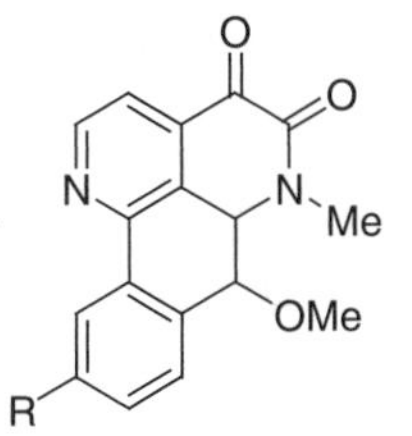

hadranthine A (**255**) R=OMe
imbiline 1 (**256**) R=H

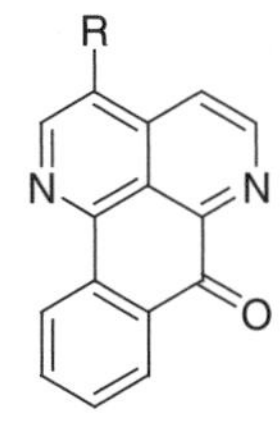

sampangine (**257**) R=H
3-methoxysampangine (**258**) R=OMe

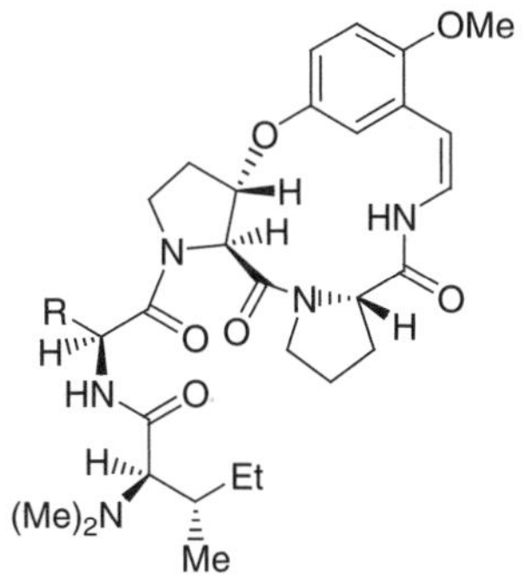

ziziphine N (**259**) R=i-Bu
ziziphine Q (**260**) R=i-Pr

An ethanolic extract of the stem bark of *Duguetia hadrantha* R.E.Fr. (Annonaceae) showed sufficient antimalarial activity to warrant bioassay-guided fractionation. This led to the isolation of antimalarial 4,5-dioxo-1-aza-aporphine alkaloids, hadranthine A (**255**) and imbiline 1 (**256**), and the copyrin alkaloids, sampangine (**257**) and 3-methoxysampangine (**258**), which were more active against the chloroquine-resistant *P. falciparum* clone W-2 (IC$_{50}$ = 0.37, 0.96, 0.29, and 0.36 μM, respectively) than the chloroquine-sensitive clone D-6 (369). Due to an absence of cytotoxicity toward VERO cells, the alkaloids hadranthine A, sampangine, and 3-methoxysampangine could be considered as potential antimalarial lead compounds.

Bioassay-guided fractionation of the EtOAc extract of the roots of *Ziziphus oenoplia* (L.) Mill. (Rhamnaceae) resulted in the isolation of some new 13-membered cyclopeptide alkaloids. Ziziphine N (**259**) and ziziphine Q (**260**) demonstrated significant antiplasmodial activity against the multidrug-resistant strain K1 of *P. falciparum* with IC$_{50}$ values of 6.41 and 5.86 μM, respectively. Both the methoxyl and the *N,N*-dimethylamino groups are seemingly crucial for the activity of these alkaloids (370).

IV. CURRENT STRATEGIES AND RECENT DEVELOPMENTS

The burden of leishmaniasis, malaria, Chagas' disease, and African trypanosomiasis is very costly in terms of human suffering, as well as contributing to poverty and underdevelopment. No vaccines are currently available for any of these diseases and therefore treatment remains a key element of disease control. The majority of available drugs have one or more of the following drawbacks: (a) insufficient efficacy or increasing loss of effectiveness due to resistance, (b) a high level of toxicity, (c) high cost, and (d) they are not readily available. Thus, new therapies for all of these protozoal diseases are urgently needed to reduce the mortality and morbidity associated with them (371). However, there is a lack of a robust pipeline of products in discovery and development to deliver drugs that meet the desired target product profiles for these diseases. It has been established that new antiprotozoal drugs should be dosed orally and curative regimens should be short. They should also be inexpensive to ensure routine availability in poverty-stricken countries where these diseases are endemic. Thus, considering these facts, new drugs must have the following characteristics (372,373):

- Malaria: For uncomplicated *P. falciparum* malaria, drugs must be orally active; be of low cost (~US$1 per full treatment course); be effective against drug-resistant parasites; have a low propensity to generate rapid resistance; be curative within 3 days; have potential for combination with other agents; have a pediatric formulation and be stable under tropical conditions (shelf life of >2 years).
- Leishmaniasis: The course of treatment must be short (≤14 days), with a single daily dose, alternate day or weekly dosing being acceptable; an oral drug is desired but can be injectable if treatment time is reduced; safety of available treatment should be improved, particularly for children and pregnant women; cost should be less than current treatments (US$200–400) and stability under standard tropical conditions is required (shelf life >2 years).
- Chagas' disease: The drugs should be active against blood and tissue forms as well as chronic forms of the disease; parental administration with reduced treatment time would be acceptable; an oral drug is desired as well as improved safety over current products (without cardiac effects), with a formulation that can be used for children and during pregnancy; they should be inexpensive and stable under tropical conditions (shelf life >2 years).
- African trypanosomiasis: Activity against both major species *T. rhodesiense* and *T. gambiense* is required as well as against known resistant strains, for example, melarsoprol failures; treatment for early-stage diseases is acceptable but efficacy against both early- and late-stage is desired, as well as parental administration in late-stage disease and oral formulation for early-stage disease; a cure should be achieved in 14 days or less; the cost should be less than the current treatment for early-stage disease ($100–140); safety for use during pregnancy and stability under tropical conditions (shelf life >2 years) are also required.

Since markets for drugs that treat tropical diseases are primarily in poor countries, marketing opportunities are generally considered to be limited. Thus, it is very important to employ a coordinated approach involving multi-disciplinary networks of academic and government institutions, researchers in several disciplines, as well as partnerships between large pharmaceutical companies and the public sector in both developed and developing countries. Fortunately, during the past 8 years there has been a dramatic increase in interest in Research and Development (R&D) directed towards producing new drugs for tropical diseases and various partnerships involving academic consortia, industry, governments, and philanthropic organizations are now dedicated to drug discovery, preclinical and late-stage development candidates and drug development in countries where most patients cannot afford to pay prices for the patented drugs they need (2,374–379).

One of the first steps in drug discovery and development is the identification of lead compounds that can be taken forward into lead optimization, or drug candidates that can be tested in human clinical trials. However, discovering lead compounds with the potential to become usable drugs is currently a key bottleneck in the pipeline for needed new and improved treatments for these tropical diseases. To improve discovery of active compounds and let them reach clinical application in the near future, different approaches have been identified and established (380–382). These approaches include: (a) *the identification of antiprotozoal active compounds from natural products* by evaluation of biological activity of plant extracts and metabolites purified from these extracts to identify products as parent compounds for the semi-synthetic or fully synthetic production of new drugs, (b) *the development of analogs of existing agents* to improve on existing drugs by chemical modifications of these compounds; this approach does not require knowledge of the mechanism of action or the biological target of the parent compound, (c) *the identification of new chemotherapeutic targets* by using the available genome sequences of human pathogens and their vectors and subsequent discovery of compounds that act on these targets (383–385), (d) *the optimization of therapy with available drugs* to improve efficacy, providing additive or synergistic antiparasitic activity, or preventing resistance development. The optimization may include new dosing and/or formulations, combination therapies, including newer agents and new combinations of older agents, (e) *the secondary use of compounds originally developed to treat other diseases*, an approach which usually involves "whole organism" screening of a known drug or compound against the disease-causing parasite. Finally, (f) *the identification of drug resistance* by combining previously effective agents with compounds that reverse parasite resistance to these agents.

All the approaches are being carried out by a number of initiatives, such as the Compound Evaluation and the Natural Products Initiative (CEN) (386), created in 2004 at the Special Programme for Research and Training in TDR. The CEN program coordinates the evaluation of compounds and natural products in validated whole organism and molecular target assays, promotes the use of natural products as starting points for the discovery of new lead compounds, and supports development of new tools and methods to accelerate drug discovery. In response to this initiative, extensive evaluations of natural products as potential new therapies for these tropical diseases are underway.

In the same direction, the Genomics and Discovery Research Committees (GDR) (387) at TDR are supporting several new networks for drug discovery involving developed and developing country researchers, institutions, and industry. These networks include the following: (a) a Compound Evaluation Network to screen thousands of compounds a year for antiparasitic activity, (b) Medicinal Chemistry and Pharmacokinetics Networks to promote the "hits" or "leads" identified through the compound evaluation network, (c) a Drug Target Portfolio Network to create a globally accessible database containing a prioritized list of drug targets across the range of disease-causing parasites that are the focus of TDR research.

In addition, new organizations have been recently created to fund research into protozoal diseases and are focused on drug development rather than drug discovery. The Tropical Disease Initiative (TDI) (388), the Institute for One World Health (IOWH) (389), and the World Bank (390) are some of the various organizations committed to the development of new antiprotozoal drugs, working in one or more stages of the drug development process. The open source TDI is asking sponsors, governments, and charities to subsidize developing country purchases at a guaranteed price, as well as creating non-profit venture-capital firms (Virtual Pharmas) that look for promising drug candidates before pushing drug development through contracts with corporate partners by charities. The IOWH, a not-for-profit pharmaceutical company, based in San Francisco and established in 2000, is focused on developing safe, effective, and affordable new medicines for people with infectious diseases in the developing world.

Similarly, the TDR Programme is also supporting drug discovery and development activities by the creation of some public-private partnerships (PPPs) to address the imbalance in funding between developed and developing world diseases (373). In these PPPs, the private partner can expand its business opportunities in return for assuming the new or expanded responsibilities and risks. Among these PPPs are Drug Neglected Drug Initiative (DNDi) (391) and Medicines for Malaria Venture (MMV) (392). The DNDi was created in 2003 by the Oswaldo Cruz Foundation (Brazil), the Indian Council for Medical Research, the Kenya Medical Research Institute, the Ministry of Health of Malaysia, Pasteur Institute (France), Médecins sans Frontières (MSF) (393), and the UNDP/World Bank/WHO/TDR (371) to improve the quality of life and the health of people suffering from neglected diseases. The MMV was established in 1999 to bring public and private sector partners together to fund and provide managerial and logistical support for the discovery, development, and delivery of new medicines to treat and prevent malaria.

Pharmaceutical companies are also increasing their participation in this process. GlaxoSmithKline, for example, is focused on discovering, developing, and making new drugs and vaccines available for treatment or prevention of diseases in the developing world, primarily malaria, but also HIV/AIDS and tuberculosis. The Novartis Institute in Singapore and the Wellcome Trust are focusing on malaria in partnership with MMV. Sanofi-Aventis has established the Impact Malaria Programme and continues collaborating with TDR and DNDi. Pfizer Inc. has also been collaborating with TDR to provide compounds and drug discovery support to identify leads for malaria, African sleeping sickness, Chagas' disease, leishmaniasis, and other tropical diseases.

Ongoing examples of drug discovery and development partnerships exploiting some of these approaches are listed below (373). Most of the drugs under study are in the discovery phase, while others are in the post-regulatory label extension phase.

Malaria: (a) *Discovery phase*: Improved 4-aminoquinoline (MMV, Glaxo-SmithKline, University of Liverpool); Farnesyltransferase inhibitors (MMV, Bristol-Myers Squibb, University of Washington); Manzamine derivatives (MMV, University of Mississippi); Cysteine protease inhibitors (MMV, University of California San Francisco, GSK); Fatty acid biosynthesis inhibition (MMV, Texas A&M, Albert Einstein/Howard Hughes, Jacobus); Pyridone (MMV, Glaxo-SmithKline); New dicationic molecules (MMV, University of North Carolina, Immtech); Dihydrofolate reductase inhibition (MMV, BIOTEC Thailand). (b) *Development phase*: Rectal artesunate (TDR); Chlorproguanil-dapsone-artesunate (MMV, TDR, GlaxoSmithKline); Pyronaridine-artesunate (MMV, TDR, Shin Poong); Amodiaquine-artesunate (TDR, European Union, DNDi); Mefloqine-artesunate (TDR, European Union, DNDi); Artemisone (MMV, Bayer, University of Hong Kong); Synthetic peroxide (MMV, Ranbaxy, University of Nebraska); i.v. artesunate (MMV, Walter Reed Army Institute of Research); and (c) *Post-regulatory label extensions phase*: Coartem in children of 5 kg weight (TDR, Novartis); Coartem pediatric form (MMV, TDR, Novartis); Chlorproguanil-dapsone (TDR, GlaxoSmithKline).

Leishmaniasis: (a) *Development phase*: Paromomycin (IOWH-TDR) and (b) *Post-regulatory label extensions phase*: Miltefosine (TDR-Zentaris).

Chagas' disease: (a) *Development phase*: Azole compounds (IOWH-Yale University).

African trypanosomiasis: (a) *Discovery phase*: Novel diamidines supported by Bill and Melinda Gates Foundation at University of North Carolina-Immtech; (b) *Development phase*: Oral eflornithine by TDR-Aventis; Novel diamidine (DB289) by Bill and Melinda Gates Foundation at University of North Carolina-Immtech.

It is noteworthy that although in the last 50 years a hundred alkaloid-containing plants traditionally used to treat parasitic diseases have proven to be biologically active against protozoan parasites in *in vitro* and *in vivo* models of infection, relatively few alkaloids have been studied further to assess their potential as lead compounds for the development of new antiprotozoal drugs. What is worse, only a few of these alkaloids have been tested in humans and none are under investigation in advanced stages of the drug development process. Moreover, despite all the analytical techniques available, a high number of alkaloid-containing plant species have not been investigated chemically or biologically in any detail and their mechanisms of action are yet to be properly determined (394). In the specific case of alkaloids active against *Leishmania* sp., *T. cruzi*, *T. brucei*, and *Plasmodium* sp. parasites, berberine (**68**) has been evaluated in human malaria and human leishmaniasis, but the available studies for berberine have not been evaluated in controlled and randomized clinical trials and therefore, more studies are required to confirm these results. Chimanine D (**27**) and 2-*n*-propylquinoline (**28**) have reached the clinical evaluation phase for the treatment of cutaneous leishmaniasis (6), while 4-aminoquinoline, manzamine derivatives and pyridone have been incorporated in the discovery phase for antimalarial drugs (392).

Although no plant-derived alkaloid is currently in the drug development phase, alkaloids continue to be useful in lead discovery because of their physicochemical and biological properties. Alkaloids display all the physicochemical properties of a typical drug (395): they have a moderate molecular weight (250–600 Da), the average number of NH and OH groups being 0.97 and 5.55, respectively (fewer than 5 and 10, according to the Lipinski rules). Alkaloids are amenable to standard techniques of purification and spectral analysis and most of them are relatively simple to synthesize, which permits a structure–activity relationship to be explored. Alkaloids can also be made more bioavailable by modulating their biological lipophobicity by substitution of certain chemical groups. Additionally, they display biological activities in the nM range and are derived from a sustainable resource (12). On the other hand, there is substantial ethnomedical and ecological information about plants that might contain alkaloids as their active metabolites, which remains uninvestigated chemically and biologically.

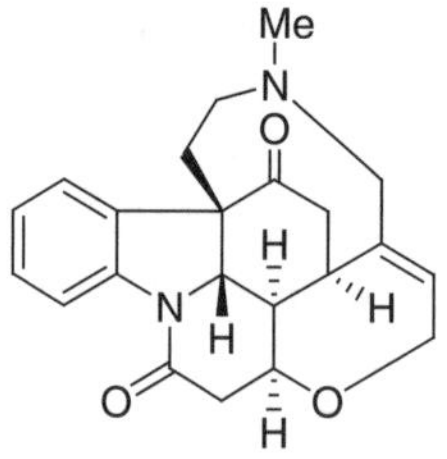

icajine **261**

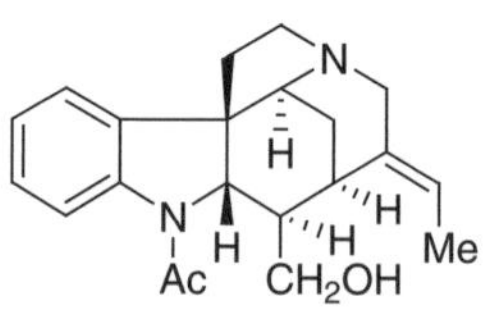

isoretuline **262**

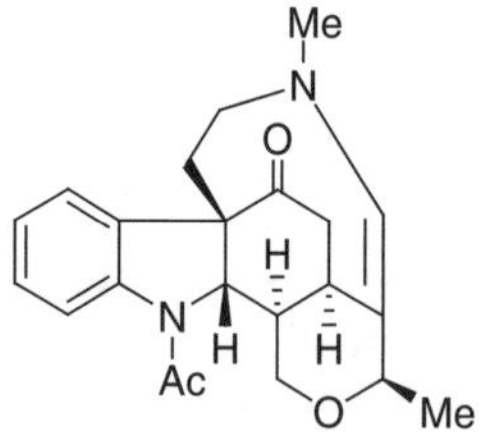

strychnobrasiline **263**

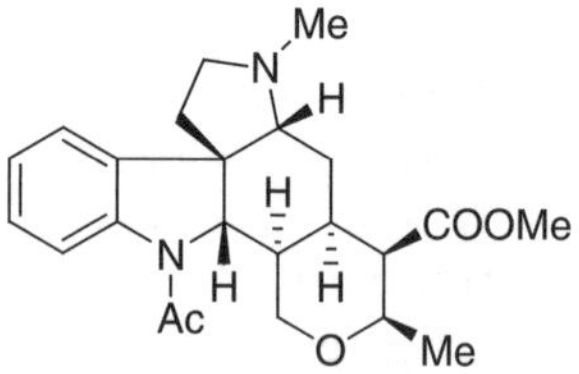

malagashanine **264**

Additionally, some plant-derived alkaloids have potential as modulators of the multidrug resistance (MDR) phenotype. Transporters of the ATP-Binding Cassette (ABC) family are known to provide the basis of MDR in mammalian cancer cells and in pathogenic yeasts, fungi, bacteria, and parasitic protozoa (396–404). One strategy to reverse the resistance of cells expressing ABC transporters is a combined and simultaneous use of chemotherapeutic agents and modulators. These modulators, also known as chemosensitizers, represent a wide range of chemical structures and can exert different cellular effects; however, they are supposed to act by the same mechanism for the reversal of MDR (405). Among potential candidates that have been screened for their ability to reverse MDR in protozoa are monoterpene indole, BBIQ, and isoquinoline alkaloids.

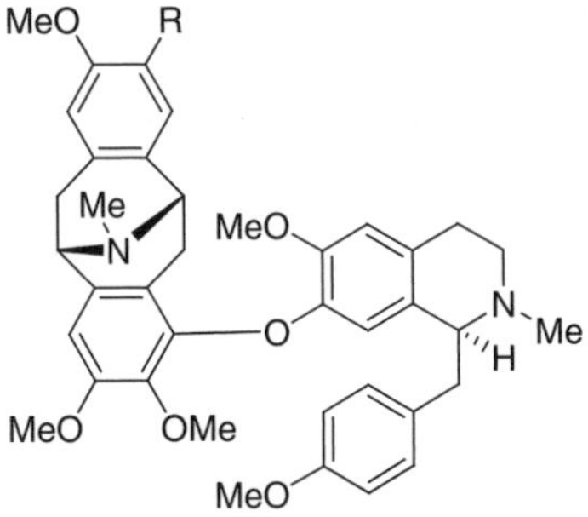

herveline B (**265**) R=H
herveline C (**266**) R=Me

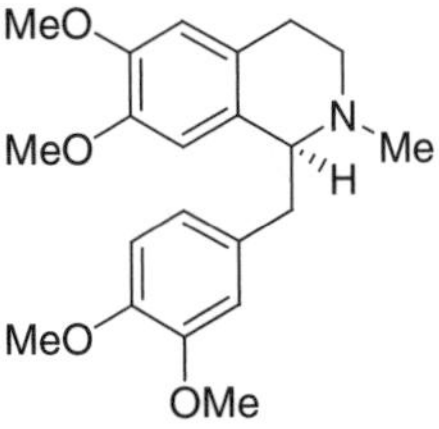

laudanosine **267**

Three alkaloids, icajine (**261**), isoretuline (**262**), and strychnobrasiline (**263**) are able to reverse chloroquine-resistance at concentrations of between 2.5 and 25 µg/mL. Icajine has also proved to be synergistic with mefloquine against a mefloquine-resistant strain of *P. falciparum* (406). Likewise, the alkaloid malagashanine (**264**) displayed a good biological profile, selectively enhancing the *in vitro* antimalarial activity of quinolines (chloroquine, quinine, and mefloquine), aminoacridines (quinacrine and pyronaridine), and a structurally unrelated drug (halofantrine) against chloroquine-resistant strains of *P. falciparum* (135,407). It was recently demonstrated that malagashanine prevents chloroquine efflux from, and stimulates chloroquine influx into, drug-resistant *P. falciparum* (408). Limacine (**82**) and fangchinoline (**86**), two isomeric BBIQs, are the most efficient potentiating drugs against the chloroquine-resistant strain FcM29 of *P. falciparum*, with fangchinoline being significantly more active than limacine (409). Moreover, hervelines B (**265**) and C (**266**), two alkaloids with moderate activity against *P. falciparum*, acted as enhancers of chloroquine activity in a dose-dependent manner (410,411). Laudanosine (**267**) also displayed *in vitro* chloroquine-potentiating action (9). These results showed that the methylation was vital for the potentiating action and that the bioactivity probably resulted mainly from the benzyltetrahydroisoquinoline moiety. Structure–activity relationship studies on MDR modulators have defined pharmacophoric substructures and physicochemical properties for anti-MDR activity. Among them are aromatic ring structures, a basic nitrogen atom, and high lipophilicity (405,412). One recurring tenet in these studies is the requirement of a nitrogen atom in the molecule, although it has since been established that the nitrogen is not an absolute parameter for activity, but exerts an influence through its contribution to hydrogen bond acceptor strength (413). Nevertheless, the derivation of structure–activity relationships for chemosensitisers should be restricted to chemically related compounds, with careful checking that MDR reversion is due to the same mechanism, and relinquishing efforts to establish general rules (414).

Alkaloids should continue to be an important part of drug development well into the future. The sequencing of the human genome opens new territory in terms of our ability to identify the proteins expressed by genes associated with the onset of diseases. These proteins can be used as molecular targets for testing thousands of compounds, including plant-derived alkaloids, in high throughput assays.

V. CONCLUSIONS AND FUTURE DIRECTIONS

A large number of antiprotozoal alkaloids with immense structural variety have been isolated from plants and marine organisms. However, despite all the analytical techniques available, considerable work is still needed to determine their mechanisms of action. Furthermore, many of these compounds have only been subjected to *in vitro* testing and *in vivo* results are still lacking. Similarly, although substantial progress has been made in identifying novel drug leads from the ocean's resources, great efforts are still needed to advance to clinical applications. In the future, the world's oceans will surely be playing an important part in alleviating the global protozoa-disease burden.

Considering the current interest in screening plants and/or marine organisms for antiprotozoal activity, and the incomplete knowledge of promising antiprotozoal natural alkaloids, potentially useful species and/or compounds should be tested principally in animals in order to determine their effectiveness in whole-organism systems, with particular emphasis on toxicity studies as well as an examination of effects on normal beneficial microbiota. In addition, methods need to be standardized to enable a more systematic search and to facilitate interpretation of results.

Major changes are occurring in R&D for drugs to treat tropical and neglected diseases, with the emergence of many organizations involved in product development. However, although this is encouraging, much more remains to be done. These organizations collaborate and form partnerships, but, in addition, some competition for projects and funding can develop. The challenge is to ensure that the products are delivered to the people who need them and to ensure that scientists in endemic countries are involved in the whole process. To guarantee the long-term sustainability of these programs, greater involvement of disease-endemic countries has to be built into the PPP organization model. More focused and result-oriented technology transfer and capacity building will support a future role of disease-endemic countries in discovering and developing the drugs they need.

Plant-derived alkaloids have provided many medicinal drugs in the past. They have contributed almost 50% of plant-derived natural products of pharmaceutical and biological significance and remain a potential source of novel therapeutic agents against infectious and non-infectious disorders. Even though there exist an important number of alkaloids from natural sources that have demonstrated potential as possible antiprotozoal agents, most of them do not meet all the requirements considered to be essential for their commercialization. Notwithstanding these problems, natural alkaloids will continue to play an important role in the development of a new generation of antiprotozoal drugs.

ACKNOWLEDGMENTS

The authors thank Ms. Lucy Brzoska for language revision, as well as Dr. Strahil H. Berkov for his critical review. Edison J. Osorio is grateful for a Fundación Carolina doctoral grant.

REFERENCES

[1] Neglected tropical diseases, hidden successes, emerging opportunities. WHO/CDS/NTD/ 2006.2.
[2] P. J. Hotez, D. H. Molyneux, A. Fenwick, J. Kumaresan, S. E. Sachs, J. D. Sachs, and L. Savioli, *N. Engl. J. Med.* **357**, 1018 (2007).
[3] D. J. Newman, G. M. Cragg, and K. M. Snader, *J. Nat. Prod.* **66**, 1022 (2003).
[4] D. J. Newman and G. M. Cragg, *J. Nat. Prod.* **70**, 461 (2007).
[5] S. Sepúlveda-Boza and B. K. Cassels, *Planta Med.* **62**, 98 (1996).
[6] M. J. Chan-Bacab and L. M. Peña-Rodríguez, *Nat. Prod. Rep.* **18**, 674 (2001).
[7] A. Fournet and V. Muñoz, *Curr. Top. Med. Chem.* **2**, 1215 (2002).
[8] C. W. Wright, *in:* "Trease and Evans Pharmacognosy" (W. C. Evans, ed.), (15th Edition), p. 407. W. B. Saunders, Edinburgh, 2002.
[9] S. Schwikkard and F. R. van Heerden, *Nat. Prod. Rep.* **19**, 675 (2002).
[10] O. Kayser, A. F. Kiderlen, and S. L. Croft, *Parasitol. Res.* **90(suppl 2)**, S55 (2003).
[11] S. Hoet, F. Opperdoes, R. Brun, and J. Quetin-Leclercq, *Nat. Prod. Rep.* **21**, 353 (2004).
[12] G. A. Cordell, M. L. Quinn-Beattie, and N. R. Farnsworth, *Phytother. Res.* **15**, 183 (2001).
[13] http://www.who.int/tdr/diseases/default.htm. Accessed on January 15, 2008.
[14] J. Alvar, S. Yactayo, and C. Bern, *Trends Parasitol.* **22**, 552 (2006).
[15] http://www.who.int/neglected_diseases/en/. Accessed on January 15, 2008.
[16] J. Orbinski, C. Beyrer, and S. Singh, *Lancet* **25**, 370 (2007).
[17] http://www.who.int/tdr/diseases/leish/default.htm. Accessed on January 15, 2008.
[18] Report of the Scientific Working Group on Leishmaniasis, TDR/SWG/04 (2004).
[19] R. Reithinger, J. C. Dujardin, H. Louzir, C. Pirmez, B. Alexander, and S. Brooker, *Lancet Infect. Dis.* **7**, 581 (2007).
[20] J. Shaw, *Mem. Inst. Oswaldo Cruz* **102**, 541 (2007).
[21] J. Mishra, A. Saxena, and S. Singh, *Curr. Med. Chem.* **14**, 1153 (2007).
[22] P. Minodier and P. Parola, *Travel Med. Infect. Dis.* **5**, 150 (2007).
[23] S. L. Croft, M. P. Barrett, and J. A. Urbina, *Trends Parasitol.* **21**, 508 (2005).
[24] S. L. Croft, S. Sundar, and A. H. Fairlamb, *Clin. Microbiol. Rev.* **19**, 111 (2006).
[25] S. Sundar, D. K. More, M. K. Singh, V. P. Singh, S. Sharma, A. Makharia, P. C. Kumar, and H. W. Murray, *Clin. Infect. Dis.* **31**, 1104 (2000).
[26] M. Roussel, M. Nacher, G. Fremont, B. Rotureau, E. Clyti, D. Sainte-Marie, B. Carme, R. Pradinaud, and P. Couppie, *Ann. Trop. Med. Parasitol.* **100**, 307 (2006).
[27] V. S. Amato, F. F. Tuon, A. M. Siqueira, A. C. Nicodemo, and V. A. Neto, *Am. J. Trop. Med. Hyg.* **77**, 266 (2007).
[28] J. Brajtburg and J. Bolard, *Clin. Microbiol. Rev.* **9**, 512 (1996).
[29] V. Yardley and S. L. Croft, *Int. J. Antimicrob. Agents* **13**, 243 (2000).
[30] N. K. Ganguly, *TDRnews* **68**, 2 (2002).
[31] J. Berman, A. D. Bryceson, S. Croft, J. Engel, W. Gutteridge, J. Karbwang, H. Sindermann, J. Soto, S. Sundar, and J. A. Urbina, *Trans. R. Soc. Trop. Med. Hyg.* **100**, S41 (2006).

[32] http://www.who.int/tdr/diseases/chagas/default.htm. Accessed on January 15, 2008.

[33] Report of the Scientific Working Group on Chagas' Disease, TDR/SWG/05 (2005).

[34] A. R. L. Teixeira, N. Nitz, M. C. Guimaro, C. Gomes, and C. A. Santos-Buch, *Postgrad. Med. J.* **82**, 788 (2006).

[35] K. Senior, *Lancet Infect. Dis.* **7**, 572 (2007).

[36] J. A. Urbina and R. Docampo, *Trends Parasitol.* **19**, 495 (2003).

[37] http://www.who.int/tdr/diseases/tryp/default.htm. Accessed on January 15, 2008.

[38] Report of the Scientific Working Group on African Trypanosomiasis, TDR/SWG/01 (2001).

[39] I. Maudlin, *Ann. Trop. Med. Parasitol.* **100**, 679 (2006).

[40] A. J. Nok, *Parasitol. Res.* **90**, 71 (2003).

[41] S. Bisser, F. X. N'Siesi, V. Lejon, P. M. Preux, S. Van Nieuwenhove, C. Miaka Mia Bilenge, and P. Buscher, *J. Infect. Dis.* **195**, 322 (2007).

[42] J. M. Kagira and N. Maina, *Onderstepoort J. Vet. Res.* **74**, 17 (2007).

[43] http://www.who.int/tdr/diseases/malaria/default.htm. Accessed on January 15, 2008.

[44] Report of the Scientific Working Group on Malaria, TDR/SWG/03 (2003).

[45] B. M. Greenwood, K. Bojang, C. J. M. Whitty, and G. A. T. Targett, *Lancet* **365**, 1487 (2005).

[46] S. Hay, C. Guerra, A. Tatem, A. Noor, and R. Snow, *Lancet Infect. Dis.* **4**, 327 (2004).

[47] J. Sachs and P. Malaney, *Nature* **415**, 680 (2002).

[48] J. van Geertruyden, F. Thomas, A. Erhart, and U. D'Alessandro, *Am. J. Trop. Med. Hyg.* **71**, 35 (2004).

[49] J. Meis, J. Verhave, P. Jap, R. Sinden, and J. Meuwissen, *Nature* **302**, 424 (1983).

[50] www.who.int/malaria/pages/performance/antimalarialmedicines.html. Accessed on January 15, 2008.

[51] E. H. Ekland and D. A. Fidock, *Curr. Opin. Microbiol.* **10**, 363 (2007).

[52] L. Tilley, T. M. Davis, and P. G. Bray, *Future Microbiol.* **1**, 127 (2006).

[53] A. Fournet, R. Hocquemiller, F. Roblot, A. Cavé, P. Richomme, and J. Bruneton, *J. Nat. Prod.* **56**, 1547 (1993).

[54] A. Fournet, A. A. Barrios, V. Muñoz, R. Hocquemiller, A. Cavé, and J. Bruneton, *Antimicrob. Agents Chemother.* **37**, 859 (1993).

[55] A. Fournet, J. C. Gantier, A. Gautheret, L. Leysalles, M. H. Muñoz, J. Mayrarge, H. Moskowitz, A. Cavé, and R. Hocquemiller, *J. Antimicrob. Chemother.* **33**, 537 (1994).

[56] M. A. Fakhfakh, A. Fournet, E. Prina, J. F. Mouscadet, X. Franck, R. Hocquemiller, and B. Figadère, *Bioorg. Med. Chem.* **11**, 5013 (2003).

[57] H. Nakayama, P. M. Loiseau, C. Bories, S. Torres De Ortiz, A. Schinini, E. Serna, A. Rojas De Arias, M. A. Fakhfakh, X. Franck, B. Figadère, R. Hocquemiller, and A. Fournet, *Antimicrob. Agents Chemother.* **49**, 4950 (2005).

[58] H. Nakayama, J. Desrivot, C. Bories, X. Franck, B. Figadère, R. Hocquemiller, A. Fournet, and P. M. Loiseau, *Biomed. Pharmacother.* **61**, 186 (2007).

[59] A. M. Belliard, B. Baunes, M. Fakhfakh, R. Hocquemiller, and R. Farinotti, *Xenobiotica* **33**, 341 (2003).

[60] J. Desrivot, P. O. Edlund, R. Svensson, P. Baranczewski, A. Fournet, B. Figadère, and C. Herrenknecht, *Toxicology* **235**, 27 (2007).

[61] I. Jacquemond-Collet, F. Benoit-Vical, Mustofa, A. Valentin, E. Stanislas, M. Mallié, and I. Fourasté, *Planta Med.* **68**, 68 (2002).

[62] C. Lavaud, G. Massiot, C. Vasquez, C. Moretti, M. Sauvain, and L. Balderrama, *Phytochemistry* **40**, 317 (1995).

[63] M. W. Muriithi, W. R. Abraham, J. Addae-Kyereme, I. Scowen, S. L. Croft, P. M. Gitu, H. Kendrick, E. N. M. Njagi, and C. W. Wright, *J. Nat. Prod.* **65**, 956 (2002).

[64] M. Mena and J. Bonjoch, *Tetrahedron* **61**, 8264 (2005).

[65] J. W. Daly, *J. Nat. Prod.* **61**, 162 (1998).

[66] R. A. Davis, A. R. Carroll, and R. J. Quinn, *J. Nat. Prod.* **65**, 454 (2002).

[67] B. Steffan, *Tetrahedron* **47**, 8729 (1991).

[68] J. Kubanek, D. E. Williams, E. Dilip de Silva, T. Allen, and R. J. Andersen, *Tetrahedron Lett.* **36**, 6189 (1995).

[69] A. D. Wright, E. Goclik, G. M. König, and R. Kaminsky, *J. Med. Chem.* **45**, 3067 (2002).
[70] M. Tori, T. Shimoji, E. Shimura, S. Takaoka, K. Nakashima, M. Sono, and W. A. Ayer, *Phytochemistry* **53**, 503 (2000).
[71] J. W. Daly, H. M. Garraffo, P. Jain, T. F. Spande, R. R. Snelling, C. Jaramillo, and A. S. Rand, *J. Chem. Ecol.* **26**, 73 (2000).
[72] T. H. Jones, J. S. T. Gorman, R. R. Snelling, J. H. C. Delabie, M. S. Blum, H. M. Garraffo, P. Jain, J. W. Daly, and T. F. Spande, *J. Chem. Ecol.* **25**, 1179 (1999).
[73] T. F. Spande, P. Jain, H. M. Garraffo, L. K. Pannell, H. J. C. Yeh, J. W. Daly, S. Fukumoto, K. Imamura, T. Tokuyama, J. A. Torres, R. R. Snelling, and T. H. Jones, *J. Nat. Prod.* **62**, 5 (1999).
[74] E. R. Watkins and S. R. Meshnick, *Semin. Pediatr. Infect. Dis.* **11**, 202 (2000).
[75] T. Frosch, M. Schmitt, and J. Popp, *J. Phys. Chem. B* **111**, 4171 (2007).
[76] P. Deloron, J. A. Ramanamirija, J. Le Bras, B. Larouze, and P. Coulanges, *Arch. Inst. Pasteur Madagascar* **51**, 69 (1984).
[77] J. L. Vennerstrom, E. O. Nuzum, R. E. Miller, A. Dorn, L. Gerena, P. A. Dande, W. Y. Ellis, R. G. Ridley, and W. K. Milhous, *Antimicrob. Agents Chemother.* **43**, 598 (1999).
[78] S. L. Croft and G. H. Coombs, *Trends Parasitol.* **19**, 502 (2003).
[79] R. Dietze, S. F. Carvalho, L. C. Valli, J. Berman, T. Brewer, W. Milhous, I. Sanchez, B. Schuster, and M. Grogl, *Am. J. Trop. Med. Hyg.* **65**, 685 (2001).
[80] M. K. Wasunna, J. R. Rashid, J. Mbui, G. Kirigi, D. Kinoti, H. Lodenyo, J. M. Felton, A. J. Sabin, and J. Horton, *Am. J. Trop. Med. Hyg.* **73**, 871 (2005).
[81] T. K. Jha, S. Sundar, C. P. Thakur, J. M. Felton, A. J. Sabin, and J. Horton, *Am. J. Trop. Med. Hyg.* **73**, 1005 (2005).
[82] A. M. Dueñas-Romero, P. M. Loiseau, and M. Saint-Pierre-Chazalet, *Biochim. Biophys. Acta* **1768**, 246 (2007).
[83] T. J. Egan, D. C. Ross, and P. A. Adams, *FEBS Lett.* **352**, 54 (1994).
[84] J. Zhang, M. Krugliak, and H. Ginsburg, *Mol. Biochem. Parasitol.* **99**, 129 (1999).
[85] T. J. Egan and K. K. Ncokazi, *J. Inorg. Biochem.* **99**, 1532 (2005).
[86] H. Ginsburg, O. Famin, F. Zhang, and M. Krugliak, *Biochem. Pharmacol.* **56**, 1305 (1998).
[87] R. Banerjee, J. Liu, W. Beatty, L. Pelosof, M. Klemba, and D. E. Goldberg, *Proc. Natl. Acad. Sci. U.S.A.* **99**, 990 (2002).
[88] P. J. Rosenthal, P. S. Sijwali, A. Singh, and B. R. Shenai, *Curr. Pharm. Des.* **8**, 1659 (2002).
[89] D. E. Goldberg, A. F. G. Slater, A. Cerami, and G. B. Henderson, *Proc. Natl. Acad. Sci. U.S.A.* **87**, 2931 (1990).
[90] P. Loria, S. Miller, M. Foley, and L. Tilley, *Biochem. J.* **339**, 363 (1999).
[91] A. F. G. Slater, W. J. Swiggard, B. R. Orton, W. D. Flitter, D. E. Goldberg, A. Cerami, and G. B. Henderson, *Proc. Natl. Acad. Sci. U.S.A.* **88**, 325 (1991).
[92] D. E. Goldberg and A. F. G. Slater, *Parasitol. Today* **8**, 280 (1992).
[93] T. J. Egan, J. M. Combrinck, J. Egan, G. R. Hearne, H. M. Marques, S. Ntenteni, B. T. Sewell, P. J. Smith, D. Taylor, D. A. van Schalkwyk, and J. C. Walden, *Biochem. J.* **365**, 343 (2002).
[94] A. Dorn, S. R. Vippagunta, H. Matile, C. Jaquet, J. L. Vennerstrom, and R. G. Ridley, *Biochem. Pharmacol.* **55**, 737 (1998).
[95] T. J. Egan and H. M. Marques, *Coord. Chem. Rev.* **192**, 493 (1999).
[96] I. Solomonov, M. Osipova, Y. Feldman, C. Baehtz, K. Kjaer, I. K. Robinson, G. T. Webster, D. McNaughton, B. R. Wood, I. Weissbuch, and L. Leiserowitz, *J. Am. Chem. Soc.* **129**, 2615 (2007).
[97] H. Guinaudeau, M. Lebœuf, and A. Cavé, *Lloydia* **38**, 275 (1975).
[98] H. Guinaudeau, M. Lebœuf, and A. Cavé, *J. Nat. Prod.* **42**, 133 (1979).
[99] H. Guinaudeau, M. Lebœuf, and A. Cavé, *J. Nat. Prod.* **46**, 761 (1983).
[100] H. Guinaudeau, M. Lebœuf, and A. Cavé, *J. Nat. Prod.* **51**, 389 (1988).
[101] H. Guinaudeau, M. Lebœuf, and A. Cavé, *J. Nat. Prod.* **57**, 1033 (1994).
[102] M. Lebœuf, A. Cavé, P. K. Bhaumik, B. Mukherjee, and R. Mukherjee, *Phytochemistry* **21**, 2783 (1980).
[103] E. F. Queiroz, F. Roblot, A. Cavé, M. Q. Paulo, and A. Fournet, *J. Nat. Prod.* **59**, 438 (1996).
[104] A. Février, M. E. Ferreira, A. Fournet, G. Yaluff, A. Inchausti, A. Rojas de Arias, R. Hocquemiller, and A. I. Waechter, *Planta Med.* **65**, 47 (1999).

[105] A. I. Waechter, A. Cavé, R. Hocquemiller, C. Bories, V. Muñoz, and A. Fournet, *Phytother. Res.* **13**, 175 (1999).

[106] B. Akendengue, E. Ngou-Milama, F. Roblot, A. Laurens, R. Hocquemiller, Ph. Grellier, and F. Frappier, *Planta Med.* **68**, 167 (2002).

[107] R. Hocquemiller, D. Cortés, G. J. Arango, S. H. Myint, A. Cavé, A. Angelo, V. Muñoz, and A. Foumet, *J. Nat. Prod.* **54**, 445 (1991).

[108] V. Mahiou, F. Roblot, R. Hocquemiller, A. Cavé, A. Rojas de Arias, A. Inchausti, G. Yaluff, A. Fournet, and A. Angelo, *J. Nat. Prod.* **57**, 890 (1994).

[109] B. Weniger, R. Aragón, E. Deharo, J. Bastida, C. Codina, A. Lobstein, and R. Anton, *Pharmazie* **55**, 867 (2000).

[110] H. Montenegro, M. Gutiérrez, L. I. Romero, E. Ortega-Barría, T. L. Capson, and L. C. Rios, *Planta Med.* **69**, 677 (2003).

[111] B. Weniger, S. Robledo, G. J. Arango, E. Deharo, R. Aragón, V. Muñoz, J. Callapa, A. Lobstein, and R. Anton, *J. Ethnopharmacol.* **78**, 193 (2001).

[112] M. R. Camacho, G. C. Kirby, D. C. Warhurst, S. L. Croft, and J. D. Phillipson, *Planta Med.* **66**, 478 (2000).

[113] V. Muñoz, M. Sauvain, P. Mollinedo, J. Callapa, I. Rojas, A. Gimenez, A. Valentin, and M. Mallié, *Planta Med.* **65**, 448 (1999).

[114] S. Hoet, C. Stévigny, S. Block, F. Opperdoes, P. Colson, B. Baldeyrou, A. Lansiaux, C. Bailly, and J. Quetin-Leclercq, *Planta Med.* **70**, 407 (2004).

[115] K. Likhitwitayawuid, S. Dej-adisai, V. Jongbunprasert, and J. Krungkrai, *Planta Med.* **65**, 754 (1999).

[116] A. Morello, I. Lipchenca, B. K. Cassels, H. Speisky, J. Aldunate, and Y. Repetto, *Comp. Biochem. Physiol. Pharmacol. Toxicol. Endocrinol.* **107**, 367 (1994).

[117] C. W. Wright, S. J. Marshall, P. F. Russell, M. M. Anderson, J. D. Phillipson, G. C. Kirby, D. C. Warhurst, and P. L. Schiff Jr., *J. Nat. Prod.* **63**, 1638 (2000).

[118] S. H. Woo, N. J. Sun, J. M. Cassady, and R. M. Snapka, *Biochem. Pharmacol.* **57**, 1141 (1999).

[119] S. H. Woo, M. C. Reynolds, N. J. Sun, J. M. Cassady, and R. M. Snapku, *Biochem. Pharmacol.* **54**, 467 (1997).

[120] M. E. Letelier, E. Rodríguez, A. Wallace, M. Lorca, Y. Repetto, A. Morello, and J. Aldunate, *Exper. Parasitol.* **71**, 357 (1990).

[121] G. L. Montoya, E. J. Osorio, N. Jiménez, and G. J. Arango, *Vitae* **11**, 51 (2004).

[122] L. Grycová, J. Dostál, and R. Marek, *Phytochemistry* **68**, 150 (2007).

[123] K. W. Bentley, *Nat. Prod. Rep.* **17**, 247 (2000).

[124] K. W. Bentley, *Nat. Prod. Rep.* **18**, 148 (2001).

[125] K. W. Bentley, *Nat. Prod. Rep.* **19**, 332 (2002).

[126] K. W. Bentley, *Nat. Prod. Rep.* **20**, 342 (2003).

[127] K. W. Bentley, *Nat. Prod. Rep.* **21**, 395 (2004).

[128] K. W. Bentley, *Nat. Prod. Rep.* **22**, 249 (2005).

[129] K. W. Bentley, *Nat. Prod. Rep.* **23**, 444 (2006).

[130] J. L. Vennerstrom, J. K. Lovelace, V. B. Waits, W. L. Hanson, and D. L. Klayman, *Antimicrob. Agents Chemother.* **34**, 918 (1990).

[131] J. D. Phillipson and C. W. Wright, *Trans. R. Soc. Trop. Med. Hyg.* **85**, 18 (1991).

[132] M. M. Iwu, J. E. Jackson, and B. G. Schuster, *Parasitol. Today* **10**, 65 (1994).

[133] C. W. Wright and J. D. Phillipson, *Phytother. Res.* **4**, 127 (1990).

[134] J. El-On, G. P. Jacobs, and L. Weinrauch, *Parasitol. Today* **4**, 76 (1988).

[135] H. Rafatro, D. Ramanitrahasimbola, P. Rasoanaivo, S. Ratsimamanga-Urverg, A. Rakoto-Ratsimamanga, and F. Frappier, *Biochem. Pharmacol.* **59**, 1053 (2000).

[136] K. Iwasa, H. S. Kim, Y. Wataya, and D. U. Lee, *Eur. J. Med. Chem.* **33**, 65 (1998).

[137] K. Iwasa, Y. Nishiyama, M. Ichimaru, M. Moriyasu, H. S. Kim, Y. Wataya, T. Yamori, T. Takashi, and D. U. Lee, *Eur. J. Med. Chem.* **34**, 1077 (1999).

[138] K. Iwasa, M. Moriyasu, Y. Tachibana, H. S. Kim, Y. Wataya, W. Wiegrebe, K. F. Bastow, L. M. Cosentino, M. Kozuka, and K. H. Lee, *Bioorg. Med. Chem.* **9**, 2871 (2001).

[139] J. Suchomelová, H. Bochořáková, H. Paulová, P. Musil, and E. Táborská, *J. Pharm. Biomed. Anal.* **44**, 283 (2007).

[140] M. Wink, T. Schmeller, and B. Latz-Bruning, *J. Chem. Ecol.* **24**, 1881 (1998).
[141] R. Verpoorte, *in:* "Alkaloids: Biochemistry, Ecology and Medicinal Applications" (M. F. Roberts and M. Wink, eds.), p. 397. Plenum Press, New York, 1998.
[142] K. Merschjohann, F. Sporer, D. Steverding, and M. Wink, *Planta Med.* **67**, 623 (2001).
[143] D. M. N. Gakunju, E. K. Mberu, S. F. Dossaji, A. I. Gray, R. D. Waigh, P. G. Waterman, and W. M. Watkins, *Antimicrob. Agents Chemother.* **39**, 2606 (1995).
[144] J. Kluza, B. Baldeyrou, P. Colson, P. Rasoanaivo, L. Mambu, F. Frappier, and C. Bailly, *Eur. J. Pharm. Sci.* **20**, 383 (2003).
[145] B. C. Ellbrd, *Parasitol. Today* **2**, 309 (1986).
[146] M. W. Davidson, I. Lopp, S. Alexander, and W. D. Wilson, *Nucleic Acids Res.* **4**, 2697 (1977).
[147] M. Maiti and K. Chaudhuri, *Indian J. Biochem. Biophys.* **18**, 245 (1981).
[148] D. Debnath, G. S. Kumar, R. Nandi, and M. Maiti, *Indian J. Biochem. Biophys.* **26**, 201 (1989).
[149] G. S. Kumar, D. Debnath, A. Sen, and M. Maiti, *Biochem. Pharmacol.* **46**, 1665 (1993).
[150] S. Mazzini, M. C. Belluci, and R. Mondelli, *Bioorg. Med. Chem. Lett.* **11**, 505 (2003).
[151] K. Bhadra, M. Maiti, and G. S. Kumar, *Biochim. Biophys. Acta* **1770**, 1071 (2007).
[152] S. D. Fang, L. K. Wang, and S. M. Hecht, *J. Org. Chem.* **58**, 5025 (1993).
[153] M. Stiborová, V. Šimánek, E. Frei, P. Hobza, and J. Ulrichová, *Chem.-Biol. Interact.* **140**, 231 (2002).
[154] S. Das, G. S. Kumar, A. Ray, and M. Maiti, *J. Biomol. Struct. Dyn.* **20**, 703 (2003).
[155] L. P. Bai, Z. Z. Zhao, Z. Cai, and Z. H. Jiang, *Bioorg. Med. Chem.* **14**, 5439 (2006).
[156] P. L. Schiff Jr., *J. Nat. Prod.* **46**, 1 (1983).
[157] P. L. Schiff Jr., *J. Nat. Prod.* **50**, 529 (1987).
[158] P. L. Schiff Jr., *J. Nat. Prod.* **54**, 645 (1991).
[159] P. L. Schiff Jr., *J. Nat. Prod.* **60**, 934 (1997).
[160] M. R. Camacho, J. D. Phillipson, S. L. Croft, P. Rock, S. J. Marshall, and P. L. Schiff Jr., *Phytother. Res.* **16**, 432 (2002).
[161] L. J. Ho, D. M. Chang, T. C. Lee, M. L. Chang, and J. H. Lai, *Eur. J. Pharmacol.* **367**, 389 (1999).
[162] N. Ivanovska, P. Nikolova, M. Hristova, S. Philipov, and R. Istatkova, *Int. J. Immunopharmacol.* **21**, 325 (1999).
[163] H. S. Kim, Y. H. Zhang, K. W. Oh, and H. Y. Ahn, *J. Ethnopharmacol.* **58**, 117 (1997).
[164] J. L. Martínez, R. Torres, and M. A. Morales, *Phytother. Res.* **11**, 246 (1997).
[165] H. S. Kim, Y. H. Zhang, and Y. P. Yun, *Planta Med.* **65**, 135 (1999).
[166] M. Okamoto, M. Ono, and M. Baba, *AIDS Res. Hum. Retroviruses* **14**, 1239 (1998).
[167] A. Fournet, V. Muñoz, A. M. Manjón, A. Angelo, R. Hocquemiller, D. Cortés, A. Cavé, and J. Bruneton, *J. Ethnopharmacol.* **24**, 327 (1988).
[168] A. Fournet, A. A. Barrios, V. Muñoz, R. Hocquemiller, and A. Cavé, *Phytother. Res.* **7**, 281 (1993).
[169] G. Dreyfuss, D. P. Allais, H. Guinaudeau, and J. Bruneton, *Ann. Pharm. Fr.* **45**, 243 (1987).
[170] V. Mahiou, F. Roblot, A. Fournet, and R. Hocquemiller, *Phytochemistry* **54**, 709 (2000).
[171] A. Fournet, A. M. Manjón, V. Muñoz, A. Angelo, J. Bruneton, R. Hocquemiller, D. Cortés, and A. Cavé, *J. Ethnopharmacol.* **24**, 337 (1988).
[172] A. Rojas de Arias, A. Inchausti, M. Ascurrat, N. Fleitas, E. Rodríguez, and A. Fournet, *Phythother. Res.* **8**, 141 (1994).
[173] A. Fournet, M. E. Ferreira, A. Rojas de Arias, A. Schinini, H. Nakayama, S. Torres, R. Sanabria, H. Guinaudeau, and J. Bruneton, *Int. J. Antimicrob. Agents* **8**, 163 (1997).
[174] A. Fournet, A. Rojas de Arias, M. E. Ferreira, H. Nakayama, S. Torres de Ortiz, A. Schinini, M. Samudio, N. Vera de Bilbao, M. Lavault, and F. Bonté, *Int. J. Antimicrob. Agents* **13**, 189 (2000).
[175] A. Schmidt and R. L. Krauth-Siegel, *Curr. Top. Med. Chem.* **2**, 1239 (2002).
[176] C. K. Angerhofer, H. Guinaudeau, V. Wongpanich, J. M. Pezzuto, and G. A. Cordell, *J. Nat. Prod.* **62**, 59 (1999).
[177] S. J. Marshall, P. F. Russell, C. W. Wright, M. M. Anderson, J. D. Phillipson, G. C. Kirby, D. C. Warhurst, and P. L. Schiff Jr., *Antimicrob. Agents Chemother.* **38**, 96 (1994).
[178] M. L. Lohombo-Ekomba, P. N. Okusa, O. Penge, C. Kabongo, M. I. Choudhary, and O. E. Kasende, *J. Ethnopharmacol.* **93**, 331 (2004).
[179] L. Z. Lin, H. L. Shieh, C. K. Angerhofer, J. M. Pezzuto, G. A. Cordell, L. Xue, M. E. Johnson, and N. Ruangrungsi, *J. Nat. Prod.* **56**, 22 (1993).

[180] A. Chea, S. Hout, S. S. Bun, N. Tabatadze, M. Gasquet, N. Azas, R. Elias, and G. Balansard, *J. Ethnopharmacol.* **112**, 132 (2007).

[181] M. Kozuka, K. Miyaji, T. Sawada, and M. Tomita, *J. Nat. Prod.* **48**, 341 (1985).

[182] T. T. Thuy, T. V. Sung, K. Franke, and L. Wessjonann, *J. Chem.* **44**, 110 (2006).

[183] T. T. Thuy, K. Franke, A. Porzel, L. Wessjohann, and T. V. Sung, *J. Chem.* **44**, 259 (2006).

[184] P. A. Tamez, D. Lantvit, E. Lim, and J. M. Pezzuto, *J. Ethnopharmacol.* **98**, 137 (2005).

[185] K. Likhitwitayawuid, C. K. Angerhofer, G. A. Cordell, J. M. Pezzuto, and N. Ruangrungsi, *J. Nat. Prod.* **56**, 30 (1993).

[186] A. L. Otshudi, A. Foriers, A. Vercruysse, A. Van Zeebroeck, and S. Lauwers, *Phytomedicine* **7**, 167 (2000).

[187] A. L. Otshudi, S. Apers, L. Pieters, M. Claeys, C. Pannecouque, E. De Clercq, A. Van Zeebroeck, S. Lauwers, M. Frédérich, and A. Foriers, *J. Ethnopharmacol.* **102**, 89 (2005).

[188] L. Mambu, M. T. Martin, D. Razafimahefa, D. Ramanitrahasimbola, P. Rasoanaivo, and F. Frappier, *Planta Med.* **66**, 537 (2000).

[189] M. Said, A. Latiff, S. J. Partridge, and J. D. Phillipson, *Planta Med.* **57**, 389 (1991).

[190] M. Böhlke, H. Guinaudeau, C. K. Angerhofer, V. Wongpanich, D. D. Soejarto, N. R. Farnsworth, G. A. Mora, and L. J. Poveda, *J. Nat. Prod.* **59**, 576 (1996).

[191] L. Z. Lin, S. F. Hu, K. Zaw, C. K. Angerhofer, H. Chai, J. M. Pezzuto, G. A. Cordell, J. Lin, and D. M. Zheng, *J. Nat. Prod.* **57**, 1430 (1994).

[192] A. Valentin, F. Benoit-Vical, C. Moulis, E. Stanislas, M. Mallié, I. Fouraste, and J. M. Bastide, *Antimicrob. Agents Chemother.* **41**, 2305 (1997).

[193] J. P. Felix, V. F. King, J. L. Shevell, M. L. Garcia, G. J. Kaczorowski, I. R. C. Bick, and R. S. Slaughter, *Biochemistry* **31**, 793 (1992).

[194] Y. Kwan, *Stem Cells* **12**, 64 (1994).

[195] Y. M. Leung, M. Berdik, C. Y. Kwan, and T. T. Loh, *Clin. Exp. Pharmacol. Physiol.* **23**, 653 (1996).

[196] R. Docampo and S. N. J. Moreno, *Parasitol. Today* **12**, 61 (1995).

[197] M. L. Dorta, A. T. Ferreira, M. E. M. Oshiro, and N. Yoshida, *Mol. Biochem. Parasitol.* **73**, 285 (1995).

[198] A. Fournet, A. Inchausti, G. Yaluff, A. Rojas de Arias, H. Guinaudeau, J. Bruneton, M. A. Breidenbach, and C. H. Faerman, *J. Enzyme Inhib.* **13**, 1 (1998).

[199] G. Bringmann, in: "Guidelines and Issue for the Discovery and Drug Development against Tropical Diseases" (A. H. Fairlamb, R. G. Ridley and H. J. Vial, eds.), p. 145. World Health Organisation, Geneva, 2003.

[200] G. Bringmann, J. Holenz, W. Saeb, L. Aké Assi, and K. Hostettmann, *Planta Med.* **64**, 485 (1998).

[201] G. François, M. Van Looveren, G. Timperman, B. Chimanuka, L. Aké Assi, J. Holenz, and G. Bringmann, *J. Ethnopharmacol.* **54**, 125 (1996).

[202] G. Bringmann, V. Hoerr, U. Holzgrabe, and A. Stich, *Pharmazie* **58**, 343 (2003).

[203] G. Bringmann, K. Messer, K. Wolf, J. Muhlbacher, M. Grüne, R. Brun, and A. M. Louis, *Phytochemistry* **60**, 389 (2002).

[204] G. Bringmann, G. François, and L. Aké Assi, *Chimia* **52**, 18 (1998).

[205] G. Bringmann, M. Wenzel, M. Rübenacker, M. Schäffer, M. Rückert, and L. Aké Assi, *Phytochemistry* **49**, 1151 (1998).

[206] G. Bringmann, W. Saeb, R. God, M. Schäffer, G. François, K. Peters, E. M. Peters, P. Proksch, K. Hostettmann, and L. Aké Assi, *Phytochemistry* **49**, 1667 (1998).

[207] G. Bringmann, A. Hamm, C. Günther, M. Michel, R. Brun, and V. Mudogo, *J. Nat. Prod.* **63**, 1465 (2000).

[208] G. Bringmann, M. Dreyer, J. H. Faber, P. W. Dalsgaard, D. Stærk, J. W. Jaroszewski, H. Ndangalasi, F. Mbago, R. Brun, M. Reichert, K. Maksimenka, and S. B. Christensen, *J. Nat. Prod.* **66**, 1159 (2003).

[209] C. P. Tang, Y. P. Yang, Y. Zhong, Q. X. Zhong, H. M. Wu, and Y. Ye, *J. Nat. Prod.* **63**, 1384 (2000).

[210] G. Bringmann, M. Dreyer, J. H. Faber, P. W. Dalsgaard, D. Stærk, J. W. Jaroszewski, H. Ndangalasi, F. Mbago, R. Brun, and S. B. Christensen, *J. Nat. Prod.* **67**, 743 (2004).

[211] G. Bringmann, K. Messer, R. Brun, and V. Mudogo, *J. Nat. Prod.* **65**, 1096 (2002).

[212] G. Bringmann, M. Dreyer, M. Michel, F. S. K. Tayman, and R. Brun, *Phytochemistry* **65**, 2903 (2004).

[213] G. Bringmann, W. Saeb, M. Rückert, J. Mies, M. Michel, V. Mudogo, and R. Brun, *Phytochemistry* **62**, 631 (2003).

[214] G. Bringmann, M. Wohlfarth, H. Rischer, J. Schlauer, and R. Brun, *Phytochemistry* **61**, 195 (2002).

[215] G. Bringmann, K. Messer, M. Wohlfarth, J. Kraus, K. Dumbuya, and M. Rückert, *Anal. Chem.* **71**, 2678 (1999).

[216] G. Bringmann, K. Messer, B. Schwöbel, R. Brun, and L. A. Assi, *Phytochemistry* **62**, 345 (2003).

[217] A. Ponte-Sucre, J. H. Faber, T. Gulder, I. Kajahn, S. E. H. Pedersen, M. Schultheis, G. Bringmann, and H. Moll, *Antimicrob. Agents Chemother.* **51**, 188 (2007).

[218] G. François, G. Bringmann, C. Dochez, C. Schneider, G. Timperman, and L. Aké Assi, *J. Ethnopharmacol.* **46**, 115 (1995).

[219] G. François, G. Bringmann, J. D. Phillipson, L. Aké Assi, C. Dochez, M. Rübenacker, C. Schneider, M. Wery, D. C. Warhurst, and G. C. Kirby, *Phytochemistry* **35**, 1461 (1994).

[220] G. Bringmann, C. Günther, W. Saeb, J. Mies, R. Brun, and L. Aké Assi, *Phytochemistry* **54**, 337 (2000).

[221] G. Bringmann, D. Koppler, B. Wiesen, G. François, A. S. S. Narayanan, M. R. Almeida, H. Schneider, and U. Zimmermann, *Phytochemistry* **43**, 1405 (1996).

[222] G. François, G. Timperman, R. D. Haller, S. Bar, M. A. Isahakia, S. A. Robertson, N. J. Zhao, L. A. De Assi, J. Holenz, and G. Bringmann, *Int. J. Pharmacog.* **35**, 55 (1997).

[223] Y. F. Hallock, K. P. Manfredi, J. W. Blunt, J. H. Cardellina, M. Schäffer, K.-P. Gulden, G. Bringmann, A. Y. Lee, J. Clardy, G. François, and M. R. Boyd, *J. Org. Chem.* **59**, 6349 (1994).

[224] Y. F. Hallock, K. P. Manfredi, J.-R. Dai, J. H. Cardellina, R. J. Gulakowski, J. B. McMahon, M. Schäffer, M. Stahl, K.-P. Gulden, G. Bringmann, G. François, and M. R. Boyd, *J. Nat. Prod.* **60**, 677 (1997).

[225] Y. F. Hallock, J. H. Cardellina II, M. Schäffer, G. Bringmann, G. François, and M. R. Boyd, *Bioorg. Med. Chem. Lett.* **8**, 1729 (1998).

[226] G. Bringmann, C. Günther, W. Saeb, J. Mies, A. Wickramasinghe, V. Mudogo, and R. Brun, *J. Nat. Prod.* **63**, 1333 (2000).

[227] G. Bringmann, F. Teltschik, M. Michel, S. Busemann, M. Rückert, R. Haller, S. Bär, S. A. Robertson, and R. Kaminsky, *Phytochemistry* **52**, 321 (1999).

[228] G. François, G. Timperman, J. Holenz, L. Aké Assi, T. Geuder, L. Maes, J. Dubois, M. Hanocq, and G. Bringmann, *Ann. Trop. Med. Parasitol.* **90**, 115 (1996).

[229] G. François, G. Timperman, T. Steenackers, L. Aké Assi, J. Holenz, and G. Bringmann, *Parasitol. Res.* **83**, 673 (1997).

[230] G. François, G. Timperman, W. Eling, L. Aké Assi, J. Holenz, and G. Bringmann, *Antimicrob. Agents Chemother.* **41**, 2533 (1997).

[231] G. François, T. Steenackers, G. Timperman, L. Aké Assi, R. D. Haller, S. Bär, M. A. Isahakia, S. A. Robertson, C. Zhao, N. J. De Souza, J. Holenz, and G. Bringmann, *Int. J. Parasitol.* **27**, 29 (1997).

[232] G. François, B. Chimanuka, G. Timperman, J. Holenz, J. Plaizier-Vercammen, L. Aké Assi, and G. Bringmann, *Parasitol. Res.* **85**, 935 (1999).

[233] G. Bringmann and C. Rummey, *J. Chem. Inf. Comput. Sci.* **43**, 304 (2003).

[234] G. Bringmann and F. Pokorny, *in:* "The Alkaloids" (G. A. Cordell, ed.), Vol. 46, p. 127. Academic Press, New York, 1995.

[235] Y. F. Hallock, J. H. Cardellina II, M. Schäffer, M. Stahl, G. Bringmann, G. François, and M. R. Boyd, *Tetrahedron* **53**, 8121 (1997).

[236] J. Bastida, R. Lavilla, and F. Viladomat, *in:* "The Alkaloids" (G. A. Cordell, ed.), Vol. 63, p. 87. Elsevier-Academic Press, New York, 2006.

[237] Dictionary of Natural Products (Net Database). Chapman & Hall/CRC Press, London, 2005.

[238] K. Likhitwitayawuid, C. K. Angerhofer, H. Chai, J. M. Pezzuto, G. A. Cordell, and N. Ruangrungsi, *J. Nat. Prod.* **56**, 1331 (1993).

[239] W. E. Campbell, J. J. Nair, D. W. Gammon, C. Codina, J. Bastida, F. Viladomat, P. J. Smith, and C. F. Albrecht, *Phytochemistry* **53**, 587 (2000).

[240] B. Şener, I. Orhan, and J. Satayavivad, *Phytother. Res.* **17**, 1220 (2003).

[241] M. R. Herrera, A. K. Machocho, R. Brun, F. Viladomat, C. Codina, and J. Bastida, *Planta Med.* **67**, 191 (2001).

[242] J. Labraña, A. K. Machocho, V. Kricsfalusy, R. Brun, C. Codina, F. Viladomat, and J. Bastida, *Phytochemistry* **60**, 847 (2002).

[243] M. R. Herrera, A. K. Machocho, J. J. Nair, W. E. Campbell, R. Brun, F. Viladomat, C. Codina, and J. Bastida, *Fitoterapia* **72**, 444 (2001).

[244] A. K. Machocho, J. Bastida, C. Codina, F. Viladomat, R. Brun, and S. C. Chhabra, *Phytochemistry* **65**, 3143 (2004).

[245] R. Cao, W. Peng, Z. Wang, and A. Xu, *Curr. Med. Chem.* **14**, 479 (2007).

[246] H. Tsuchiya, M. Sato, and I. Watanabe, *J. Agri. Food. Chem.* **47**, 4167 (1999).

[247] C. C. Shi, S. Y. Chen, G. J. Wang, J. F. Liao, and C. F. Chen, *Eur. J. Pharmacol.* **390**, 319 (2000).

[248] C. C. Shi, J. F. Liao, and C. F. Chen, *Pharmacol. Toxicol.* **89**, 259 (2001).

[249] J. Ishida, H. K. Wang, O. Masayoshi, C. L. Cosentino, C. Q. Hu, and K. H. Lee, *J. Nat. Prod.* **64**, 958 (2001).

[250] H. Aassila, M. L. Bourguet-Kondracki, S. Rifai, A. Fassouane, and M. Guyot, *Marine Biotechnol.* **5**, 163 (2003).

[251] A. T. Evans and S. L. Croft, *Phytother. Res.* **1**, 25 (1987).

[252] C. Di Giorgio, F. Delmas, E. Ollivier, R. Elias, G. Balansard, and P. Timon-David, *Exp. Parasitol.* **106**, 67 (2004).

[253] T. S. Kam, K. M. Sim, T. Koyano, and K. Komiyama, *Phytochemistry* **50**, 75 (1999).

[254] P. Rivas, B. K. Cassels, A. Morello, and Y. Repetto, *Comp. Biochem. Physiol. Part C* **122**, 27 (1999).

[255] F. Freiburghaus, R. Kamisky, M. H. Nkunya, and R. Brun, *J. Ethnopharmacol.* **55**, 1 (1996).

[256] J. C. Cavin, S. M. Krassner, and E. Rodríguez, *J. Ethnopharmacol.* **19**, 89 (1987).

[257] E. V. Costa, M. L. B. Pinheiro, C. M. Xavier, J. R. A. Silva, A. C. F. Amaral, A. D. L. Souza, A. Barison, F. R. Campos, A. G. Ferreira, G. M. C. Machado, and L. L. P. Leon, *J. Nat. Prod.* **69**, 292 (2006).

[258] W. Milliken, *Econ. Bot.* **51**, 212 (1997).

[259] J. C. P. Steele, N. C. Veitch, G. C. Kite, M. S. J. Simmonds, and D. C. Warhurst, *J. Nat. Prod.* **65**, 85 (2002).

[260] M. Beljanski and M. S. Beljanski, *Exp. Cell Biol.* **50**, 79 (1982).

[261] C. W. Wright, J. D. Phillipson, S. O. Awe, G. C. Kirby, D. C. Warhurst, J. Quetin-Leclercq, and L. Angenot, *Phytother. Res.* **10**, 361 (1996).

[262] M. Sauvain, C. Moretti, J. A. Bravo, J. Callapa, M. H. Muñoz, E. Ruiz, B. Richard, and L. Le Men-Olivier, *Phytother. Res.* **10**, 198 (1996).

[263] F. Leteurtre, J. Madalengoitia, A. Orr, T. Guzi, E. Lehnert, T. Macdonald, and Y. Pommier, *Cancer. Res.* **52**, 4478 (1992).

[264] A. Codoñer, I. Monzó, F. Tomás, and R. Valero, *Spectrochim. Acta* **42A**, 765 (1986).

[265] A. Codoñer, I. Monzó, C. Ortiz, and A. Olba, *J. Chem. Soc. Perkin Trans. II* **1989**, 107 (1989).

[266] J. N. Picada, K. V. da Silva, B. Erdtmann, A. T. Henriques, and J. A. Henriques, *Mutation Res.* **379**, 135 (1997).

[267] T. Herraiz and C. Chaparro, *Life Sci.* **78**, 795 (2006).

[268] P. Sharma and G. A. Cordell, *J. Nat. Prod.* **51**, 528 (1988).

[269] S. M. K. Rates, E. E. S. Schapoval, I. A. Souza, and A. T. Henriques, *Int. J. Pharmacog.* **31**, 288 (1993).

[270] A. T. Henriques, A. A. Melo, P. R. Moreno, L. L. Ene, J. A. Henriques, and E. E. Schapoval, *J. Ethnopharmacol.* **50**, 19 (1996).

[271] J. C. Delorenzi, L. Freire-de-Lima, C. R. Gattass, D. Andrade Costa, L. He, M. E. Kuehne, and E. M. B. Saraiva, *Antimicrob. Agents Chemother.* **46**, 2111 (2002).

[272] E. M. Silva, C. C. Cirne-Santos, I. C. P. P. Frugulhetti, B. Galvao-Castro, E. M. B. Saraiva, M. E. Kuehne, and D. C. Bou-Habib, *Planta Med.* **70**, 808 (2004).

[273] T. Kam, K. Sim, H. Pang, T. Koyano, M. Hayashi, and K. Komiyama, *Bioorg. Med. Chem. Lett.* **14**, 4487 (2004).

[274] M. T. Andrade, J. A. Lima, A. C. Pinto, C. M. Rezende, M. P. Carvalho, and R. A. Epifanio, *Bioorg. Med. Chem.* **13**, 4092 (2005).

[275] J. C. Delorenzi, M. Attias, C. R. Gattass, M. T. Andrade, C. Rezende, C. A. Pinto, A. T. Henriques, D. C. Bou-Habib, and E. M. B. Saraiva, *Antimicrob. Agents Chemother.* **45**, 1349 (2001).

[276] M. You, X. Ma, R. Mukherjee, N. R. Farnsworth, G. A. Cordell, A. D. Kinghorn, and J. M. Pezzuto, *J. Nat. Prod.* **57**, 1517 (1994).

[277] V. Muñoz, C. Moretti, M. Sauvain, C. Caron, A. Porzel, G. Massiot, B. Richard, and L. L. Men-Olivier, *Planta Med.* **60**, 455 (1994).

[278] M. Azoug, A. Loukaci, B. Richard, J. M. Nuzillard, C. Moretti, M. Zeches-Hanrot, and L. L. Men-Olivier, *Phytochemistry* **39**, 1223 (1995).

[279] A. L. Bodley and T. A. Shapiro, *Proc. Natl. Acad. Sci. U.S.A.* **92**, 3726 (1995).

[280] A. Lorence and C. L. Nessler, *Phytochemistry* **65**, 2735 (2004).

[281] M. E. Wall, M. C. Wani, C. E. Cook, K. H. Palmer, A. T. McPhail, and G. A. Sim, *J. Am. Chem. Soc.* **88**, 3888 (1966).

[282] E. J. Corey, D. N. Crouse, and J. E. Anderson, *J. Org. Chem.* **40**, 2140 (1975).

[283] A. L. Bodley, J. N. Cumming, and T. A. Shapiro, *Biochem. Pharmacol.* **55**, 709 (1998).

[284] G. Capranico, F. Ferri, M. V. Fogli, A. Russo, L. Lotito, and L. Baranello, *Biochimie* **89**, 482 (2007).

[285] D. Stærk, E. Lemmich, J. Christensen, A. Kharazmi, C. E. Olsen, and J. W. Jaroszewski, *Planta Med.* **66**, 531 (2000).

[286] C. W. Wright, D. H. Bray, M. J. O'Neill, D. C. Warhurst, J. D. Phillipson, J. Quetin-Leclercq, and L. Angenot, *Planta Med.* **57**, 337 (1991).

[287] C. W. Wright, D. Allen, Y. Cai, Z. P. Chen, J. D. Phillipson, G. C. Kirby, D. C. Warhurst, M. Tits, and L. Angenot, *Phytother. Res.* **8**, 149 (1994).

[288] M. Frederich, M. P. Hayette, M. Tits, P. De Mol, and L. Angenot, *Antimicrob. Agents Chemother.* **43**, 2328 (1999).

[289] G. Philippe, P. De Mol, M. Zèches-Hanrot, J. M. Nuzillard, M. H. Tits, L. Angenot, and M. Frédérich, *Phytochemistry* **62**, 623 (2003).

[290] V. Brandt, M. Tits, J. Penelle, M. Frédérich, and L. Angenot, *Phytochemistry* **57**, 653 (2001).

[291] P. Thongphasuk, R. Suttisri, R. Bavovada, and R. Verpoorte, *Phytochemistry* **64**, 897 (2003).

[292] A. C. Mitaine, K. Mesbah, C. Petermann, S. Arrazola, C. Moretti, M. Zèches-Hanrot, and L. Le Men-Olivier, *Planta Med.* **62**, 458 (1996).

[293] A. C. Mitaine, B. Weniger, M. Sauvain, E. Lucumi, R. Aragón, and M. Zèches-Hanrot, *Planta Med.* **64**, 487 (1998).

[294] A. C. Mitaine-Offer, M. Sauvain, A. Valentin, J. Callapa, M. Mallié, and M. Zèches-Hanrot, *Phytomedicine* **9**, 142 (2002).

[295] M. M. Iwu, J. E. Jackson, J. D. Tally, and D. L. Klayman, *Planta Med.* **58**, 436 (1992).

[296] M. M. Iwu and D. L. Klayman, *J. Ethnopharmacol.* **36**, 133 (1992).

[297] G. François, L. Aké Assi, J. Holenz, and G. Bringmann, *J. Ethnopharmacol.* **54**, 113 (1996).

[298] J. E. Okokon, B. S. Antia, A. C. Igboasoiyi, E. E. Essien, and H. O. C. Mbagwu, *J. Ethnopharmacol.* **111**, 464 (2007).

[299] R. Ansa-Asamoah, G. J. Kapadia, H. A. Lloyd, and E. A. Sokoloski, *J. Nat. Prod.* **53**, 975 (1990).

[300] G. J. Kapadia, C. K. Angerhofer, and R. Ansa-Asamoah, *Planta Med.* **59**, 565 (1993).

[301] T. A. van Beek, R. Verpoorte, and A. Baerheim-Svendsen, *Planta Med.* **50**, 180 (1984).

[302] T. Sevenet, L. Allorge, B. David, K. Awang, A. H. A. Hadi, J. C. Quirion, F. Remy, H. Schaller, and L. E. Teo, *J. Ethnopharmacol.* **41**, 147 (1994).

[303] T. S. Kam and K. M. Sim, *Phytochemistry* **47**, 145 (1998).

[304] E. Federici, G. Palazzino, M. Nicoletti, and C. Galeffi, *Planta Med.* **66**, 93 (2000).

[305] C. W. Wright, D. Allen, J. D. Phillipson, G. C. Kirby, D. C. Warhurst, G. Massiot, and L. Le Men-Olivier, *J. Ethnopharmacol.* **40**, 41 (1993).

[306] C. W. Wright, D. Allen, Y. Cai, J. D. Phillipson, I. M. Said, G. C. Kirby, and D. C. Warhurst, *Phytother. Res.* **6**, 121 (1992).

[307] N. Keawpradub, G. C. Kirby, J. C. P. Steele, and P. J. Houghton, *Planta Med.* **65**, 690 (1999).

[308] I. C. Ferreira, M. V. Lonardoni, G. M. Machado, L. L. Leon, L. Gobbi Filho, L. H. Pinto, and A. J. de Oliveira, *Mem. Inst. Oswaldo Cruz* **99**, 325 (2004).

[309] J. C. Tanaka, C. C. da Silva, I. C. P. Ferreira, G. M. C. Machado, L. L. Leon, and A. J. B. Oliveira, *Phytomedicine* **14**, 377 (2007).

[310] K. Bonjean, M. C. De Pauw-Gellet, M. P. Defresne, P. Colson, C. Houssier, L. Dassonneville, C. Bailly, R. Greimers, C. Wright, J. Quentin-Leclercq, M. Tits, and L. Angenot, *Biochemistry* **37**, 5136 (1998).

[311] G. Philippe, L. Angenot, M. Tits, and M. Frédérich, *Toxicon* **44**, 405 (2004).

[312] M. Frédérich, M. Tits, E. Goffin, G. Philippe, P. Grellier, P. De Mol, M. P. Hayette, and L. Angenot, *Planta Med.* **70**, 520 (2004).

[313] M. Frédérich, M. J. Jacquier, P. Thépenier, P. De Mol, M. Tits, G. Philippe, C. Delaude, L. Angenot, and M. Zèches-Hanrot, *J. Nat. Prod.* **65**, 1381 (2002).

[314] M. Frédérich, M. C. Depauw-Gillet, G. Llabrés, M. Tits, M. P. Hayette, V. Brandt, J. Penelle, P. De Mol, and L. Angenot, *Planta Med.* **66**, 262 (2000).

[315] M. Frédérich, M. C. De Pauw, C. Prosperi, M. Tits, V. Brandt, J. Penelle, M. Hayette, P. De Mol, and L. Angenot, *J. Nat. Prod.* **64**, 12 (2001).

[316] G. Philippe, L. Nguyen, L. Angenot, M. Frédérich, G. Moonen, M. Tits, and J. M. Rigo, *Eur. J. Pharmacol.* **530**, 15 (2006).

[317] F. Sandberg and K. Kristianson, *Acta Pharm. Suec.* **7**, 329 (1970).

[318] J. N. Lisgarten, M. Coll, J. Portugal, C. W. Wright, and J. Aymami, *Nat. Struct. Biol.* **9**, 57 (2002).

[319] S. Van Miert, S. Hostyn, B. U. W. Maes, K. Cimanga, R. Brun, M. Kaiser, P. Mátyus, R. Dommisse, G. Lemière, A. Vlietinck, and L. Pieters, *J. Nat. Prod.* **68**, 674 (2005).

[320] G. C. Kirby, A. Paine, D. C. Warhurst, B. K. Noamese, and J. D. Phillipson, *Phytother. Res.* **9**, 359 (1995).

[321] K. Cimanga, T. De Bruyne, L. Pieters, A. Vlietinck, and C. A. Turger, *J. Nat. Prod.* **60**, 688 (1997).

[322] P. Grellier, L. Ramiaramanana, V. Millerioux, E. Deharo, J. Schrével, F. Frappier, F. Trigalo, B. Bodo, and J. L. Pousset, *Phytother. Res.* **10**, 317 (1996).

[323] T. H. M. Jonckers, S. Van Miert, K. Cimanga, C. Bailly, P. Colson, M. C. De Pauw-Gillet, H. Van Den Heuvel, M. Claeys, F. Lemière, E. L. Esmans, J. Rozenski, L. Quirijnen, L. Maes, R. Dommisse, G. L. F. Lemière, A. Vlietinck, and L. Pieters, *J. Med. Chem.* **45**, 3497 (2002).

[324] L. Dassonneville, K. Bonjean, M. C. De Pauw-Gillet, P. Colson, C. Houssier, J. Quetin-Leclercq, L. Angenot, and C. Bailly, *Biochemistry* **38**, 7719 (1999).

[325] C. W. Wright, J. Addae-Kyereme, A. G. Breen, J. E. Brown, M. F. Cox, S. L. Croft, Y. Gökçek, H. Kendrick, R. M. Phillips, and P. L. Pollet, *J. Med. Chem.* **44**, 3187 (2001).

[326] B. Chataing, J. L. Concepción, R. Lobatón, and A. Usubillaga, *Planta Med.* **64**, 31 (1998).

[327] C. A. Hall, T. Hobby, and M. Cipollini, *J. Chem. Ecol.* **32**, 2405 (2006).

[328] C. Moretti, M. Sauvain, C. Lavaud, G. Massiot, J. A. Bravo, and V. Muñoz, *J. Nat. Prod.* **61**, 1390 (1998).

[329] K. P. Devkota, M. I. Choudhary, R. Ranjit, Samreen, and N. Sewald, *Nat. Prod. Res.* **21**, 292 (2007).

[330] T. S. Kam, K. M. Sim, T. Koyano, M. Toyoshima, M. Hayashi, and K. Komiyama, *J. Nat. Prod.* **61**, 1332 (1998).

[331] G. N. Zirihi, P. Grellier, F. Guédé-Guina, B. Bodo, and L. Mambu, *Bioorg. Med. Chem. Lett.* **15**, 2637 (2005).

[332] P. González, C. Marín, I. Rodríguez-González, A. B. Hitos, M. J. Rosales, M. Reina, J. G. Díaz, A. González-Coloma, and M. S. Moreno, *Int. J. Antimicrob. Agents* **25**, 136 (2005).

[333] A. M. Mayer and K. R. Gustafson, *Int. J. Cancer* **105**, 291 (2003).

[334] B. Haefner, *Drug Discov. Today* **8**, 536 (2003).

[335] K. A. El Sayed, M. Kelly, U. A. K. Kara, K. K. H. Ang, I. Katsuyama, D. C. Dunbar, A. A. Khan, and M. T. Hamann, *J. Am. Chem. Soc.* **123**, 1804 (2001).

[336] R. Sakai, T. Higa, C. W. Jefford, and G. Bernardinelli, *J. Am. Chem. Soc.* **108**, 6404 (1986).

[337] Y. C. Shen, H. R. Tai, and C. Y. Duh, *Chin. Pharm. J.* **48**, 1 (1996).

[338] M. Tsuda and J. Kobayashi, *Heterocycles* **46**, 765 (1997).

[339] E. Magnier and Y. Langlois, *Tetrahedron* **54**, 6201 (1998).

[340] U. K. Pandit, *Farmaco* **53**, 749 (1995).

[341] R. A. Edrada, P. Proksch, V. Wray, L. Witte, W. E. Muller, and R. W. Van Soest, *J. Nat. Prod.* **59**, 1056 (1996).

[342] K. V. Rao, M. S. Donia, J. Peng, E. Garcia-Palomero, D. Alonso, A. Martinez, M. Medina, S. G. Franzblau, B. L. Tekwani, S. I. Khan, S. Wahyuono, K. L. Willett, and M. T. Hamann, *J. Nat. Prod.* **69**, 1034 (2006).

[343] K. K. Ang, M. J. Holmes, T. Higa, M. T. Hamann, and U. A. Kara, *Antimicrob. Agents Chemother.* **44**, 1645 (2000).
[344] M. Yousaf, N. L. Hammond, J. Peng, S. Wahyuono, K. A. McIntosh, W. N. Charman, A. M. S. Mayer, and M. T. Hamann, *J. Med. Chem.* **47**, 3512 (2004).
[345] M. Donia and M. T. Hamann, *Lancet Infect. Dis.* **3**, 338 (2003).
[346] W. Gul and M. T. Hamann, *Life Sci.* **78**, 442 (2005).
[347] G. Kirsch, G. M. Kong, A. D. Wright, and R. Kaminsky, *J. Nat. Prod.* **63**, 825 (2000).
[348] L. Heys, C. G. Moore, and P. Murphy, *J. Chem. Soc. Rev.* **29**, 57 (2000).
[349] H. Hua, J. Peng, F. R. Fronczek, M. Nelly, and M. T. Hamann, *Bioorg. Med. Chem.* **12**, 6461 (2004).
[350] Y. Nakao, T. Shiroiwa, S. Murayama, S. Matsunaga, Y. Goto, Y. Matsumoto, and N. Fusetani, *Mar. Drugs* **2**, 55 (2004).
[351] Y. Mizuno, A. Makioka, S. Kawazu, S. Kano, S. Kawai, M. Akaki, M. Aikawa, and H. Ohtomo, *Parasitol. Res.* **88**, 844 (2002).
[352] C. Osterhage, R. Kaminsky, G. M. Konig, and A. D. Wright, *J. Org. Chem.* **65**, 6412 (2000).
[353] J. E. Lazaro, J. Nitcheu, R. Z. Predicala, G. C. Mangalindan, F. Nesslany, D. Marzin, G. P. Concepcion, and B. Diquet, *J. Nat. Toxins* **11**, 367 (2002).
[354] A. M. S. Mayera and M. T. Hamann, *Comp. Biochem. Physiol. C. Toxicol. Pharmacol.* **140**, 265 (2005).
[355] C. Larksarp and H. Alper, *J. Org. Chem.* **65**, 2773 (2000).
[356] A. Witt and J. Bergman, *Curr. Org. Chem.* **7**, 659 (2003).
[357] S. R. Padala, P. R. Padi, and V. Thipireddy, *Heterocycles* **60**, 183 (2003).
[358] S. B. Mhaske and N. P. Argade, *Tetrahedron* **62**, 9787 (2006).
[359] E. Christopher, E. Bedir, C. Dunbar, I. A. Khan, C. O. Okunji, B. M. Schuster, and M. M. Iwu, *Helv. Chim. Acta* **86**, 2914 (2003).
[360] J. Scovill, E. Blank, M. Konnick, E. Nenortas, and T. Shapiro, *Antimicrob. Agents Chemother.* **46**, 882 (2002).
[361] V. F. de Paula, L. C. D. Barbosa, A. J. Demuner, D. Pilo-Veloso, and M. C. Picanco, *Pest. Manag. Sci.* **56**, 168 (2000).
[362] F. Kiuchi, N. Nakamura, Y. Tsuda, K. Kondo, and H. Yoshimura, *Chem. Pharm. Bull.* **36**, 1452 (1988).
[363] S. Koul, J. L. Koul, S. C. Taneja, K. L. Dhar, D. S. Jamwal, K. Singh, R. K. Reen, and J. Sing, *Bioorg. Med. Chem.* **8**, 251 (2000).
[364] A. Kapil, *Planta Med.* **59**, 474 (1993).
[365] T. S. Ribeiro, L. Freire-de-Lima, J. O. Previato, L. Mendonça-Previato, N. Heiseb, and M. E. Freire de Lima, *Bioorg. Med. Chem. Lett.* **14**, 3555 (2004).
[366] B. Raay, S. Medda, S. Mukhopadhyay, and M. K. Basu, *Indian J. Biochem. Biophys.* **36**, 248 (1999).
[367] P. R. Veerareddy, V. Vobalaboina, and A. Nahid, *Pharmazie* **59**, 194 (2004).
[368] I. Muhammad, D. C. Dunbar, S. I. Khan, B. L. Tekwani, E. Bedir, S. Takamatsu, D. Ferreira, and L. A. Walker, *J. Nat. Prod.* **66**, 962 (2003).
[369] I. Muhammad, D. C. Dunbar, S. Takamatsu, L. A. Walker, and A. M. Clark, *J. Nat. Prod.* **64**, 559 (2001).
[370] S. Suksamrarn, N. Suwannapoch, N. Aunchai, M. Kuno, P. Ratananukul, R. Haritakun, C. Jansakulc, and S. Ruchirawat, *Tetrahedron* **61**, 1175 (2005).
[371] http://www.who.int/tdr. Accessed on January 22, 2008.
[372] S. Nwaka and A. Hudson, *Nat. Rev. Drug Discov.* **5**, 941 (2006).
[373] S. Nwaka and R. G. Ridley, *Nat. Rev. Drug Discov.* **2**, 919 (2003).
[374] P. Trouiller, P. Olliaro, E. Torreele, J. Orbinski, R. Laing, and N. Ford, *Lancet* **359**, 2188 (2002).
[375] R. Pink, A. Hudson, M. A. Mouriès, and M. Bendig, *Nat. Rev. Drug Discov.* **4**, 727 (2005).
[376] P. J. Hotez, E. Ottesen, A. Fenwick, and D. Molyneux, *Adv. Exp. Biol. Med.* **582**, 22 (2006).
[377] S. L. Croft, *Curr. Opin. Investig. Drugs* **8**, 103 (2007).
[378] S. M. Maurer, A. Rai, and A. Sali, *PLoS Med.* **1**, e56 (2004).
[379] M. F. Mrazek and E. Mossialos, *Health Policy* **64**, 75 (2003).
[380] M. J. Witty, *Int. J. Parasitol.* **29**, 95 (1999).
[381] F. Modabber, P. A. Buffet, E. Torreele, G. Milon, and S. L. Croft, *Kinetoplastid Biol. Dis.* **6**, 3 (2007).
[382] P. J. Rosenthal, *J. Exp. Biol.* **206**, 3735 (2003).

[383] J. A. Gutiérrez, *Int. J. Parasitol.* **30**, 247 (2000).

[384] M. J. Gardner, N. Hall, E. Fung, O. White, M. Berriman, R. W. Hyman, J. M. Carlton, A. Pain, K. E. Nelson, S. Bowman, I. T. Paulsen, K. James, J. A. Eisen, K. Rutherford, S. L. Salzberg, A. Craig, S. Kyes, M. S. Chan, V. Nene, S. J. Shallom, B. Suh, J. Peterson, S. Angiuoli, M. Pertea, J. Allen, J. Selengut, D. Haft, M. W. Mather, A. B. Vahadilla, D. M. Martin, A. H. Fairlamb, M. J. Fraunholz, D. S. Roos, S. A. Ralph, G. I. McFadden, L. M. Cummings, G. M. Subramanian, C. Mungall, J. C. Venter, D. J. Carucci, S. L. Hoffman, C. Newbold, R. W. Davis, C. M. Fraser, and B. Barrell, *Nature* **419**, 498 (2002).

[385] N. M. El-Sayed, P. J. Myler, G. Blandin, M. Berriman, J. Crabtree, G. Aggarwal, E. Caler, H. Renauld, E. A. Worthey, C. Hertz-Fowler, E. Ghedin, C. Peacock, D. C. Bartholomeu, B. J. Haas, A. N. Tran, J. R. Wortman, U. C. Alsmark, S. Angiuoli, A. Anupama, J. Badger, F. Bringaud, E. Cadag, J. M. Carlton, G. C. Cerqueira, T. Creasy, A. L. Delcher, A. Djikeng, T. M. Embley, C. Hauser, A. C. Ivens, S. K. Kummerfeld, J. B. Pereira-Leal, D. Nilsson, J. Peterson, S. L. Salzberg, J. Shallom, J. C. Silva, J. Sundaram, S. Westenberger, O. White, S. E. Melville, J. E. Donelson, B. Andersson, K. D. Stuart, and N. Hall, *Science* **309**, 404 (2005).

[386] http://www.who.int/tdr/grants/workplans/cen.htm. Accessed on January 22, 2008.

[387] http://www.who.int/tdr/grants/workplans/gdr.htm. Accessed on January 22, 2008.

[388] http://www.tropicaldisease.org. Accessed on January 22, 2008.

[389] http://www.oneworldhealth.org. Accessed on January 22, 2008.

[390] http://www.worldbank.org. Accessed on January 22, 2008.

[391] http://www.dndi.org. Accessed on January 22, 2008.

[392] http://www.mmv.org. Accessed on January 22, 2008.

[393] http://www.msf.org. Accessed on January 22, 2008.

[394] J. D. Phillipson, *Planta Med.* **69**, 491 (2003).

[395] C. A. Lipinski, F. Lombardo, B. W. Dominy, and P. J. Freeney, *Drug. Deliv. Rev.* **23**, 3 (1997).

[396] H. W. van Veen and W. N. Konings, *Semin. Cancer Biol.* **8**, 183 (1997).

[397] R. G. Deeley and S. P. C. Cole, *Semin. Cancer Biol.* **8**, 193 (1997).

[398] L. Chow and S. Volkman, *Exp. Parasitol.* **90**, 135 (1998).

[399] P. Borst, R. Evers, M. Kool, and J. Wijnholds, *Biochim. Biophys. Acta Biomembr.* **1461**, 347 (1999).

[400] C. G. Blackmore, P. A. McNaughton, and H. W. van Veen, *Mol. Membr. Biol.* **18**, 97 (2001).

[401] P. Borst and R. O. Elferink, *Annu. Rev. Biochem.* **71**, 537 (2002).

[402] A. Klokouzas, S. Shahi, S. B. Hladky, M. A. Barrand, and H. W. van Veen, *Int. J. Antimicrob. Agents* **22**, 301 (2003).

[403] L. R. Emerson, B. C. Skillman, H. Wolfger, K. Kuchler, and D. F. Wirth, *Ann. Trop. Med. Parasitol.* **98**, 643 (2004).

[404] E. J. Osorio, S. M. Robledo, G. J. Arango, and C. E. Muskus, *Biomedica* **25**, 242 (2005).

[405] I. Pajeva and M. Wiese, *J. Med. Chem.* **41**, 1815 (1998).

[406] M. Frédérich, M. P. Hayette, M. Tits, P. De Mol, and L. Angenot, *Planta Med.* **67**, 523 (2001).

[407] P. Rasoanaivo, S. Ratsimamanga-Urverg, R. Milijaona, H. Rafatro, A. Rakoto-Ratsimamanga, C. Galeffi, and M. Nicoletti, *Planta Med.* **60**, 13 (1994).

[408] D. Ramanitrahasimbola, P. Rasoanaivo, S. Ratsimamanga, and H. Vial, *Mol. Biochem. Parasitol.* **146**, 58 (2006).

[409] F. Frappier, A. Jossang, J. Soudon, F. Calvo, P. Rasoanaivo, S. Ratsimamanga-Urverg, J. Saez, J. Schrevel, and P. Grellier, *Antimicrob. Agents Chemother.* **40**, 1476 (1996).

[410] P. Rasoanaivo, S. Ratsimamanga-Urverg, H. Rafatro, D. Ramanitrahasimbola, G. Palazzino, C. Galeffi, and M. Nicoletti, *Planta Med.* **64**, 58 (1998).

[411] S. Ratsimamanga-Urverg, P. Rasoanaivo, H. Rafatro, B. Robijaona, and A. Rakoto-Ratsimamanga, *Ann. Trop. Med. Parasitol.* **88**, 271 (1994).

[412] J. M. Zamora, H. L. Pearce, and W. T. Beck, *Mol. Pharmacol.* **33**, 454 (1988).

[413] G. Ecker, M. Hubber, D. Schmid, and P. Chiba, *Mol. Pharmacol.* **56**, 791 (1999).

[414] E. Teodori, S. Dei, S. Scapecchi, and F. Gualtieri, *Il Farmaco* **57**, 385 (2002).

## CHAPTER 3

Buxus Steroidal Alkaloids: Chemistry and Biology

Athar Ata* and **Brad J. Andersh**

Contents		

I. INTRODUCTION

The genus *Buxus* comprises various species including *Buxus sempervirens*, *B. papillosa*, *B. microphylla*, *B. hildebrandtii*, and *B. hyrcana*. The crude extracts of these plants have been reported to have applications in traditional medicine to cure fatigue, rheumatism, malaria, depression, and skin infections (1). The ethanolic extract of *B. sempervirens* has been reported to exhibit anti-HIV activity,

Department of Chemistry and Biochemistry, Bradley University, Peoria, IL 61625-0208, USA

* Corresponding author.
E-mail address: aata@bumail.bradley.edu (A. Ata).

The Alkaloids, Volume 66
ISSN: 1099-4831, DOI 10.1016/S1099-4831(08)00203-4

and also exhibited delayed progression in HIV-infected asymptomatic patients (2,3). Phytochemical studies on various plants of the genus *Buxus* have resulted in the isolation of over 200 new steroidal alkaloids (4). A few of these alkaloids have shown interesting biological activities including antibacterial, antimyco-bacterial, antimalarial, and acetylcholinesterase (AChE) inhibitory activities (5–9).

Buxus alkaloids have a unique triterpenoid-steroidal pregnane type structure with C-4 methyl groups, a $9\beta,10\beta$-cycloartenol system, and a degraded C-20 side chain. Structurally, *Buxus* alkaloids are of the following two types:

(1)   Derivatives of $9\beta,10\beta$-*cyclo*-4,4,14α-trimethyl-5α-pregnane system (**1**)
(2)   Derivatives of $9(10 \rightarrow 19)$ *abeo* 4,4,14α-trimethyl-5α-pregnane system (**2**)

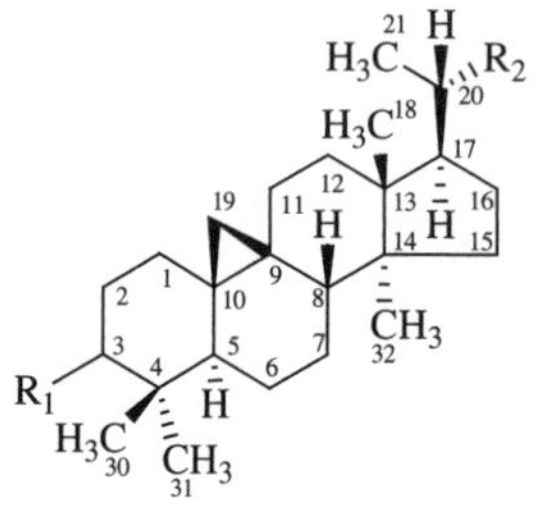

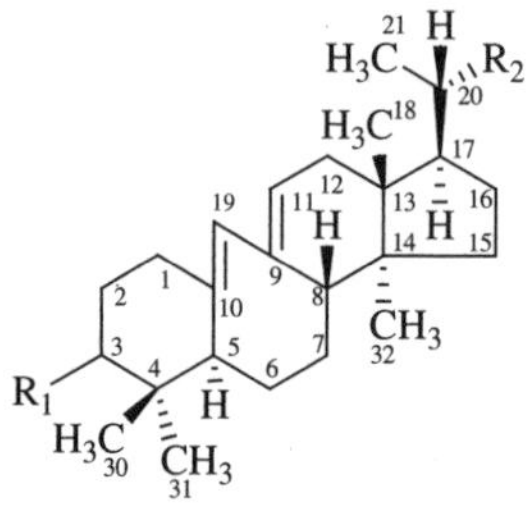

R_1 and R_2 in structures of **1** and **2** represent different amino or keto functionalities.

Both of these classes can easily be differentiated with the aid of ^{1}H NMR spectral studies. Alkaloids belonging to skeleton **1** exhibit a pair of AB doublets at δ 0.1–0.5 ($J = 4.0$ Hz) in their ^{1}H NMR spectra. Alkaloids having basic structure **2** show the lack of these signals in their ^{1}H NMR spectra, but do exhibit signals for the C-11 and C-19 vinylic protons in the olefinic range of the spectrum. Alkaloids of type **2** possessing an $9(10 \rightarrow 19)$ *abeo*-diene system also show absorption maxima at 238 and 245 nm with shoulders at 228 and 252 nm in the UV spectrum. These spectral features help in identifying two different classes of *Buxus* alkaloids. In each case, certain structural modifications are also observed, either due to the absence of one or both methyls, or due to the presence of a methylene group at C-4, and also the presence of different oxygen functions and points of unsaturation. This chapter exclusively describes the structures of alkaloids and their biological activities, isolated from various plants of the genus *Buxus* during the period 1992–2008. Additionally, the semisynthesis and structural modifications of these alkaloids are also discussed.

II. PHYTOCHEMICAL STUDIES ON *BUXUS* SPECIES

During the reporting period of 1992–2008, phytochemical studies on various *Buxus* species have been performed and these efforts resulted in the isolation of several new steroidal alkaloids. Atta-ur-Rahman and Choudhary's research group contributed more than half of these new compounds to the chemical literature. Other research groups have also worked in this area and communicated their results. Structures of new compounds, and reported compounds, isolated for the first time from various species of *Buxus*, as well as their reported biological activities, are described as follows.

A. Steroidal Alkaloids from *B. bodinieri*

B. bodinieri H. Lév. is a common ornamental plant in South-China. Qiu *et al.* carried out phytochemical studies on the crude methanolic extract of this plant and reported the isolation and structure elucidation of five new steroidal alkaloids: buxbodine A (**3**), buxbodine B (**4**), buxbodine C (**5**), buxbodine D (**6**), and buxbodine E (**7**) (10). These authors did not report any biological activities for these alkaloids. Alkaloid **3** has a unique structure due to the lack of a keto or an amino functionality at the C-3 position. Typically, all *Buxus* alkaloids have a C-3 amino or carbonyl group, except for irehine (**43**), which has a hydroxyl group substituted at C-3.

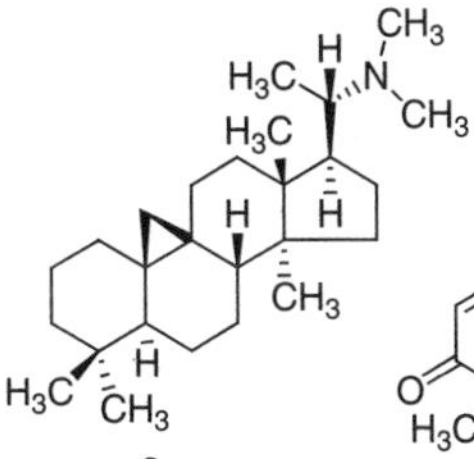

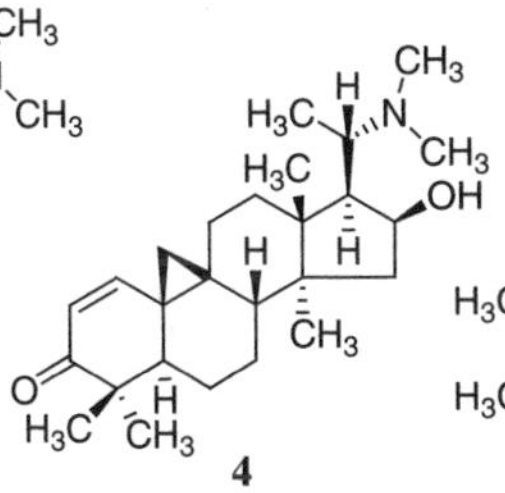

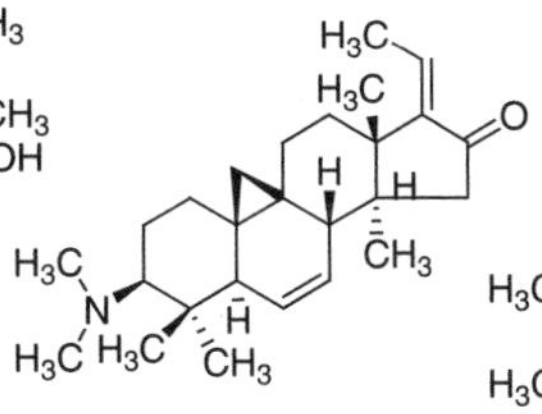

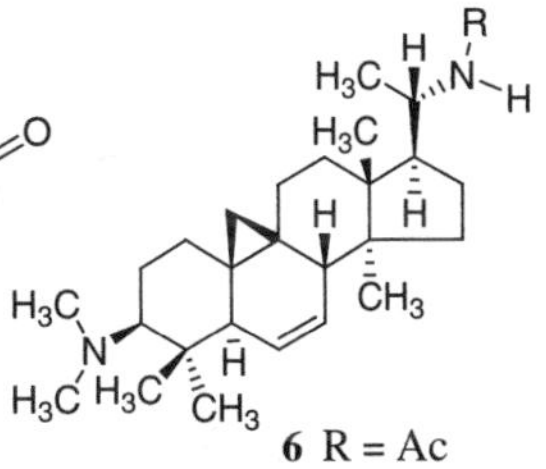

B. Steroidal Alkaloids from *B. hildebrandtii*

B. hildebrandtii Baill. is abundant in Africa. Its aqueous extract is used to treat malaria and venereal diseases. The crude extract of this plant has also been reported to exhibit antifungal and antiviral activities (11). Atta-ur-Rahman *et al.* performed chemical investigations on two different collections of this plant from Ethiopia and reported the isolation of seven new steroidal bases: 30-*O*-benzoyl-16-deoxybuxidienine C (**8**), 30-hydroxybuxamine A (**9**), 30-norbuxamine A (**10**), O^2-buxafuranamine (**11**), 6-hydroxy-O^2-buxafuranamine (**12**), O^{10}-buxafurana-mine (**13**), and 6-deoxy-O^{10}-buxafuranamine (**14**) (12,13). Additionally, the identification of six known steroidal bases, isolated for the first time from this *Buxus* species, has been reported by this research group. These alkaloids are cyclomicrobuxamine (**15**), buxamine A (**16**), cyclobuxoviridine (**17**), moenjodar-mine (**18**), buxamine C (**19**), and cyclorolfeine (**20**) (12). Alkaloids **11–14** belong to the rarely occurring class of *Buxus* alkaloids having a tetrahydrofuran ring incorporated in their structures. The C-4α methyl in *Buxus* alkaloids undergoes oxidation, and a number of alkaloids containing aldehyde or hydroxyl-bearing methylene functionalities at C-4 have been reported in the

literature (14,15). Based on these arguments and Drieding model studies, the
α-stereochemistry was assigned for the ether linkage in **11–14**. The tetrahydrofuran ring in alkaloids **11** and **12** may be biosynthesized by attack of the C-31
hydroxyl group on the C-1/C-2 double bond having a leaving group at C-10,
while the ether linkage between C-31 and C-10 in alkaloids **13** and **14** may result
from attack of the C-31 hydroxyl group on a C-1/C-10 double bond having a
leaving group at C-2 (7). The presence of a hydroxyl group or an ether
functionality at C-31 can easily be differentiated by recording the ^{1}H NMR
spectrum in pyridine, as the protons geminal to the hydroxyl group exhibit
induced paramagnetic shift of ∼1.0 ppm. An induced paramagnetic shift of
0.02 ppm is observed for protons geminal to the ether linkage (16,17).

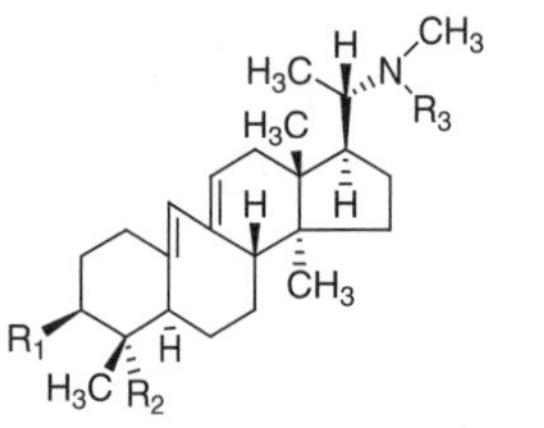

8 R$_1$ = H, R$_2$ = CH$_2$-OBz, R$_3$ = CH$_3$
9 R$_1$, R$_3$ = CH$_3$, R$_2$ = CH$_2$OH
10 R$_1$, R$_3$ = CH$_3$, R$_2$ = H

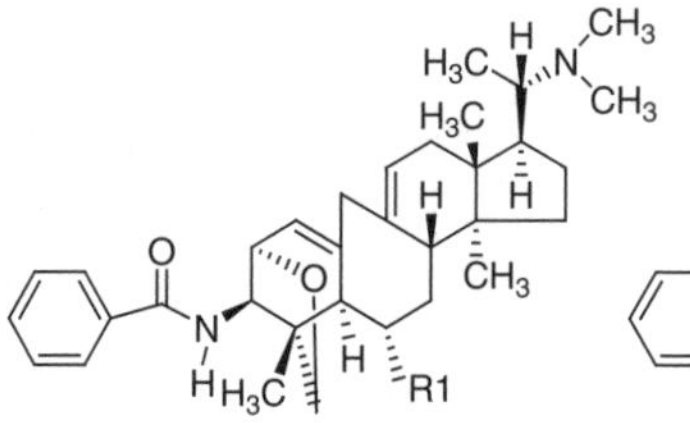

11 R$_1$ = H
12 R$_1$ = OH

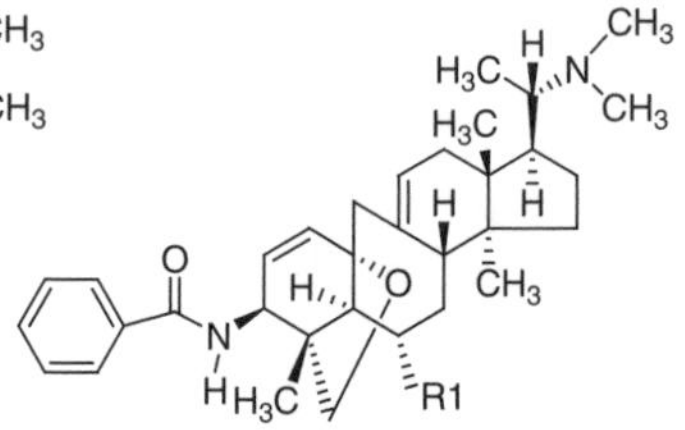

13 R$_1$ = OH
14 R$_1$ = H

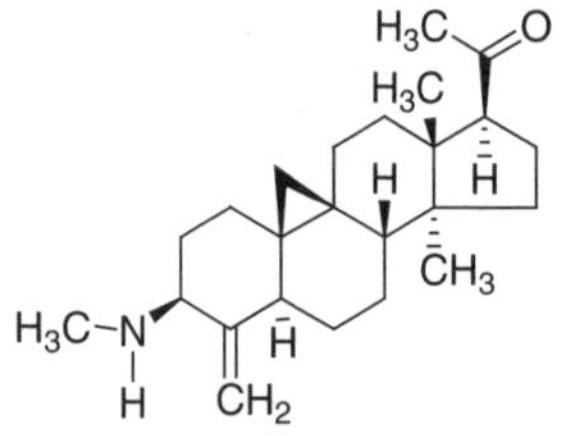

15

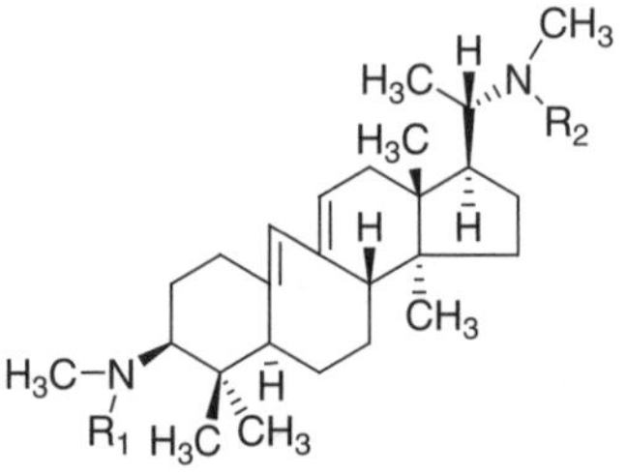

16 R$_1$, R$_2$ = CH$_3$
19 R$_1$ = H, R$_2$ = CH$_3$

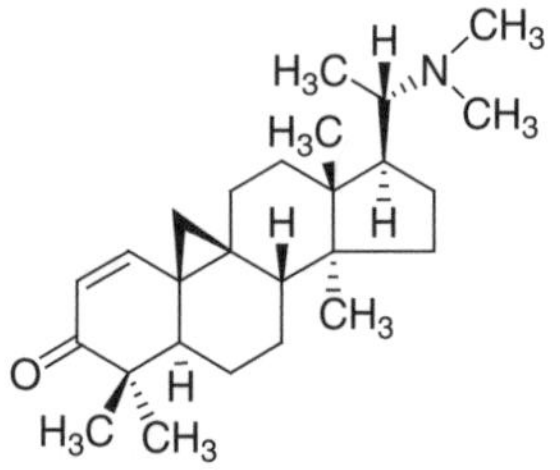

17

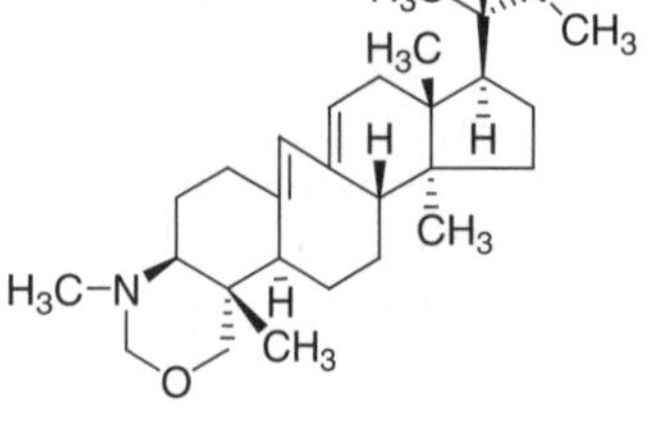

18

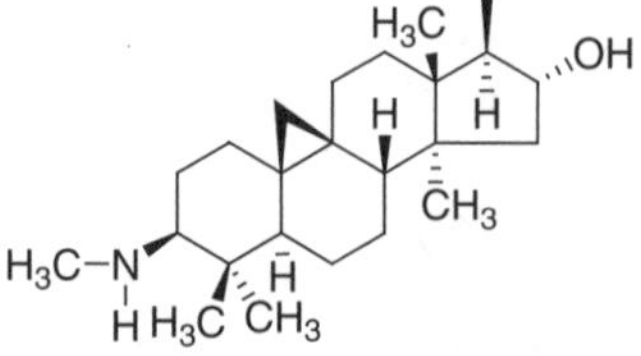

20

C. Alkaloids of *B. hyrcana*

B. hyrcana Pojark is a tree-like plant and is widely distributed in Iran. This plant was chemically studied by Choudhary's research group, and the investigations led to the isolation of five new alkaloids: hyrcanine (**21**), homomoenjodaramine (**22**), *N*-benzoylbuxahyrcanine (**23**), *N*-tigloylbuxahyrcanine (**24**), and *N*-isobutyroyl-buxahyrcanine (**25**), along with one known alkaloid, moenjodarmine (**18**) (18,19). In 2006, this plant was re-investigated by Choudhary and collaborators, resulting in the isolation of five more new alkaloids: hyrcanone (**26**), hyrcanol (**27**), hyrcatrienine (**28**), N_b-demethylcyclobuxoviricine (**29**), and hyrcamine (**30**) (20). These studies also yielded five known alkaloids: buxidine (**31**), buxandrine (**32**), buxabenzacinine (**33**), buxippine-K (**34**), and *E*-buxenone (**35**) (20).

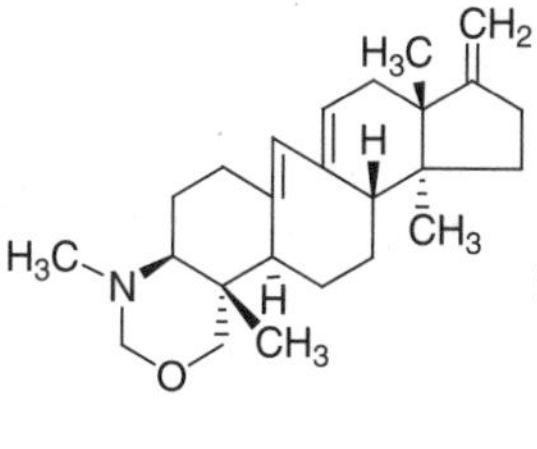

21

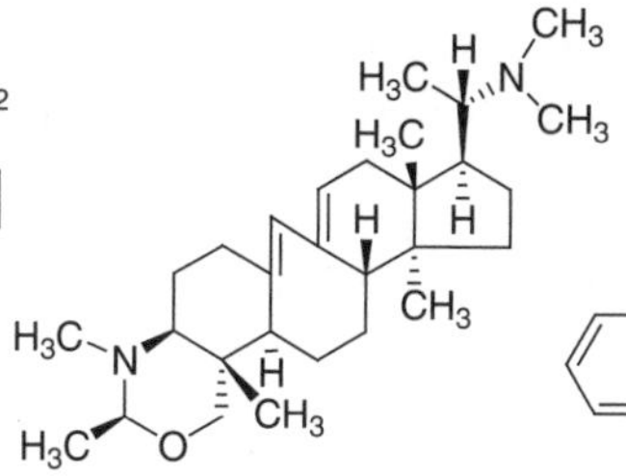

22

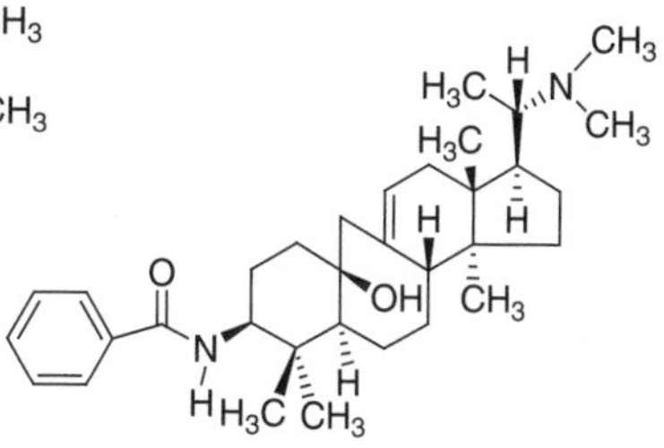

23

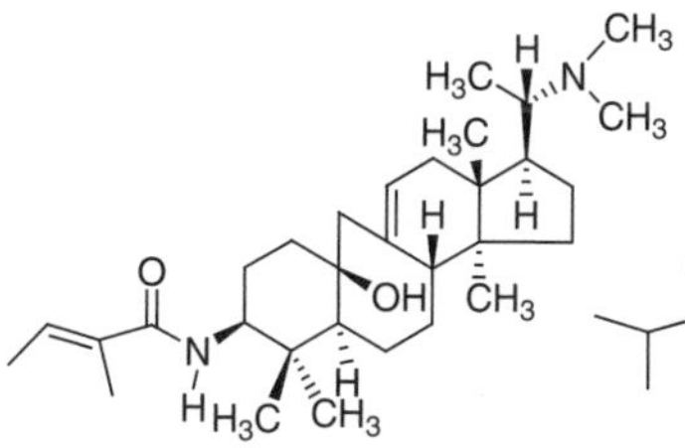

24

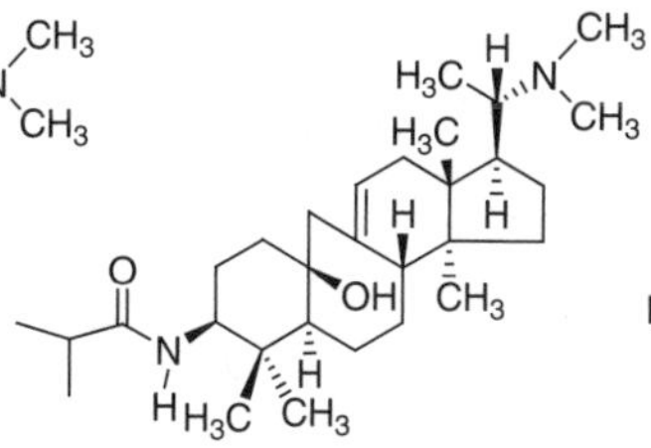

25

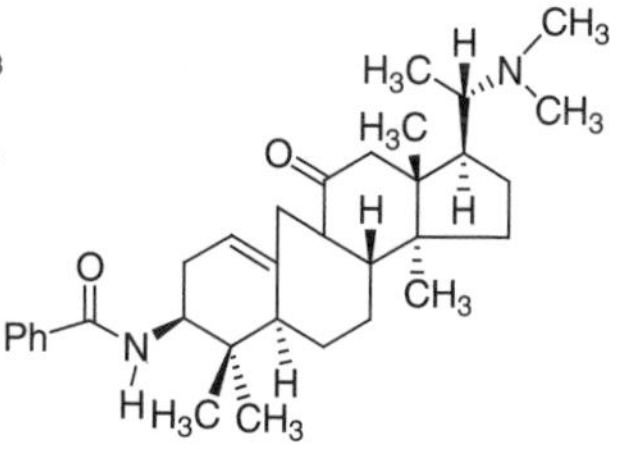

26

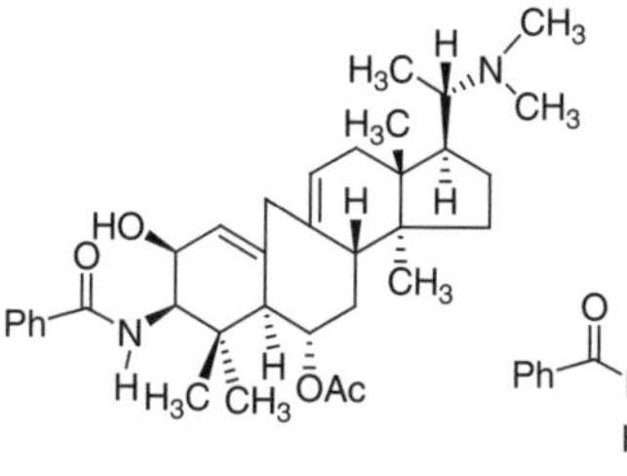

27

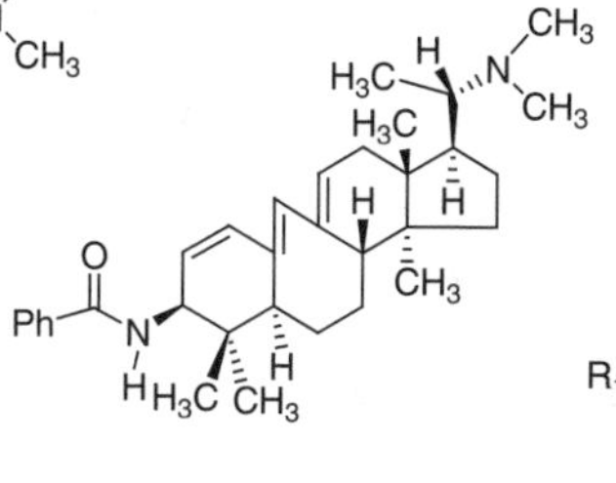

28

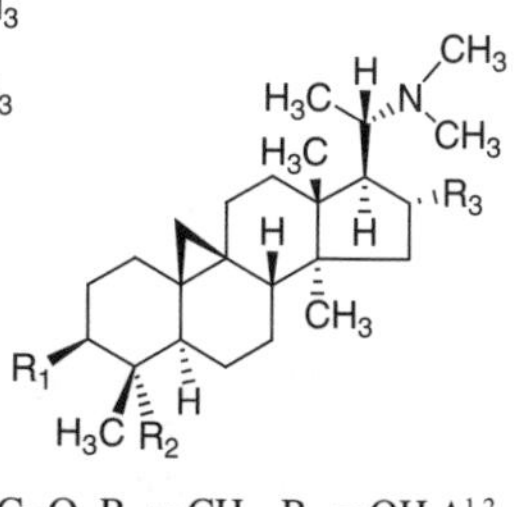

29 $R_1 = C=O$, $R_2 = CH_3$, $R_3 = OH$ $\Delta^{1,2}$
30 $R_1 = NH$-tigloyl, $R_2 = CH_2$-OH, R_3 OAc
31 $R_1 = NH$-benzoyl, $R = CH_2$-OH, $R_3 = OH$, $\Delta^{6,7}$
32 $R_1 = NH$-benzoyl, $R = CH_2$-OH, $R_3 = OAc$ $\Delta^{6,7}$

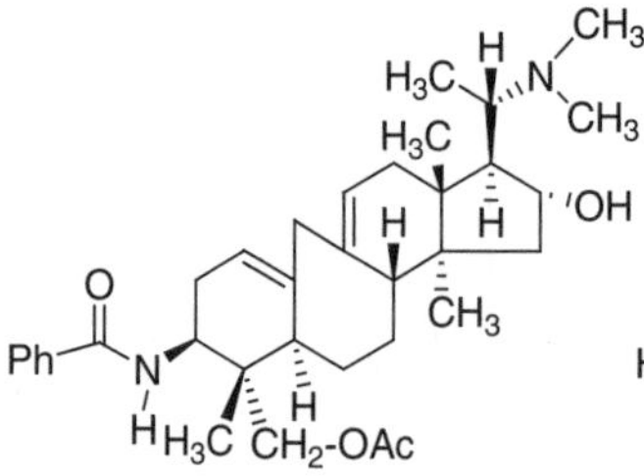

33

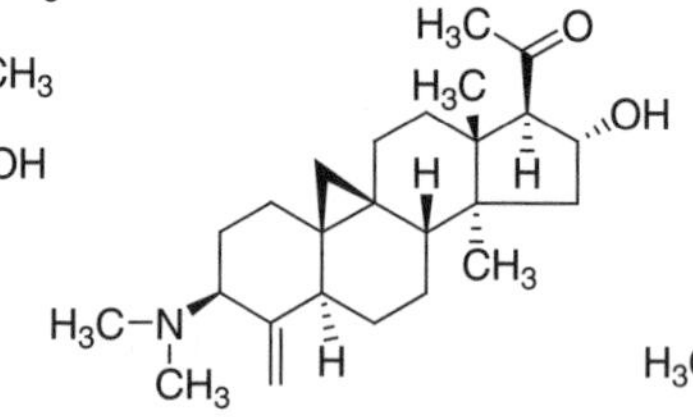

34

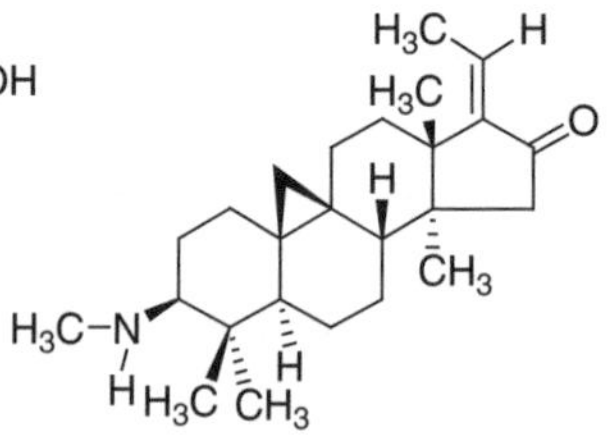

35

Recently, our research group reported the isolation of four new steroidal alkaloids: 2α,16α,31-triacetylbuxiran (36), 2α,16α,31-triacetyl-9,11-dihydrobuxiran (37), O^6-buxafurandiene (38), and 7-deoxy-O^6-buxafurandiene (39) from this plant, as well as four known steroidal bases: benzoylbuxidienine (40), buxapapillinine (41), buxaquamarine (42), and irehine (43) (21,22). Alkaloids 18, 31–35, 41, and 42 were isolated for the first time from this plant. Another important feature of this plant is that it is known to produce alkaloids having N-tigloyl, tetrahydrooxazine, and ether moieties in their structures (18–22).

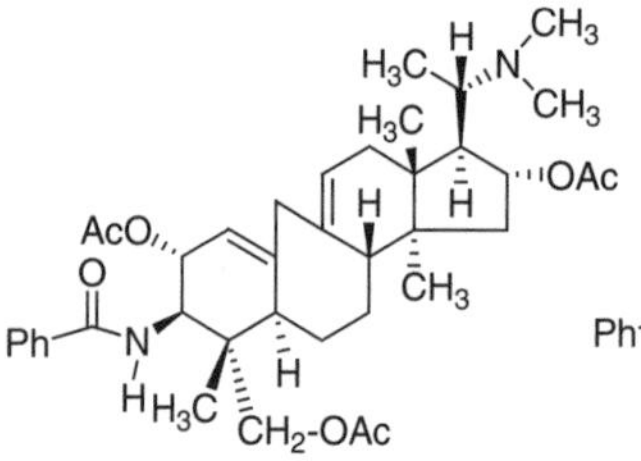

36

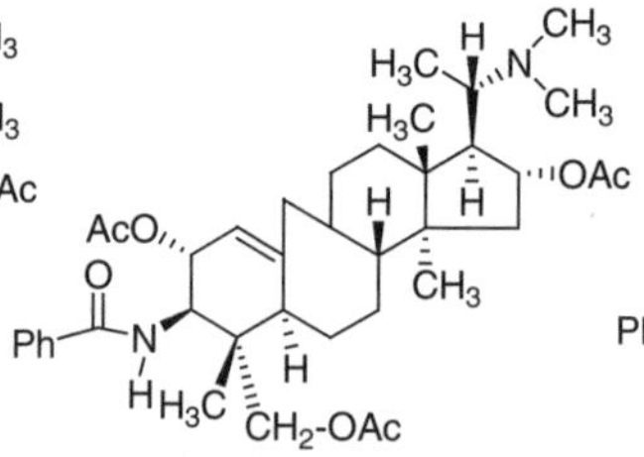

37

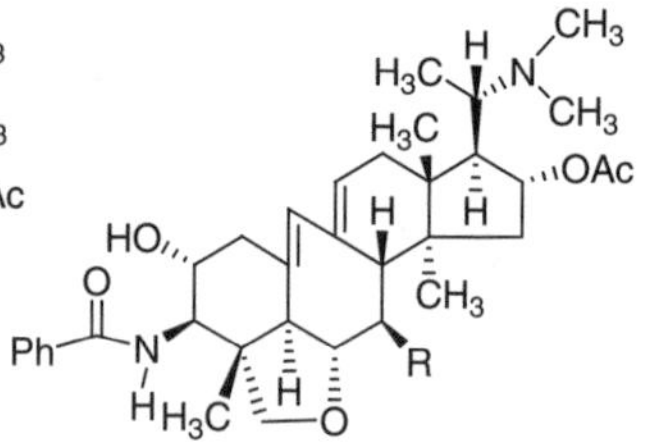

38 R = OH
39 R = H

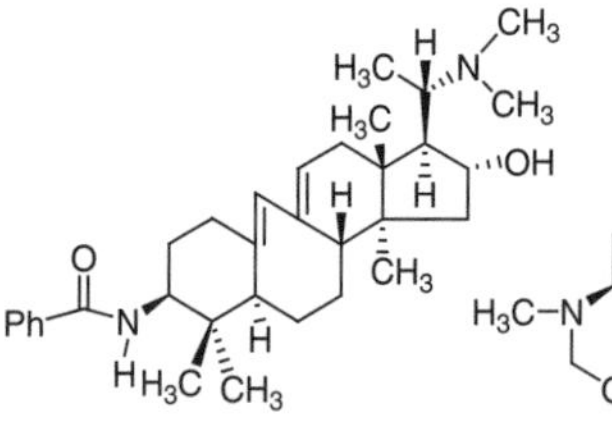

40

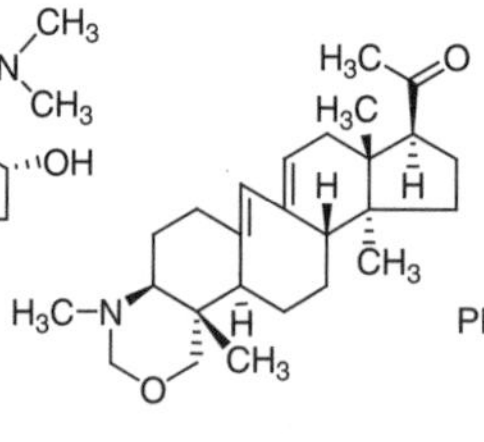

41

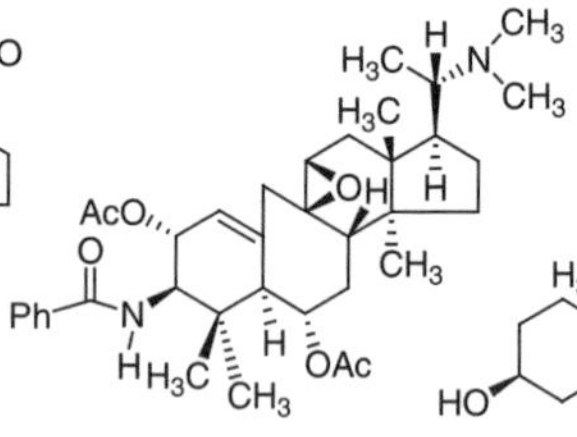

42

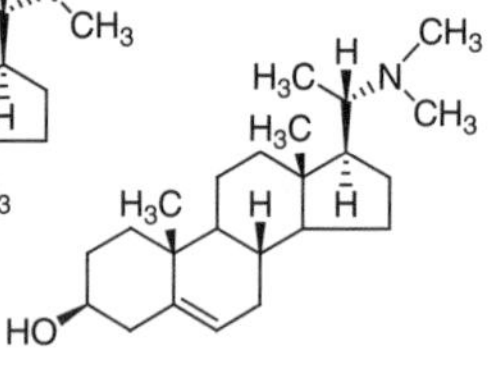

43

D. Steroidal Alkaloids from *B. longifolia*

B. longifolia, commonly known as "English Box," is widely distributed in Europe and Turkey. The crude extract of this plant has shown inhibition against serum cholinesterase of human and horse blood (23). Atta-ur-Rahman's research group performed phytochemical studies on the ethanolic extract of *B. longifolia* of

Turkish origin, and these studies resulted in the isolation of seven new steroidal bases: cyclovirobuxeine F (**44**), *N*-benzoyl-*O*-acetylbuxalongifoline (**45**), buxasamarine (**46**), cyclobuxamidine (**47**), buxalongifolamidine (**48**), buxabenzacinine (**49**), and cyclobuxomicreinine (**50**) (24–26). Two known alkaloids, 16α-acetoxybuxabenzamidienine (**51**) and *trans*-cyclosuffrobuxinine (**52**), were also isolated from this plant (24).

44

45

46

47

48 R = OH
49 R = H

50

51

52

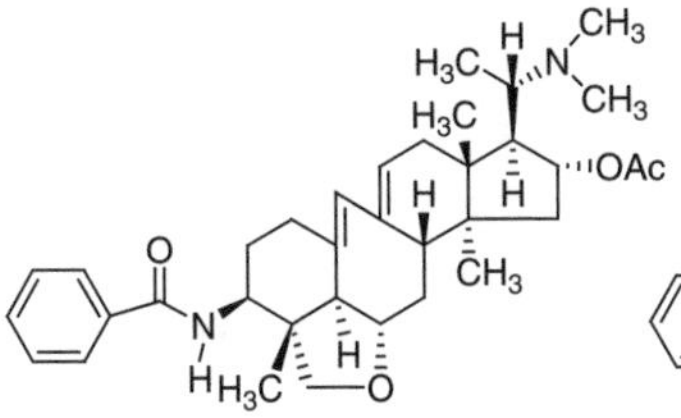

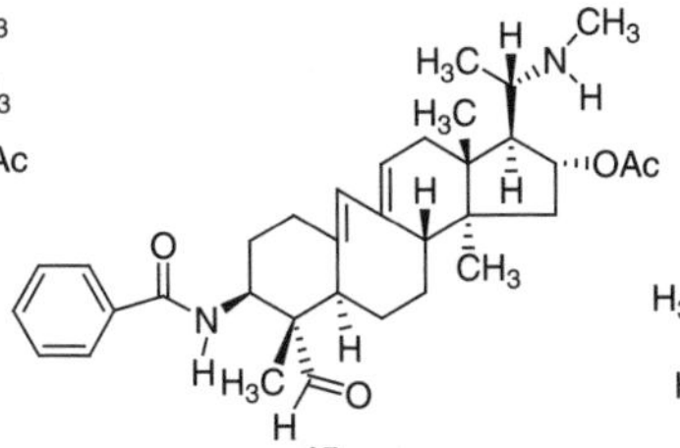

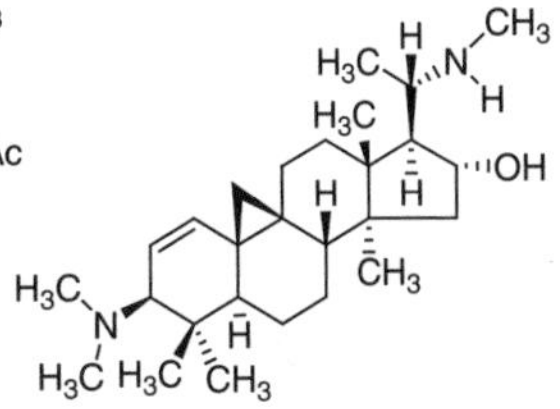

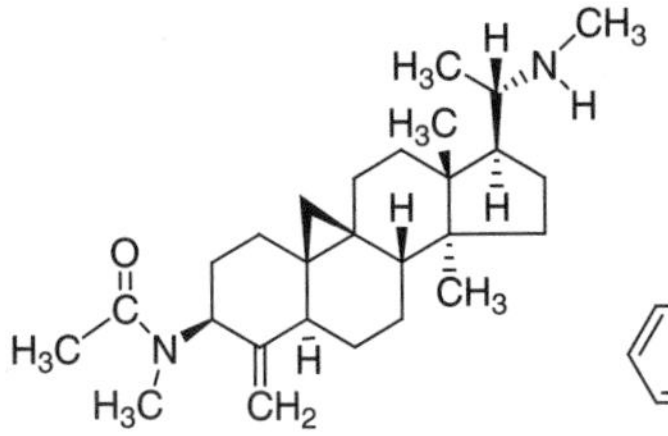

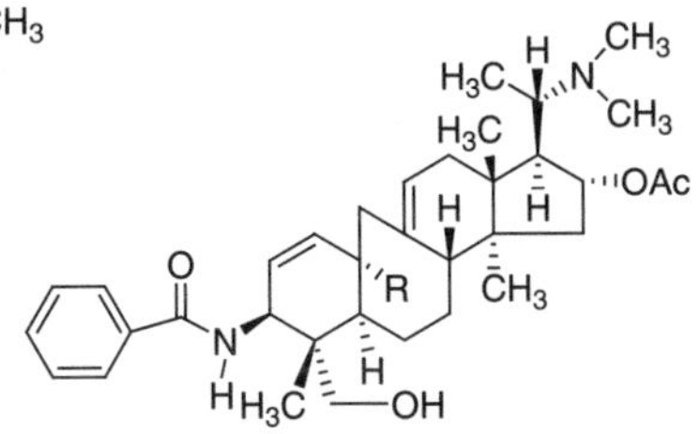

Alkaloid **48** contains a hydroxyl group at C-10 and may act as a biosynthetic precursor to O^2-buxafuranamine (**12**). The natural existence of **48** further supports the plausible biosynthesis of the ether linkage in these alkaloids.

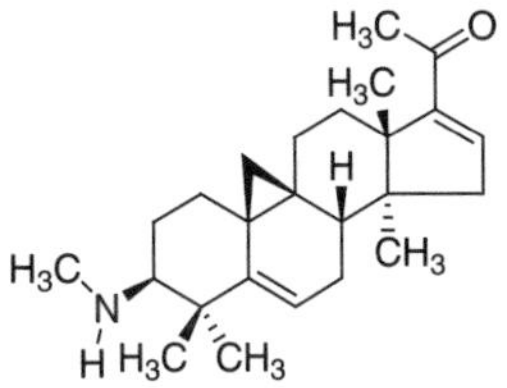

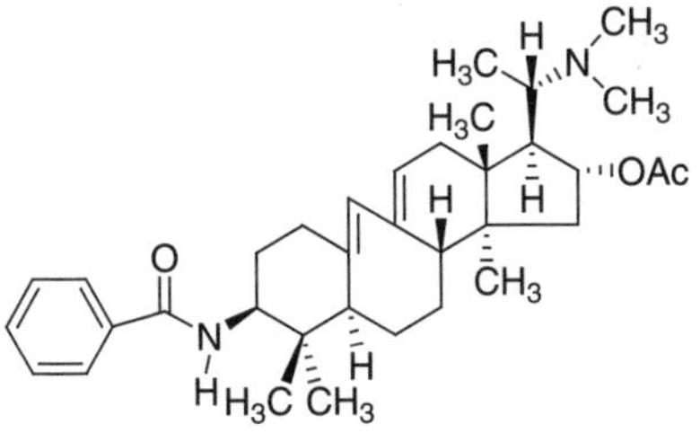

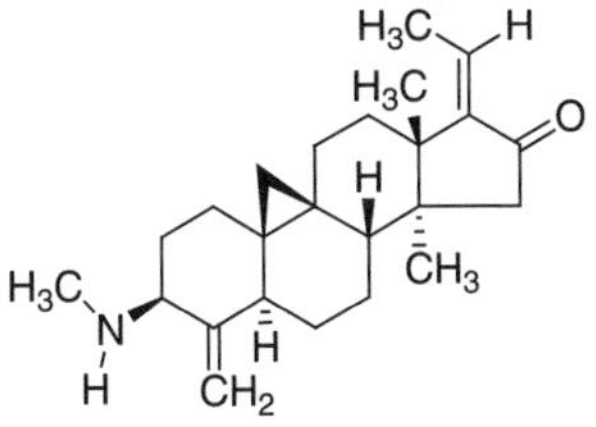

E. Steroidal Bases of *B. papillosa*

B. papillosa C.K. Schneid. is a shrub and is abundant in northern areas of Pakistan and India. This plant is locally known as *Shamshad*. The crude extract of this plant finds extensive use as a traditional medicine for the treatment of various diseases, especially malaria, rheumatism, and skin infections (27). This plant has been extensively investigated by Atta-ur-Rahman's research group, which has isolated over 40 new steroidal bases (35). During the reporting period in this chapter, Atta-ur-Rahman and collaborators have reported the isolation of 10 new steroidal alkaloids, namely, papillotrienine (**53**), N_b-demethylpapilliotrienine (**54**), N_b-demethylharapamine (**55**), N_a,N_b-dimethylbuxupapine (**56**), 16α-hydroxypapillamidine (**57**), 2α,16α-diacetoxy-9β,11β-epoxybuxamidine (**58**), papillozine C (**59**), buxakashmiramine (**60**), buxakarachiamine (**61**), and buxahejramine (**62**) (28–31). Six known steroidal bases, cycloprotobuxine C (**63**), cyclovirobuxeine A (**64**), cyclomicrophylline A (**65**), semperviraminol (**66**), sempervirone (**67**), and buxozine C (**68**), were isolated for the first time from this plant (9,30,31).

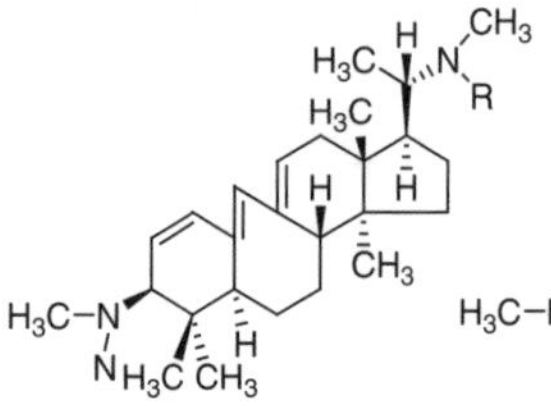

53 R = CH₃
54 R = H

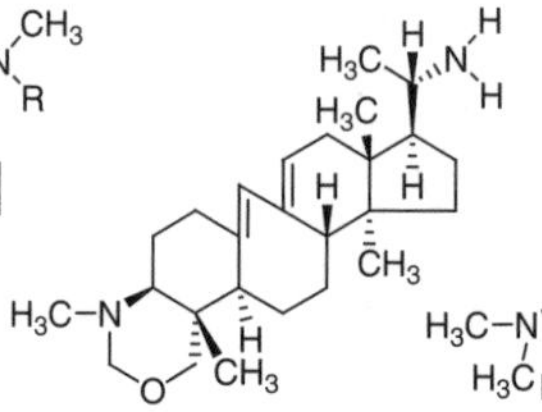

55

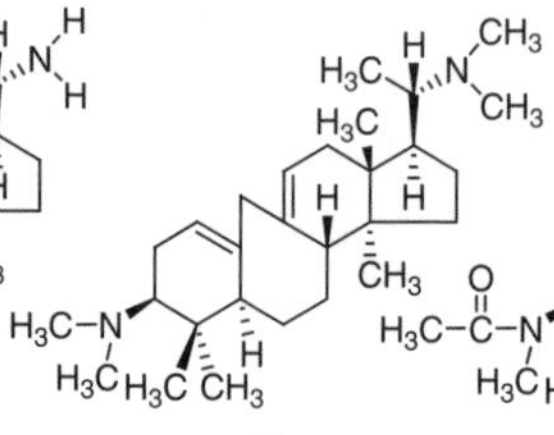

56

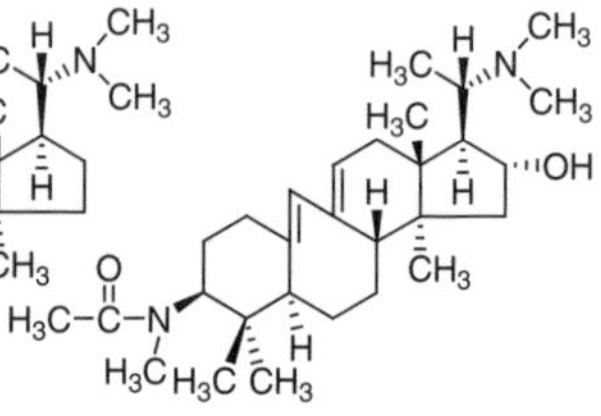

57

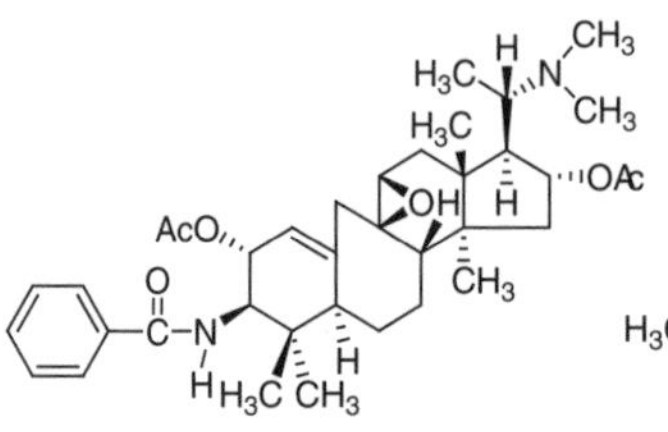

58

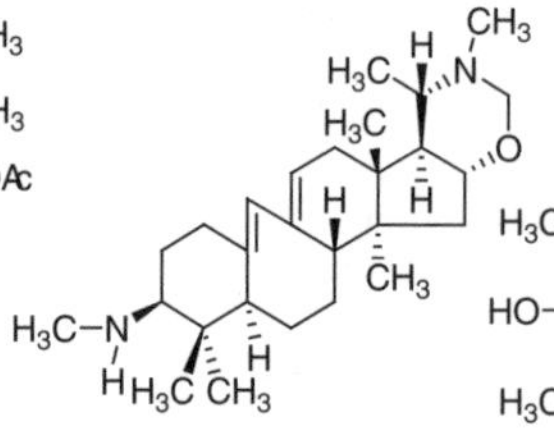

59

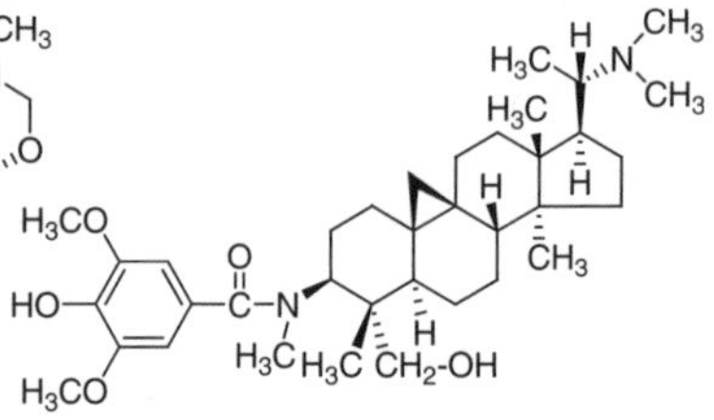

60

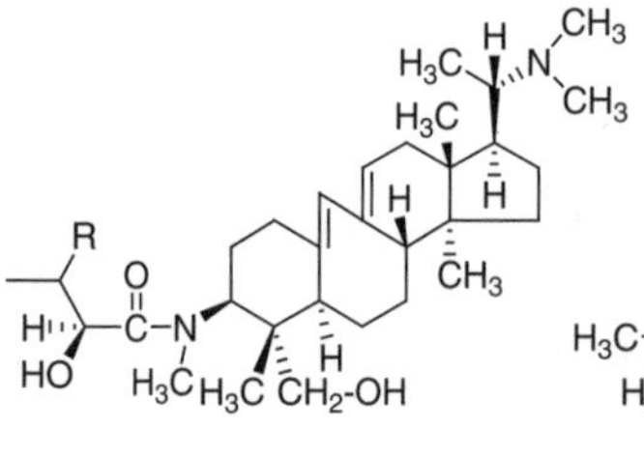

61 R = CH₃
62 R = CH₃-CH₂

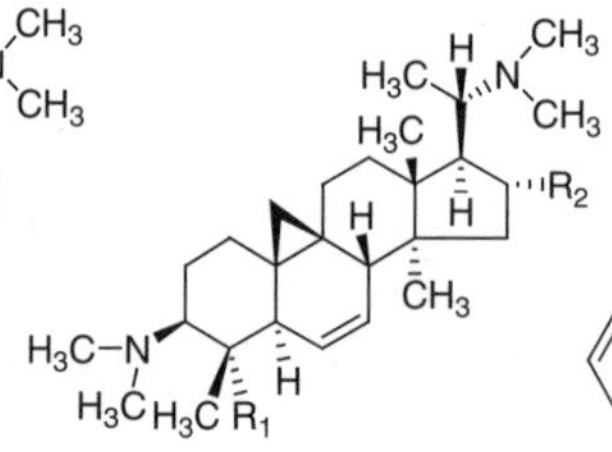

63 R₁ = CH₃, R₂ = H
64 R₁ = CH₃, R₂ = OH
65 R₁ = CH₂OH, R₂ = OH

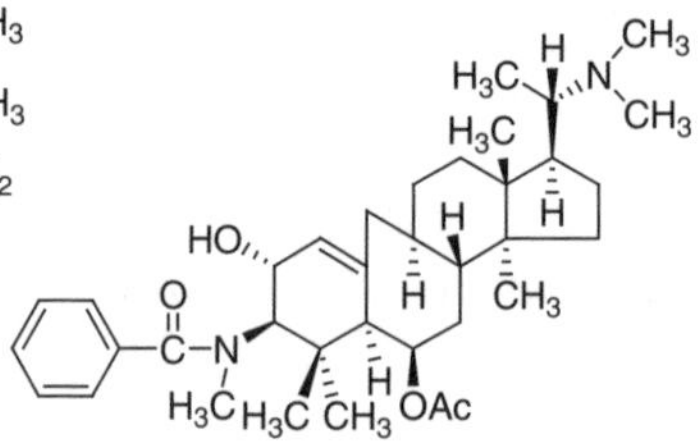

66

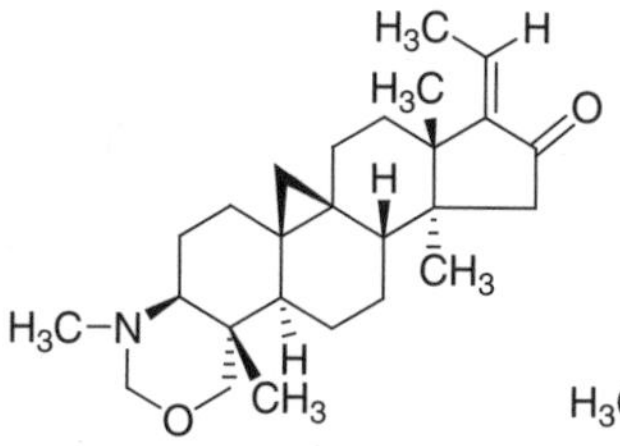

67 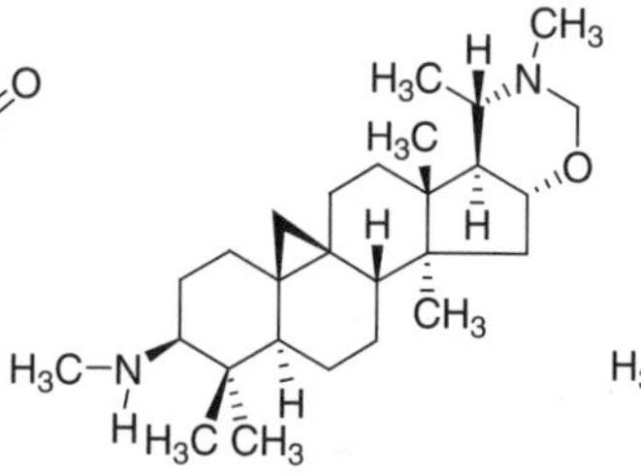   68 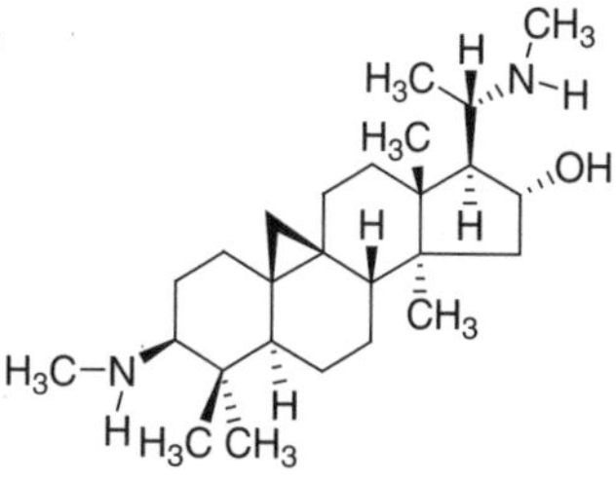   69

Alkaloids **53** and **54** belong to a unique class of *Buxus* bases having a conjugated triene system. The biosynthesis of these compounds may occur by the elimination of a suitable leaving group at C-1 or C-2 of the precursor *abeo*-diene alkaloids. Alkaloids **55** and **67** are members of the class of *Buxus* alkaloids having a tetrahydrooxazine moiety incorporated in ring A of their structures. The presence of this ring in *Buxus* alkaloids can be easily recognized by the ^{1}H NMR spectrum exhibiting the presence of two pairs of AB doublets at δ 3.20–4.50. The tetrahydrooxazine ring may be produced in nature by the condensation of formaldehyde or acetaldehyde with the C-3 amino group. The addition of the C-4β hydroxymethylene to the corresponding ketimine could then result in the formation of a tetrahydrooxazine ring. Alkaloids **59** and **68** contain a tetrahydrooxazine moiety in ring D, and **59** is the second alkaloid reported in this series, as buxozine C (**68**) represents the first example of this class of *Buxus* alkaloids (32). The mass spectra of alkaloids **59** and **68** exhibit ions at *m/z* 127 and 113 due to cleavage of ring D, and these ions serve as diagnostic features to determine the presence of a tetrahydrooxazine ring in ring D (30,32,33).

Recently, Ji and coworkers reported the short and efficient synthesis of alkaloid **68** using compound **69** as a precursor and following the proposed biosynthetic pathway. This involves the reaction of **69** with an excess of formaldehyde in the presence of ethanol as a solvent at room temperature to afford alkaloid **68** in 91% yield (34). This ring closure reaction further supports the proposed tetrahydrooxazine ring formation at ring D, as a number of N_b-formyl containing alkaloids have been reported in the literature (35).

F. Steroidal Bases from *B. sempervirens*

B. sempervirens L., commonly known as boxwood, is abundant in Eurasia. The ethanolic extract of this plant has been reported to exhibit anti-HIV properties (3). During the last three decades, over 50 new steroidal bases have been reported in the literature (35). Cyclovirobuxine D, one of the constituents of *B. sempevirens*, is reported to be active in the treatment of heart disorders (36). Chemical studies carried out by four research groups (Atta-ur-Rahman, Khodzhaev, Fourneau, and Loru) during 1992–2008 resulted in the isolation of 18 new steroidal bases. These new bases include 16-hydroxybuxaminone (**70**), isodihydrocyclomicrophylline A

(71), spriofornabuxine (72), 16α-hydroxy-N_a-benzoylbuxadine (73), sempervir-aminone (74), N_a-demethylsemperviraminone (75), buxaminol C (76), 30-hydroxycyclomicobuxene (77), sempervirooxazolidine (78), semperviraminol (66), buxamine F (79), 17-oxocycloprotobuxine (80), N_{20}-formylbuxaminol E (81), O_{16}-syringylbuxaminol E (82), N_{20}-acetylbuxamine G (83), N_{20}-acetylbux-amine E (84), 16α,31-diacetylbuxadine (85), and N_b-demethylcyclomikurane (86) (26,37–42). Additionally, nine known steroidal bases, E-cyclobuxaphylamine (87), Z-cyclobuxaphylamine (88), cyclomicrobuxeine (89), papilamine (90), buxoxy-benzamine (91), buxapapillinine (40), cyclobuxaphylline (92), cyclimikuranine (93), and buxaqumarine (41), were also isolated for the first time from this species.

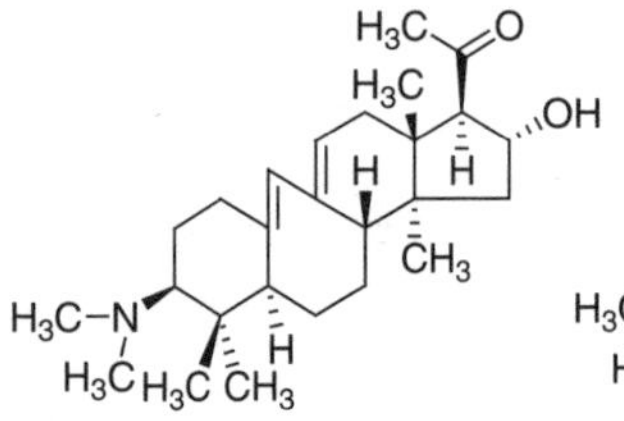

70 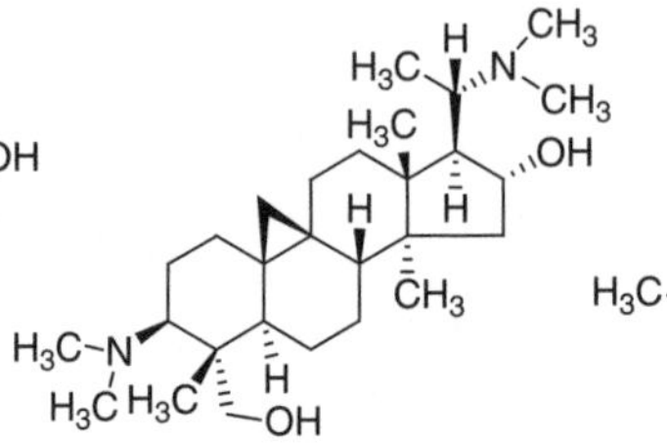   71 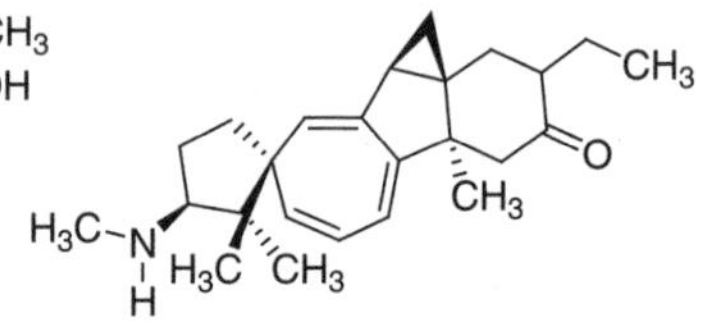   72

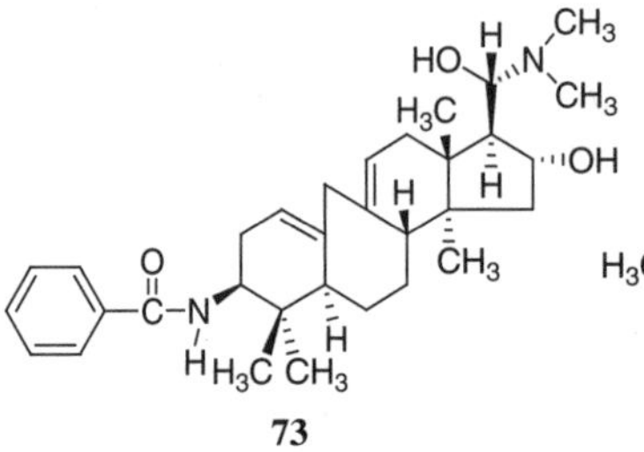

73 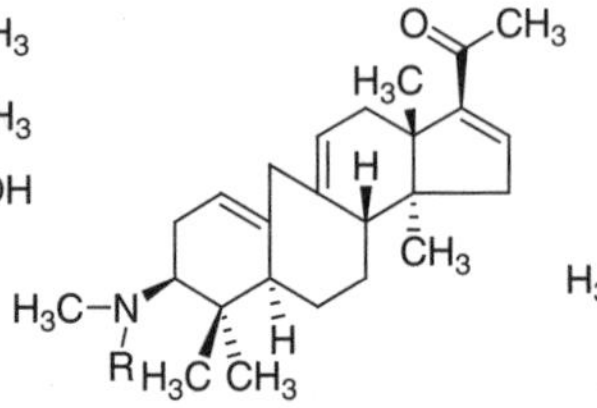   74 R = CH₃   76
75 R = H

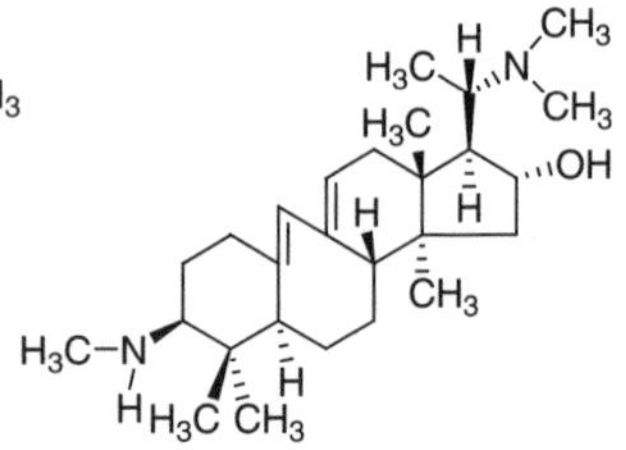

Buxus alkaloids, with a cycloheptane ring B, representing the 9(10→19) *abeo* system, usually contain one or two double bonds. Alkaloid 72, containing a spiro junction at C-10 and a cycloheptatriene ring, represents a new class of *Buxus* alkaloids. It might by produced in nature by the same biogenetic pathways as those of *Buxus* alkaloids, but through a series of complex pathways, as no plausible biosynthetic pathways have been proposed. Sempervirooxazolidine (78) also represents a novel structure having an oxazolidine moiety incorporated in its structure at C-2 and C-3. Biogenetically, this moiety may be produced by the reaction of the C-3 amino group with formaldehyde to afford a ketimine that, on further reaction with the C-2 hydroxyl group, results in the formation of an oxazolidine ring. It is also important to note that it is well documented in the literature that the C-31 methyl group (C-4α) undergoes oxidation to afford a hydroxymethylene group (14,15), and in light of this literature precedent, the name 30-hydroxycyclomicobuxene needs to be revised to 31-hydroxycyclomico-buxene for alkaloid **104**.

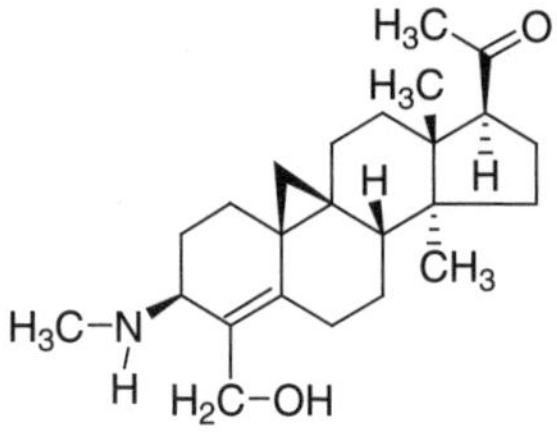

77

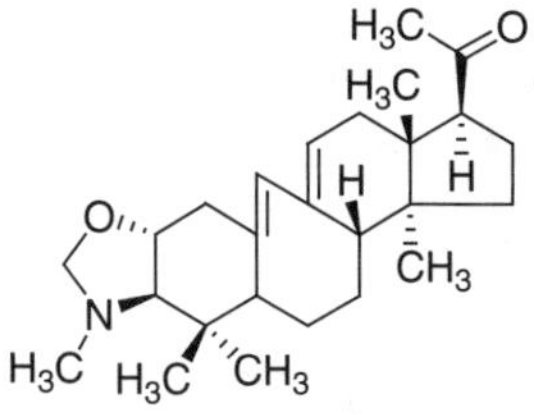

78

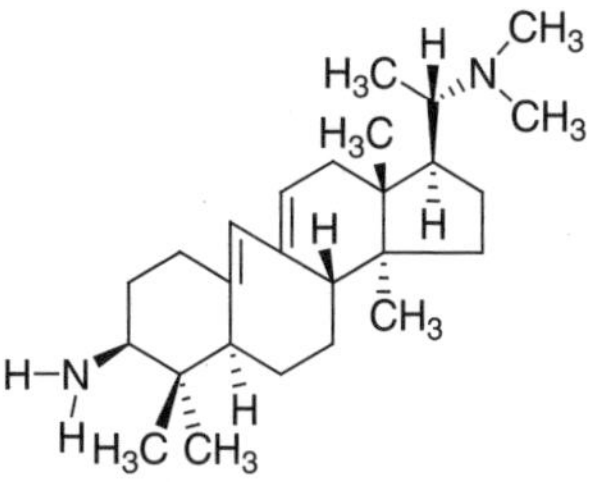

79

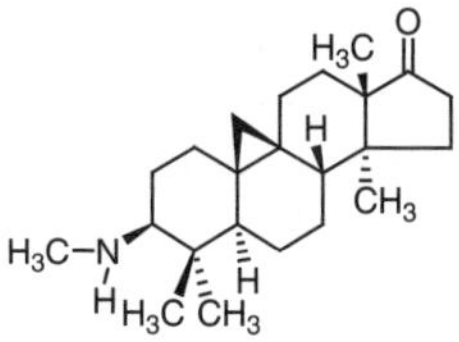

80

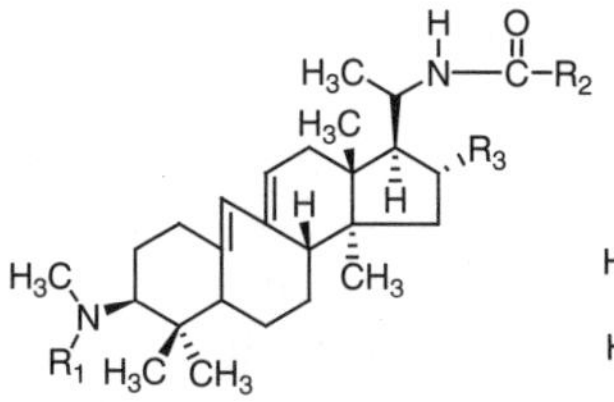

81 R$_1$ = CH$_3$, R$_2$ = H R$_3$ = OH
83 R$_1$ = H, R$_2$ = CH$_3$, R$_3$ = H
84 R$_1$ =CH$_3$, R$_2$ = CH$_3$, R$_3$ = H

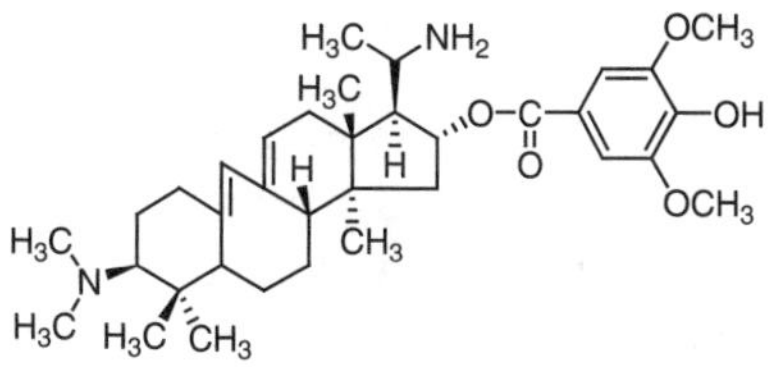

82

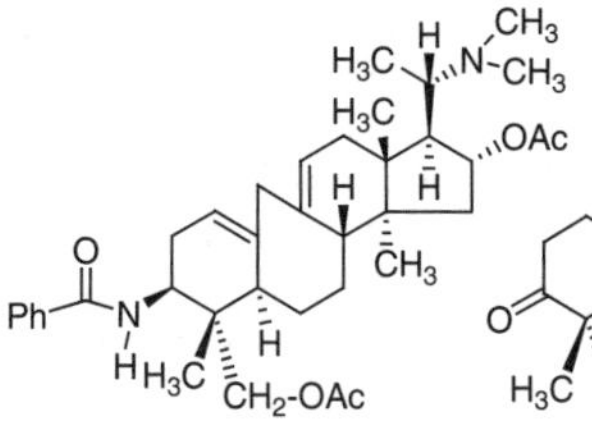

85

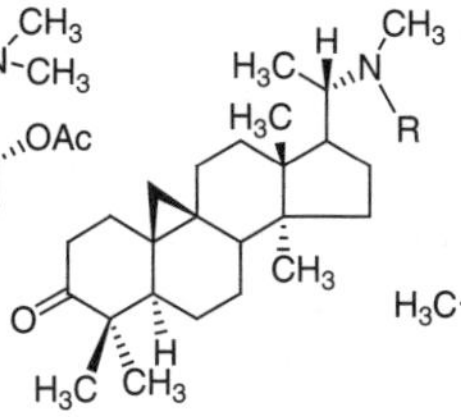

86 R = H
93 R = CH$_3$

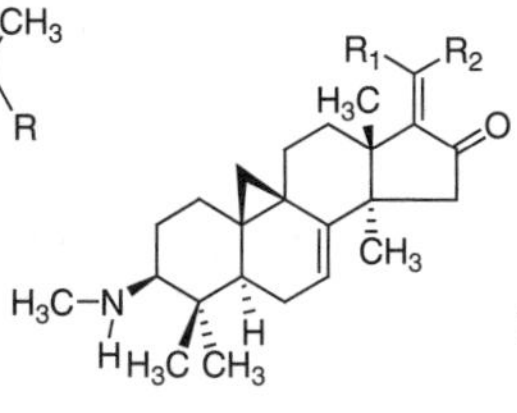

87 R$_1$ = CH$_3$, R$_2$ = H
88 R$_1$ = H, R$_2$ = CH$_3$

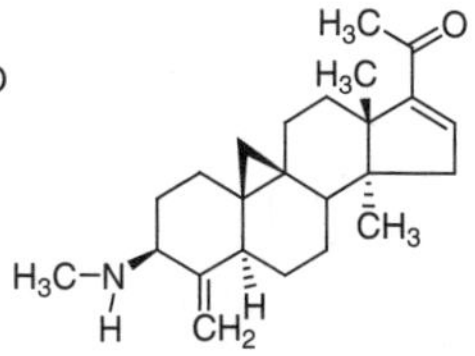

89

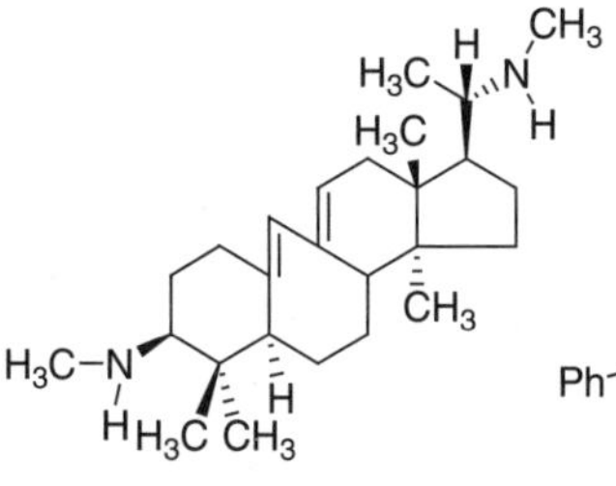

90

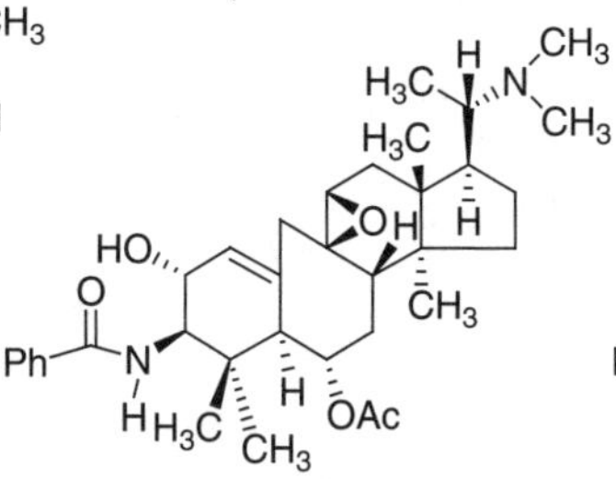

91

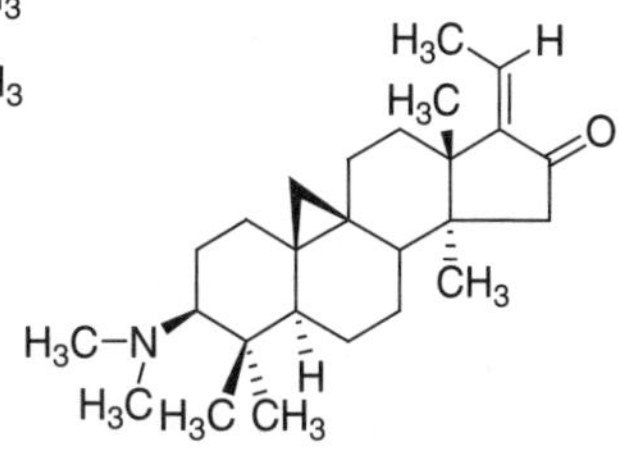

92

III. BIOLOGICAL ACTIVITIES OF *BUXUS* ALKALOIDS AND THEIR SEMISYNTHETIC ANALOGUES

Buxus alkaloids exhibit various biological effects including antimalarial, antibacterial, antifungal, antiviral, and enzyme inhibitory activities. One of the important aspects of a drug discovery program is to discover new naturally occurring enzyme inhibitors. These enzyme inhibitors may be used either as adjuvants to improve chemotherapy or to treat various diseases. *Buxus* alkaloids have shown significant potential in the area of enzyme inhibition activities.

A. Glutathione *S*-Transferase Inhibitory Activity

Presently, substantial concern has been raised about anticancer drug resistance. Several research investigations have been conducted in the past in an effort to explain the mechanisms of acquired drug resistance in the treatment of cancer and parasitic diseases. These studies have indicated that the cystosolic detoxification enzymes, glutathione *S*-transferases (GSTs; E.C. 2.1.5.18), play an active role in this process (43). GSTs are phase II detoxification isozymes that function in conjugation with a wide variety of exogenous and endogenous electrophilic substances to make a glutathione adduct. Glutathione is a tripeptide having γ-glutamyl-cysteinyl-glycine amino acids in its backbone. The glutathione adduct formed is less toxic and has a very high solubility in water, and can therefore easily be excreted from the body (44). Anticancer drugs with electrophilic centers can readily form this adduct in the presence of GST, and can be excreted from the body, thus lowering the efficiency of the chemotherapeutic agent. To increase the effectiveness of cancer chemotherapeutic agents, it might be necessary to use GST inhibitors as adjuvants during chemotherapy. Natural products for GST inhibition have not been explored extensively. It is therefore a worthwhile effort to discover natural products exhibiting GST inhibitory activity with a view to use them as an adjuvant during cancer chemotherapy. Our research group evaluated a number of *Buxus* alkaloids for anti-GST activity using a random approach, and surprisingly found alkaloids **38–43** to be active in this assay (45). The IC_{50} values of **38–43** in the GST inhibition assay were 23.1, 22.0, 33.4, 19.0, and 45.1 µM, respectively.

B. Antiacetylcholinesterase and Antibutylcholinesterase Activities

Acetylcholine serves as a neurotransmitter in the central and peripheral nervous system. Alzheimer's disease (AD) is a progressive, degenerative, neurological disorder and may result from a deficit of acetylcholine in the brain (46). Cholinesterase is a family of enzymes that consists of AChE and butylcholinesterase (BChE), and both of these enzymes are considered responsible for the hydrolytic destruction of acetylcholine into acetic acid and choline (47,48). It is well documented that AD could be cured by increasing the level of acetylcholine in the brain (49). The use of potent inhibitors of AChE and BChE activity is one of the approaches to cure AD (48,49). *Buxus* alkaloids have shown potential as anti-AChE and anti-BChE inhibitors. Most of the steroidal bases isolated from *B. hyrcana* have shown activity in AChE and BChE assays. The IC_{50} values of *Buxus* alkaloids active in these assays are listed in Table I (18,20,22,50). Anti-AChE and anti-BChE activities might be due to the presence of dimethylamino moieties at C-3 and C-20, or the presence of the ether linkage in these alkaloids. However, it is difficult to comment on the structure–activity relationships of these bioactive compounds from the data obtained thus far.

Guillou and collaborators have recently reported the discovery of the semisynthetic compound (**96**) as a highly selective inhibitor of AChE (50). These authors also report on an improved isolation method for *N*-3-isobutyrylcyclo-buxidine-F (**94**), which involves extraction of the plant material with dichloro-methane/methanol, followed by separation of the alkaloids by solvent–solvent fractionation at different pH values. Fractions obtained at pH 8 were triturated with acetone to afford alkaloid **94** in bulk quantity. This method is very straightforward as compared to the previously reported methods, which involve long and tedious chromatographic techniques and countercurrent separations (51). This research group used alkaloid **94** as a synthetic template to generate 28 different analogues in order to study their structure–activity relationships for anticholesterase activity. These authors reported a two-step synthesis of alkaloid **96** using *N*-3-isobutyrylcyclobuxidine-F (**94**) as the main precursor. This involves the pyrolysis of **94** at 240°C under 0.05 mmHg to afford a mixture of **95** and **96**. The second step involves the pyrolysis of this mixture in the presence of tetraethylammonium hydroxide using similar reaction conditions as those of the first step to afford **96** (Scheme 1).

Compounds **100a–i** are new 1,3-dihydrooxazine derivatives, and their synthesis involves the hydrolysis of *N*-3-isobutyrylcyclobuxidine-F (**94**) to afford cyclobuxidine-F (**97**). Amide derivatives (**99a–i**) are prepared by coupling of the appropriate anhydrides (**98a–i**). These authors used the reported synthetic approach to synthesize anhydrides (**98a–i**), except **98a–c** and **98f**, as these are commercially available (52). Pyrolysis of compounds **99a–i** under reduced pressure affords the 1,3-dihydrooxazines **100a–i** and **99a–i** having the 9(10→19) *abeo* system as major and minor products, respectively. This mixture is converted to compounds **100a–i** by heating under reduced pressure in the presence of tetraethylammonium hydroxide (Scheme 2). These authors further reported the purification of **99a–i** from the mixture of **99a–i** and **100a–i** using alumina.

The syntheses of 1,3-oxazinane derivatives with cyclopropane and open cyclopropane rings are shown in Schemes 3 and 4. The synthesis of an isomer of compound **106** containing an α,β-ethylenic ketone function is also outlined in Scheme 5. Further modification of ring D is described in Schemes 6 and 7. Schemes 8 and 9 show the synthesis of compounds **117** and **124**, respectively. All of the synthesized compounds and intermediates exhibit different levels of bioactivity against AChE and BChE. Structure–activity relationship studies on these newly synthesized compounds indicate that oxazines (**100d–h**) and pyrimidines (**119a,b** and **124**) are more active than their corresponding open chain precursors (**99d–f**, **118a,b**, and **123**). The inhibitory potency decreases when

Table I Inhibition of AChE and BuChE activities (IC$_{50}$ values) for *Buxus* alkaloids and their synthetic analogues

Compounds	IC$_{50}$ values	
	AChE	BuChE
18	50.8 µM	–
21	312.0 µM	–
22	19.2 mM	–
23	>1000 µM	310.6 µM
24	443.6 µM	31.2 µM
25	>1000 µM	63.7 µM
26	145.0 µM	20.0 µM
28	–	1.71 µM
29	310.0 µM	1.12 µM
30	83.0 µM	–
31	210.6 µM	58.6 µM
32	175.4 µM	37.1 µM
33	468.0 µM	350.0 µM
34	>1500 µM	210.0 µM
38	17.0 µM	–
39	13.0 µM	–
40	35.0 µM	–
41	76.0 µM	–
42	80.0 µM	–
43	100 µM	–
94	>1000 nM	>1000 nM
95	938.0 nM	747.7 mM
96	31.0 nM	1000 nM
99d	105 nM	>10,000 nM
99e	120 nM	1926.0 nM
100a	680 nM	>10,000 nM
100b	697 nM	>10,000 nM
100c	225 nM	>10,000 nM
100d	13.0 nM	>10,000 nM
100e	27.0 nM	>10,000 nM
100f	102.0 nM	>1535.0 nM
100g	7650 nM	>10,000 nM
100h	400.0 nM	>375.0 nM
100i	4205 nM	380.0 nM
101	820.0 nM	>10,000 nM
102	2380 nM	>10,000 nM
106	110.0 nM	1460 nM
107	380.0 nM	–
108	110.0 nM	–
110	110.0 nM	1440 nM

Table I (*Continued*)

Compounds	IC$_{50}$ values	
	AChE	BuChE
111	106.0 nM	–
118a	291.0 nM	–
118b	1341 nM	–
119a	18.0 nM	> 10,000 nM
119b	14.0 nM	–
123	780 nM	> 10,000 nM
124	368.0 nM	> 10,000 nM
125	28.0 nM	> 10,000 nM

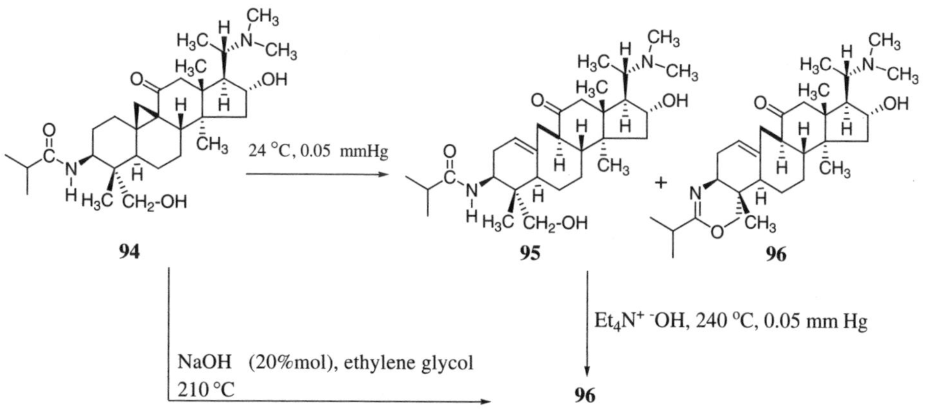

Scheme 1

the oxazine ring is reduced to an oxazinane **104**. These data suggest that the oxazine and the amidine moieties are required for anti-AChE activity. It is further observed that the presence of a 9β,10β-cyclopropyl functionality in the oxazinene **102** and pyrimidine series also increases the enzyme inhibitory activity. The IC$_{50}$ values of these compounds are listed in Table I. These bioactivity data suggest that these compounds represent the first example of steroidal alkaloids exhibiting potent anti-AChE properties.

C. Antibacterial Activity

Antibiotic resistance among bacterial pathogens threatens the treatment of infectious diseases. New antibiotics are needed to combat this situation (53). *Buxus* alkaloids have also exhibited antibacterial activities against different

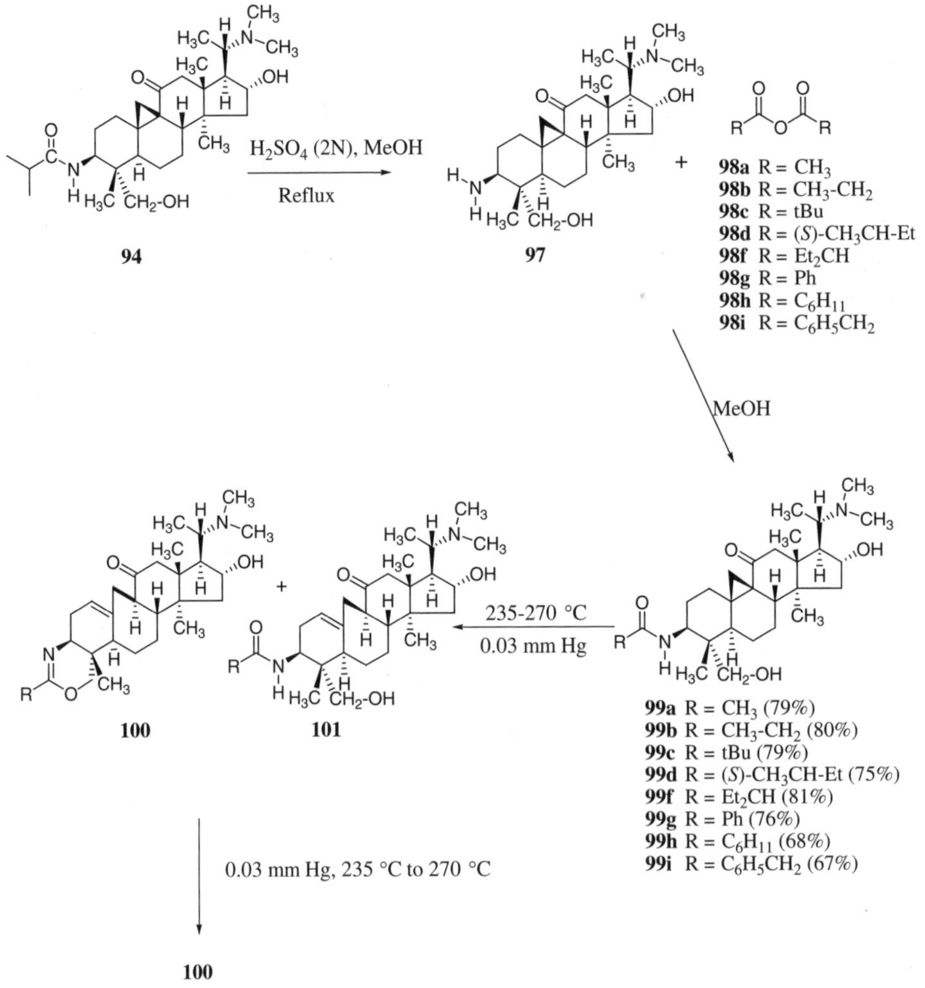

Scheme 2

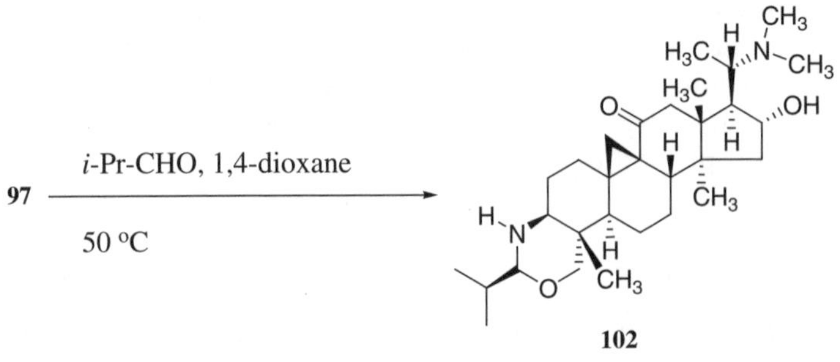

Scheme 3

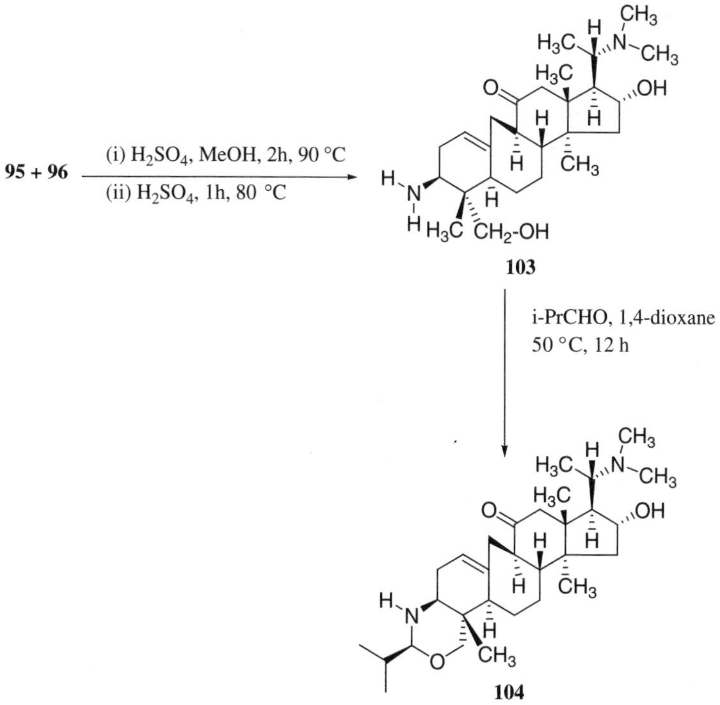

$95 + 96$ — (i) H$_2$SO$_4$, MeOH, 2h, 90 °C / (ii) H$_2$SO$_4$, 1h, 80 °C → **103**

i-PrCHO, 1,4-dioxane
50 °C, 12 h

104

Scheme 4

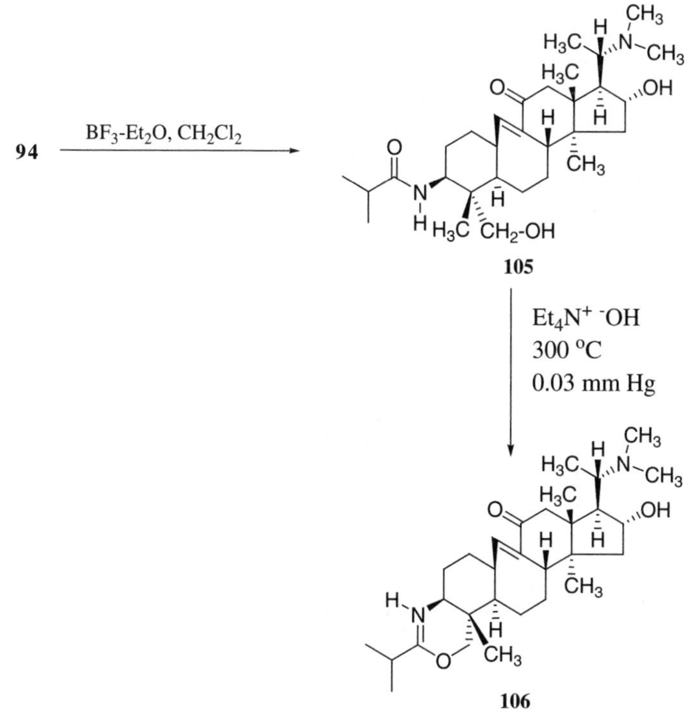

94 — BF$_3$-Et$_2$O, CH$_2$Cl$_2$ → **105**

Et$_4$N$^+$ $^-$OH
300 °C
0.03 mm Hg

106

Scheme 5

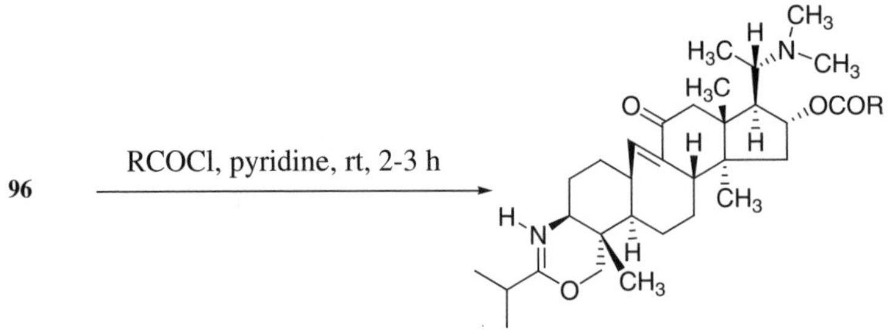

107 R = CH_3
108 R = tBu

Scheme 6

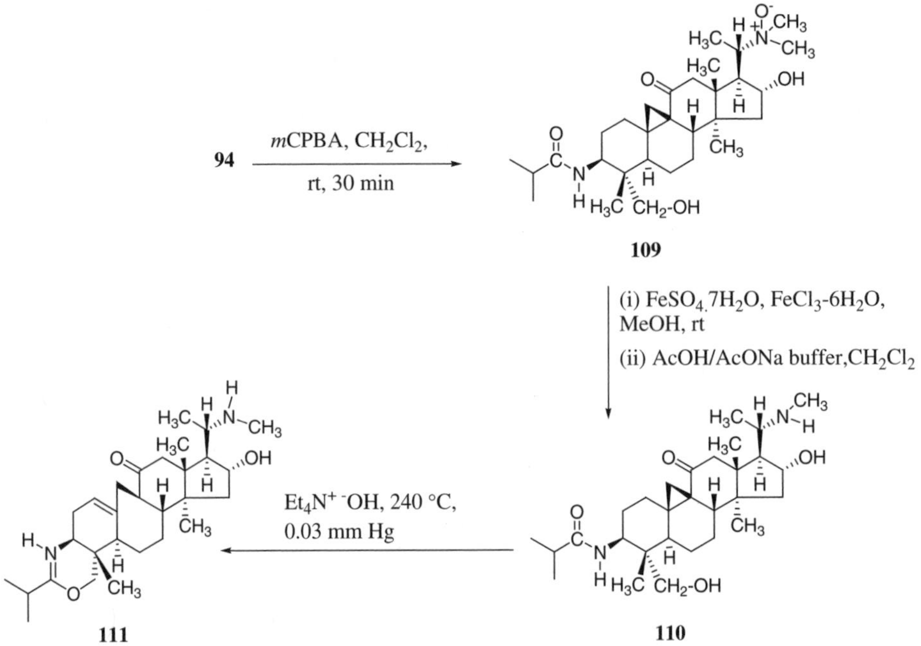

Scheme 7

human pathogenic bacteria. For instance, alkaloids **44–47** exhibit significant antibacterial activity against *Salmonella typhi, Shigella flexneri, Pseudomonas aeruginosa,* and *Escherichia coli*. The minimum inhibitory concentrations (MICs) of *Buxus* alkaloids are listed in Table II.

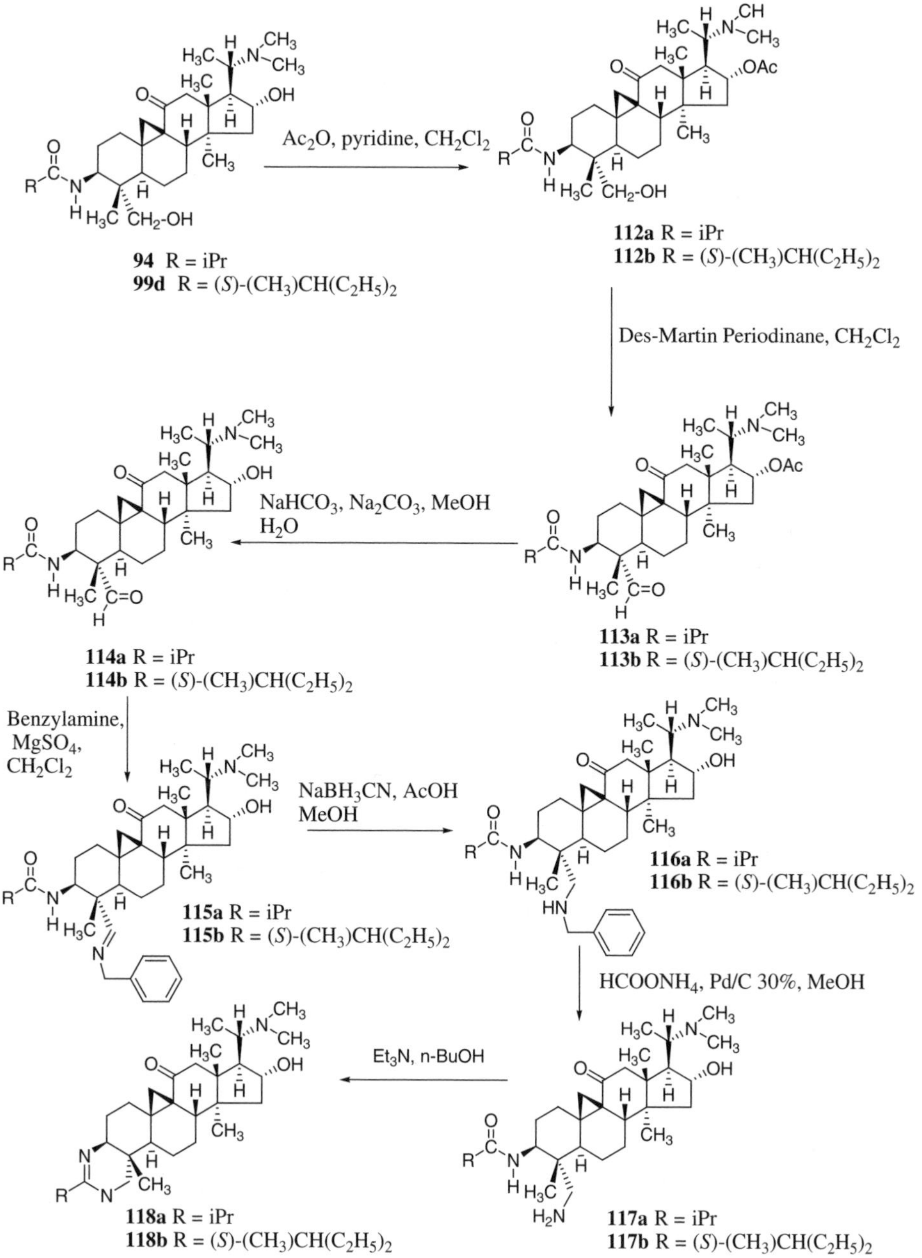

94 R = iPr
99d R = (S)-(CH₃)CH(C₂H₅)₂

112a R = iPr
112b R = (S)-(CH₃)CH(C₂H₅)₂

Ac₂O, pyridine, CH₂Cl₂

Des-Martin Periodinane, CH₂Cl₂

NaHCO₃, Na₂CO₃, MeOH
H₂O

114a R = iPr
114b R = (S)-(CH₃)CH(C₂H₅)₂

113a R = iPr
113b R = (S)-(CH₃)CH(C₂H₅)₂

Benzylamine,
MgSO₄,
CH₂Cl₂

NaBH₃CN, AcOH
MeOH

115a R = iPr
115b R = (S)-(CH₃)CH(C₂H₅)₂

116a R = iPr
116b R = (S)-(CH₃)CH(C₂H₅)₂

HCOONH₄, Pd/C 30%, MeOH

Et₃N, n-BuOH

118a R = iPr
118b R = (S)-(CH₃)CH(C₂H₅)₂

117a R = iPr
117b R = (S)-(CH₃)CH(C₂H₅)₂

Scheme 8

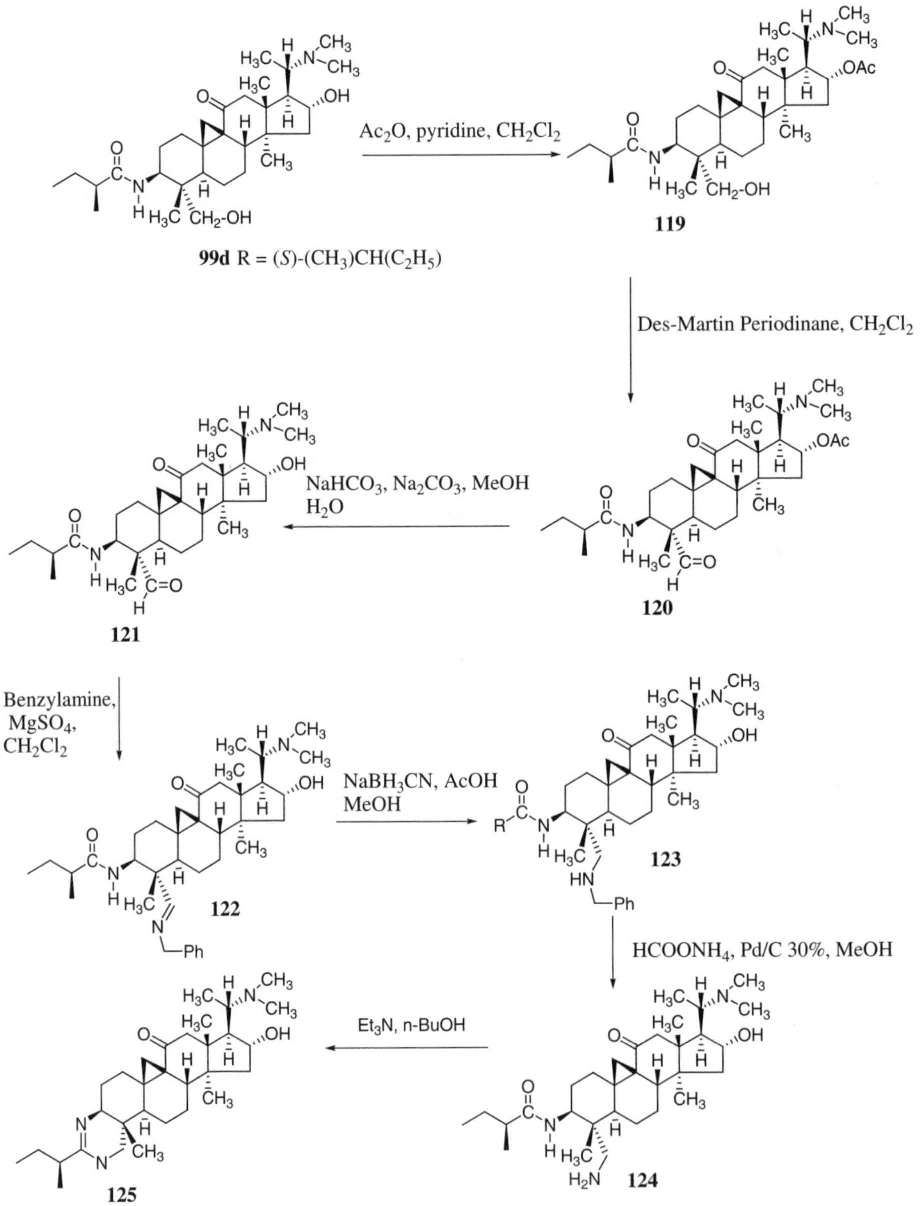

Scheme 9

Table II Minimum inhibition concentration (μg/100 μL) for the antibacterial activity of alkaloids 44–47

Alkaloids	Organisms			
	S. typhi	*S. flexneri*	*P. aeruginosa*	*E. coli*
44	100	100	10	–
45	100	200	50	–
46	100	150	50	–
47	150	–	–	200

IV. CONCLUSIONS

In summary, during the reporting period, a number of new steroidal alkaloids have been isolated from various *Buxus* species, and a few of them have a novel carbon skeleton. Steroidal bases isolated from *B. hyrcana* exhibit enzyme inhibitory activities, including GST, AChE, and BChE. One of the known steroidal bases **94** was used as a template to generate a library of compounds to evaluate for anti-AChE and anti-BChE properties. These studies resulted in the discovery of compound **96**, which represents the first example of steroidal alkaloids as a potent AChE inhibitor. This indicates that studies on structure–activity relationships of moderately bioactive compounds can be worthwhile in the discovery of new pharmaceuticals.

ACKNOWLEDGMENTS

We are thankful to Bradley University for supporting our research program. Athar Ata is also grateful to all of his collaborators and his students who helped in the area of steroidal alkaloids. Funding from NSERC is also appreciated.

REFERENCES

[1] G. A. Cordell, "Introduction to Alkaloids: A Biogenetic Approach," Wiley, New York, 1981, p. 890.
[2] J. Durant, P. Dellamonica, and A. Rebouillat, "Boxwood Extract for Treating Human Immunodeficiency Virus Infection," PCT Int. Appl., WO9300916, p. 20 1993.
[3] J. Durant, P. Chantre, G. Gonzalez, J. Vandermander, P. Halfon, B. Rousse, D. Guedon, V. Rahelinirina, S. Chamaret, L. Montagnier, and P. Dellamonica, *Phytomedicine* **5**, 1 (1998).
[4] Atta-ur-Rahman and M. I. Choudhary, *Nat. Prod. Rep.* **71**, 619 (1999).
[5] A. Ata, S. A. Van Den Bosch, D. J. Harwanik, and G. E. Pidwinski, *Pure Appl. Chem.* **79**, 2269 (2007).
[6] Atta-ur-Rahman and M. I. Choudhary, *in:* "The Alkaloids: Chemistry and Biology" (G. A. Cordell, ed.), Vol. 52, p. 233. Academic Press, San Diego, 1999.
[7] A. Ata, Phytochemical and Structural Studies on the Chemical Constituents of *Buxus hildebrandtii* and *B. sempervirens*, Ph.D. Thesis, University of Karachi, Karachi, Pakistan, 1995.
[8] Atta-ur-Rahman and M. I. Choudhary, *Pure Appl. Chem.* **71**, 1–79 (1999).

[9] S. Naz, Phytochemical and Structural Studies on the Chemical Constituents of *Buxus sempervirens* and *B. papillosa*, Ph.D. Thesis, University of Karachi, Karachi, Pakistan, 1995.

[10] M. Qiu, W. Yang, and R. Nie, *Yunnan Zhiwu Yanjiu* **23**, 357 (2001).

[11] R. A. A. Mothana, R. Grunert, U. Lindequist, and P. J. Bednarski, *Pharmazie* **62**, 305 (2007).

[12] Atta-ur-Rahman, M. Alam, H. Nasir, E. Dagne, and A. Yenesew, *Phytochemistry* **29**, 1293 (1990).

[13] Atta-ur-Rahman, M. I. Choudhary, and A. Ata, *Heterocycles* **34**, 157 (1992).

[14] M. Sangare, F. Khuong Huu Laine, D. Herlem, A. Milliet, B. Septe, G. Berenger, and G. Lukacs, *Tetrahedron Lett.* **16**, 1791 (1975).

[15] J. Guilhem, *Tetrahedron Lett.* **16**, 2937 (1975).

[16] F. Noor-e-Ain, Studies on the Steroidal Alkaloids from *Buxus longifolia* and other Related Plants, Ph.D. Thesis, University of Karachi, Karachi, Pakistan, 1996.

[17] P. V. Demarco, E. Farkas, D. Doddrell, B. L. Mylari, and E. Wenkert, *J. Am. Chem. Soc.* **90**, 5480 (1968).

[18] Atta-ur-Rahman, S. Parveen, A. Khalid, A. Farooq, S. A. M. Ayatollahi, and M. I. Choudhary, *Heterocycles* **49**, 481 (1998).

[19] M. I. Choudhary, S. Shahnaz, S. Parveen, A. Khalid, S. A. M. Ayatollahi, Atta-ur-Rahman, and M. Parvez, *J. Nat. Prod.* **66**, 739 (2003).

[20] M. I. Choudhary, S. Shahnaz, S. Parveen, A. Khalid, A. M. Mesaik, S. A. M. Ayatollahi, and Atta-ur-Rahman, *Chem. Biodivers.* **3**, 1039 (2006).

[21] M. H. Meshkatalsadat, A. Mollataghi, and A. Ata, *Z. Naturforsch. B: Chem. Sci.* **61**, 201 (2006).

[22] Z. U. Babar, A. Ata, and M. H. Meshkatalsadat, *Steroids* **71**, 1045 (2006).

[23] B. Sener and Atta-ur-Rahman, *Rec. Trav. Chim. Pays-Bas* **115**, 103 (1996).

[24] Atta-ur-Rahman, F. Noor-e-Ain, M. I. Choudhary, Z. Parveen, S. Turkoz, and B. Sener, *J. Nat. Prod.* **60**, 976 (1997).

[25] Atta-ur-Rahman, F. Noor-e-Ain, R. A. Ali, M. I. Choudhary, A. Pervin, S. Turkoz, and B. Sener, *Phytochemistry* **32**, 1059 (1993).

[26] Atta-ur-Rahman, S. Naz, F. Noor-e-Ain, R. A. Ali, M. I. Choudhary, B. Sener, and S. Turkoz, *Phytochemistry* **31**, 2933 (1992).

[27] The Wealth of India, Council of Scientific and Industrial Research, Delhi, India, p. 252, 1948.

[28] Atta-ur-Rahman, E. Asif, S. S. Ali, H. Nasir, S. A. Jamal, A. Ata, A. Farooq, M. I. Choudhary, B. Sener, and S. Turkoz, *J. Nat. Prod.* **55**, 1063 (1992).

[29] Atta-ur-Rahman, M. I. Choudhary, S. Naz, A. Ata, S. Parveen, and M. Parvez, *Nat. Prod. Lett.* **11**, 111 (1998).

[30] Atta-ur-Rahman, S. Naz, A. Ata, and M. I. Choudhary, *Heterocycles* **48**, 519 (1998).

[31] Atta-ur-Rahman, S. Parveen, A. Khalid, A. Farooq, and M. I. Choudhary, *Phytochemistry* **58**, 963 (2001).

[32] Z. Voticky, O. Bauerova, V. Paulik, and L. Dolejs, *Phytochemistry* **16**, 1860 (1977).

[33] Z. Voticky, L. Dolejis, O. Bauerova, and V. Paulik, *Colln. Czech. Commun.* **42**, 2549 (1977).

[34] H. Liu, M. Ji, and J. Cai, *Arch. Pharm. Chem. Life Sci.* **339**, 675 (2006).

[35] Atta-ur-Rahman, "Handbook of Natural Products Data, Diterpenoid and Steroidal Alkaloids", vol. 1, Elsevier, Amsterdam, 1990.

[36] Y. X. Wang, Y. W. Liu, Y. H. Tan, and B. H. Sheng, *Acta Pharmacol. Sin.* **10**, 226 (1989).

[37] B. U. Khodzhaev, M. R. Khodzhaeva, and K. Ubaev, *Khim. Prir. Soedin.* **6**, 907 (1993).

[38] C. Fourneau, R. Hocquemiller, D. Guedon, and A. Cavé, *Tetrahedron Lett.* **38**, 2965 (1997).

[39] Atta-ur-Rahman, M. I. Choudhary, S. Naz, A. Ata, B. Sener, and S. Turkoz, *J. Nat. Prod.* **60**, 770 (1997).

[40] Atta-ur-Rahman, S. Naz, A. Ata, M. I. Choudhary, B. Sener, and S. Turkoz, *Nat. Prod. Lett.* **12**, 299 (1998).

[41] Atta-ur-Rahman, A. Ata, S. Naz, M. I. Choudhary, B. Sener, and S. Turkoz, *J. Nat. Prod.* **62**, 665 (1999).

[42] F. Loru, D. Duval, A. Aumelas, F. Akeb, D. Guedon, and R. Guedj, *Phytochemistry* **54**, 951 (2000).

[43] C. C. Udenigwe, A. Ata, and R. Samarasekera, *Chem. Pharm. Bull.* **55**, 442 (2007).

[44] A. Ata and C. C. Udenigwe, *Curr. Bioact. Compd.* **4**, 41 (2008).

[45] A. Ata and C. Iverson, *Nat. Prod. Res.* (2008).

[46] A. Enz, R. Amstutz, H. Boddeke, G. Gmelin, and J. B. Malonowski, *Prog. Brain Res.* **98**, 431 (1993).

[47] T. L. Rosenberry, *Adv. Enzymol. Relat. Areas Mol. Biol.* **43**, 103 (1975).

[48] K. S. Kosmulalage, S. Zahid, C. C. Udenigwe, S. Akhtar, A. Ata, and R. Samarasekera, *Z. Naturforsch. B: Chem. Sci.* **62**, 580 (2007).

[49] A. Ata, *in:* "Innovations in Chemical Biology" (B. Sener, ed.), p. 175. Springer, New York, 2008.

[50] T. Sauvaitre, M. Barlier, D. Herlem, N. Gresh, A. Chiaroni, D. Guenard, and C. Guillou, *J. Med. Chem.* **50**, 5311 (2007).

[51] D. Herlem, Alkaloids of *Buxus balearica* Wild. (Buxaceae) – Isolation and Structure Determination, Ph.D. Thesis, Science Faculty of the University of Paris, Paris, 1967.

[52] J. T. Shaw, W. L. Corbett, D. Layman, G. D. Cuny, and J. Kerschner, *J. Heterocycl. Chem.* **25**, 1837 (1988).

[53] A. Ata, Y. H. Win, D. Holt, P. Holloway, E. P. Segstro, and G. S. Jayatilake, *Helv. Chim. Acta* **87**, 1090 (2004).

CHAPTER **4**

Pandanus Alkaloids: Chemistry and Biology

Maribel G. Nonato[1,*], Hiromitsu Takayama[2] and
Mary J. Garson[3]

Contents

I. INTRODUCTION

The plant family Pandanaceae, otherwise known as the screw pine family, is comprised of four genera, *Freycinetia* (*ca.* 200 species), *Sararanga* (two species),

[1] Research Center for the Natural Sciences, College of Science, Graduate School, University of Santo Tomas, España, Manila 1015, Philippines

[2] Graduate School of Pharmaceutical Sciences, Chiba University, Chiba 263-8522, Japan

[3] School of Molecular and Microbial Sciences, The University of Queensland, Brisbane, Qld 4072, Australia

* Corresponding author.
E-mail address: pandans2001@yahoo.com (M.G. Nonato).

The Alkaloids, Volume 66
ISSN: 1099-4831, DOI 10.1016/S1099-4831(08)00204-6

Pandanus (1,2), and the latest addition to the family, the genus *Martellindendron* (seven species) (3–5). The king of the family, the genus *Pandanus*, is made up of about 700 known species distributed mainly in tropical and subtropical regions; 52 species of the genus are found in the Philippines. The word "pandans" comes from Malay, and is applied to any member of the family Pandanaceae, a family of arborescent or lianoid dioecious monocotyledons.

A certain degree of confusion is found in the nomenclature of *Pandanus* species, since several names are known for the same species (2). At the commencement of our research program on the genus *Pandanus*, little was known regarding the chemistry and biological activities of this genus, with only four species reported in the literature. From the constituents of *Pandanus* plants, essential oils from *Pandanus latifolius* (6), lignans and ionones (7) and essential oils (8) from *Pandanus tectorius*, lignans and benzofurans from *Pandanus odoratissimus* (9), as well as 4-hydroxybenzoic acid from *Pandanus odorus* (10) were previously characterized.

P. tectorius (11) and *P. latifolius* (6) were found to contain sterols and the terpene, linalool, respectively. Alkaloids were detected in *P. veitchii* (12) and *P. amaryllifolius* (13). *P. boninesis* Warb. is indigenous to Bonin island in Japan, and has been used as a roadside tree with the fruits used as food (14). Two novel triterpenoids were isolated from the leaves of *P. boninesis* and their structures were elucidated as (24S)-24-methyl-25,32-cyclo-5α-lanost-9(11)-en-3β-ol and (24S)-24-methyl-25,32-cyclo-cycloartan-3β-ol (15).

The genus *Pandanus* is present throughout Southeast Asia and Northern Australia. Fiber from the leaves is used to make mats and baskets throughout the region, while local tribes and native animals eat the nut kernel in the fruits (16). The most studied plant in the genus is *P. amaryllifolius* Roxb.

P. amaryllifolius (Pandanaceae) is the only reported *Pandanus* species with scented leaves (Figure 1) (2). This plant is also known as fragrant screw pine, toei hom (Thailand), pandan mabango (Philippines), pandan wangi (Malay), and daun pandan (Indonesia). The leaves are used as a food flavoring, and in traditional medicine in the Philippines, Thailand, and Indonesia. *P. amaryllifolius* is used popularly as a flavoring for rice because it emits a peculiar odor similar to "ambermohor" rice (17). The alkaloid 2-acetyl-1-pyrroline was identified as the flavoring agent (13,18). The leaves are used medicinally in Southeast Asia to refresh the body, reduce fever, and relieve indigestion and flatulence (19). The oil of the leaf is described as a purgative, as a treatment for leprosy, and as a stimulant and antispasmodic. It is also reported to be effective against headaches, rheumatism, and epilepsy, and as a cure for sore throat (20). The seeds are reported to strengthen the heart and liver, while the roots are used as a diuretic and an aphrodisiac (20,21). In Indonesia, the volatile oil of *P. amaryllifolius*, known as "pandan wangi," is used as a remedy for toothache, rheumatism, and as a tranquilizer (22). Hot water extracts of the root of this plant (reported as *P. odorus* Ridl.) show hypoglycemic activity, and 4-hydroxybenzoic acid has been isolated as the active principle (23,24).

The leaves contain essential oils, carotenoids, tocopherols and tocotrienols (25), quercetin (26), and non-specific lipid transfer proteins (27). A lectin,

Figure 1 *Pandanus amaryllifolius* Roxb. plant.

pandanin, was recently isolated from the saline extract of the leaves of *P. amaryllifolius* using ammonium sulfate precipitation affinity chromatography on mannose–agarose and molecular size exclusion by gel filtration. The unglycosylated protein pandanin exhibits hemagglutinating activity toward rabbit erythrocytes, and its activity could be reversed exclusively by mannose and mannan. Pandanin also possesses antiviral activities against the human viruses herpes simplex virus type-1 (HSV-1) and influenza virus (H1N1) with 3-day EC_{50} values of 2.94 and 15.63 µM, respectively (28).

Two other non-specific lipid transfer proteins were also isolated from the saline extract of the mature leaves of *P. amaryllifolius* using affinity chromatography on fetuin-agarose and Affi-gel Blue gel anion exchange chromatography as well as gel filtration. The proteins were demonstrated as non-glycoproteins, with a molecular mass of 18 and 13 kDa, respectively, comprising peptide subunits from 6.5 to 9 kDa in the forms of a heterodimer and a homodimer. The proteins exhibit weak to moderate hemagglutinating activity toward rabbit erythrocytes, however, this activity could not be reversed by mannose. They thus could be easily differentiated from the mannose-binding lectin even though all three proteins have subunits with similar molecular weight (27).

II. ISOLATION AND STRUCTURE ELUCIDATION

A. (±)-Pandamarine from *Pandanus amaryllifolius* (29)

Phytochemical screening of Philippine medicinal plants revealed the presence of alkaloids in the leaves of *P. amaryllifolius* as detected by the Culvenor–Fitzgerald test. A specimen collected from Isabela, Philippines was extracted with ethanol, and the 5% aq. H_2SO_4 fraction basified with aq. NH_3 and extracted into chloroform, yielding a gum, which crystallized on standing. The crystals were identified as ($\pm$)-pandamarine (**1**); this component was the major alkaloid isolated from the leaves of the plant sample. In the ^1H NMR spectrum, signals at δ 6.60 (bs), 6.72 (bs), and 5.19 (t, J=8 Hz) identified three trisubstituted alkenes, as did ^{13}C signals at 117.1 (d), 133.6 (s), 134.4 (d), 135.4 (s), 138.9 (s), and 148.0 (d) ppm. The compound also possessed two methyl substituents (^1H: δ 1.89 (bs) and 1.78 (d, J=1.5 Hz)) and two amide groups (^{13}C: δ 175.2 (s) and 175.0 (s)). A signal at δ 80.0 was assigned to the spiro carbon center. After recrystallization from a mixture of MeOH, EtOAc, and ether, the complete structure of pandamarine (m.p. 210–211°C) was determined by an X-ray diffraction study. Diffractometer data at 295 K were refined by least squares techniques to a residual of 0.049 2244 "observed" reflections. Crystals of ($\pm$)-pandamarine were triclinic, and of space group P 1, a=13.077(2), b=9.857(5), c=7.214(2) Å, α=106.91(3), β=96.22(2), γ=100.01(2)°, Z=2 (29). From the X-ray structure, it was shown that pandamarine contains two γ-alkylidene-α,β-unsaturated lactam moieties, with a piperidine ring in a chair conformation and perpendicular to one of the lactam rings. The C-5/C-6 alkene bond has a Z-configuration. Full ^1H and ^{13}C assignments were not provided for this compound since HMBC data were not available (29).

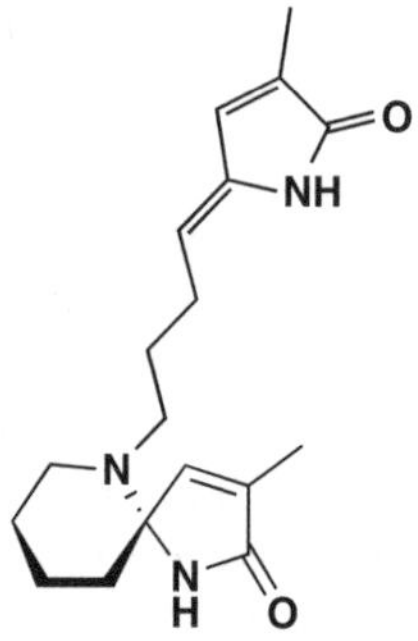

($\pm$) Pandamarine (**1**)

B. Pandamarilactones (30)

Attempts to identify other alkaloids in the leaves of *P. amaryllifolius* afforded three novel alkaloids known as pandamarilactones. The alcohol extract of *P. amaryllifolius* leaves collected from Manila, Philippines, on concentration under vacuum, yielded a dark green resinous material. Partitioning between Et_2O and 5% H_2SO_4 gave a green resinous organic extract. Alkalinization of the green resinous extract, and subsequent extraction with $CHCl_3$, gave a brownish-green resinous material. A series of chromatographic purifications using $CHCl_3$ with increasing concentrations of MeOH, and reverse-phase HPLC eluting isocratically in 80% MeCN in H_2O, gave a yellow amorphous solid, which was identified as pandamarilactone-1 (**2**). Two other alkaloids pandamarilactone-32 (**3**) and pandamarilactone-31 (**4**) were purified by reverse-phase HPLC using a linear gradient from H_2O to MeCN with UV detection at 254 nm (30).

Pandamarilactone-1 (**2**) was optically active with $[\alpha]_D^{23} = -35$ (*c*. 1.00, MeOH). High-resolution mass spectrometry gave an m/z of 317.1635 and established the molecular formula as $C_{18}H_{23}NO_4$, which corresponds to pandamarine ($C_{18}H_{25}N_3O_2$), but with two nitrogen atoms instead of two oxygen atoms. The IR spectrum of **2** showed absorption bands indicative of the α,β-unsaturated five-membered ring lactone at 1764 cm^{-1}, and the enol ester structure at 1695 cm^{-1}. These IR data indicated that pandamarilactone-1 has the structure **2**, in which the two lactam rings of **1** were instead replaced by lactone rings. The ^1H NMR spectrum was in agreement with the proposed structure for **2** and was similar to that of pandamarine (**1**). All other 2D NMR data supported the structure of **2**. ROESY data revealed a correlation between H-4 at δ 6.98 and H-6 at δ 5.04, confirming the Z geometry of the C-5/C-6 double bond.

Pandamarilactone-32 (**3**), for which an $[\alpha]_D$ was not recorded, was obtained as a white amorphous solid and was the major alkaloid based on its yield. The alkaloid purified by reverse-phase HPLC gave an m/z of 299.1521 by high-resolution EI mass spectrometry, consistent with the molecular formula $C_{18}H_{21}NO_3$ for nine double bond equivalents. The ^1H and ^{13}C NMR spectra revealed identical signals to those of **2** for the upper part of the structure, but that there were different signals for the lower portion. The UV absorption at λ_{max} 325 nm was in agreement with an α,β-unsaturated five-membered ring carbonyl with a nitrogen auxochrome at the β-position (31), while the absorption at 279 nm was in accordance with a five-membered lactone, as in **2**. The IR spectrum revealed the presence of carbonyl signals at 1764 cm^{-1} and 1710 cm^{-1} for a vinyl ester and an enol ester, respectively. A signal at 1667 cm^{-1} was suggestive of a vinylogous amide. The lower ring system had four quaternary carbons (three olefinic carbons and one carbonyl carbon), and five methylenes, of which the one at δ 110.8 (C-19) was an alkene based on the APT ^{13}C spectrum. Important ^{13}C signals at δ 168.0 and 188.0, were assigned to the β-carbon and the carbonyl carbon of a β-substituted vinylogous amide, respectively (32). HMBC data provided the structural information that established the lower part of Pandamarilactone-32. The exo-methylene signals for H-19 at δ 5.89 and 5.14 and the two-proton singlet at δ 3.16 (H-16) all correlated with the signals at δ 188.0 (C-18) and δ 141.0 (C-17). The H-16 signal also correlated to carbons at δ 113.0 (s) and 168.0 (s) assigned to C-14 and C-15, respectively, while the H-19 protons correlated to C-16 at δ 30.9. Pandamarilactone-32 thus contained a N-alkyl-6-methylene-hexahydro-5H-pyrindin-5-one unit. A Z geometry of the C-5/C-6 alkene was revealed by the ROESY correlation between H-4 at δ 6.98 and H-6 at δ 5.09.

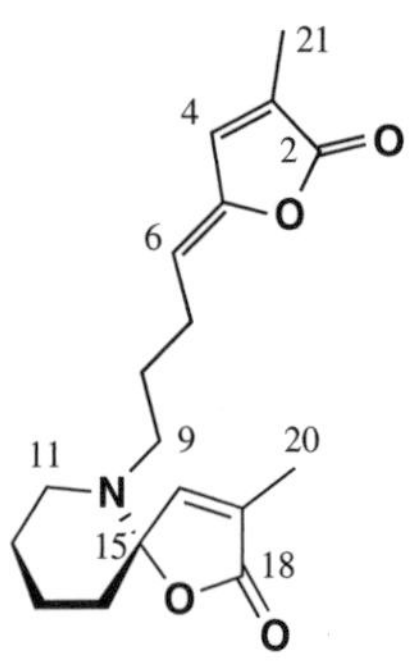

Pandamarilactone-1 (2) 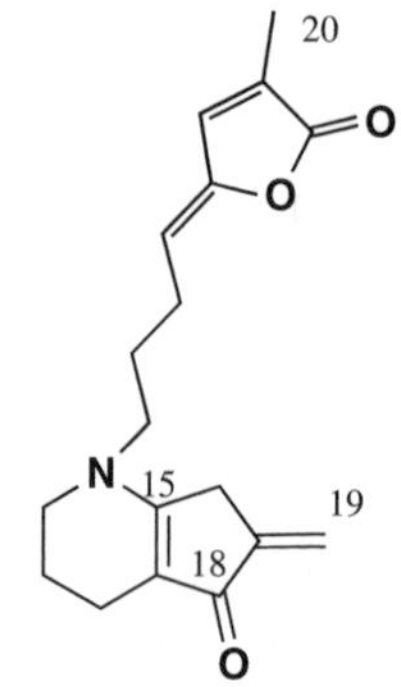   Pandamarilactone-32 (3) 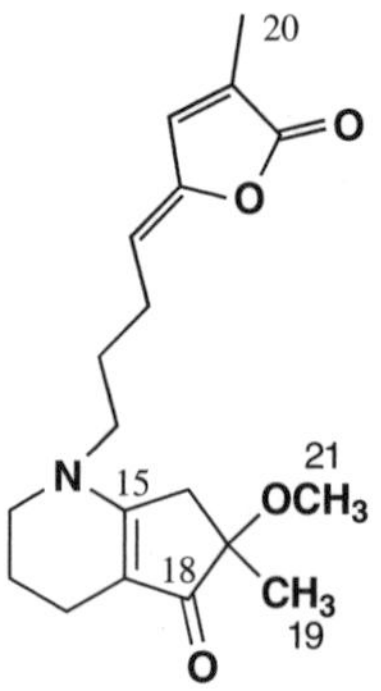   Pandamarilactone-31 (4)

Pandamarilactone-31 (4), $[\alpha]_D^{23}=-2.0$ (c. 1.0, CHCl$_3$) showed the presence of a vinyl lactone from its UV absorption at 295 nm. From its HREI mass spectrum of m/z 331.1940 for $C_{19}H_{25}NO_4$, the addition of methanol to the molecular formula of 3 was apparent, resulting in eight double bond equivalents. The ^{1}H NMR spectrum showed similar resonances with 3, but lacked the exo-methylene proton signals corresponding to H-19. Instead, four new proton resonances were observed. These were the two signals at δ 2.69 (d, J=17.0 Hz) and 2.45 (d, J=17.0 Hz) for the non-equivalent geminal protons, the methoxy signal at δ 3.20 (s, 3H) and a methyl singlet at δ 1.37. Comparison of the NMR data of 4 with 3 together with 2D NMR experiments confirmed the structure of pandamarilactone-31. In particular there were correlations between the H-16 protons and carbons at δ 188.0 (C-18) and at δ 23.0 (C-19), and between the methoxy or methyl proton signals and the quaternary carbon at δ 80.0 (C-17). Based on ROESY data, the stereochemical assignment at C-5/C-6 was again Z as in alkaloids 1–3. The absolute configuration at C-17 was not defined. Tables I and II present the ^{1}H and ^{13}C NMR assignments for 2–4.

Pandamarilactonine-31 (4) was unlikely to be an isolation artifact since an alternative orientation for the addition of methanol to the reactive enone moiety of pandamarilactonine-32 (3) would have been expected on electronic grounds.

All three pandamarilactones, together with pandamarine (1), derive from a C9–N–C9 precursor with modifications around the lower portion of the molecules. The skeletal structure of alkaloids 3 and 4 had not previously been reported in the literature. The five-membered lactone ring in the alkaloids 2–4, or the lactam ring of alkaloid 1, are most likely to be derived from 4-hydroxy-4-methyl glutamic acid. This amino acid has been identified in *P. veitchii* (33) and its biosynthesis has been studied by Peterson and Fowden (34). Further discussion of the biosynthetic pathway that may lead to alkaloids 3 and 4 is provided in Section IV.

C. Pyrrolidinones (35)

The isolation of novel piperidine alkaloids from the leaves of *P. amaryllifolius* collected in various parts of the Philippines prompted Garson et al. (35) to look at

Table I ¹H NMR spectral data of alkaloids **2–4**[a]

#	**2**[b]	**3**	**4**
4	6.98, ddd, 1.2, 1.0	6.98, ddd, 2.0, 1.0	6.99, ddd, 1.2, 1.0
6	5.04, ddd, 8.0, 1.0	5.09, ddd, 8.0, 1.0	5.10, dd, 8.0, 1.0
7	2.31, ddd, 8.0, 7.2	2.39, ddd, 8.0, 8.0	2.39, ddd, 8.0, 7.6
8	1.54, dddd, 8.0, 7.2	1.77, dddd, 8.0, 8.0	1.76, dddd, 7.6, 7.6
9	2.45, dd, 1.0	3.30, m	3.26 dd, 7.6
11	2.79, dd, 7.0	3.27, m	3.25, m
12	1.72, m	1.85, m	1.80–1.90, m
13	1.72, m	2.33, dd, 6.0	2.25, br dd, 4.8, 1.2
14	1.72, m		
16	6.68, dd, 1.2	3.16, br s	2.69, d, 17.0
			2.45, d, 17.0
19		5.89, ddd, 1.0, 1.0	1.37, s
		5.14, ddd, 1.0, 1.0	
20	1.86, br s	2.00, d, 2.0	2.00, d, 1.2
21	2.00, br s		3.20, s

[a]400 MHz; solution in CDCl₃ referenced at δ 7.25.
[b]*J* values in Hz.

Table II ¹³C NMR spectral data of alkaloids **2–4**[a]

#	**2**	**3**	HMBC correlations	**4**
2	171.0, s	171.0, s	H20, H4	171.0, s
3	129.2, s	129.0, s	H20, H4	130.0, s
4	137.6, d	137.0, d	H20, H6	137.0, d
5	148.6, s	149.0, s	H4, H6	149.0, s
6	113.7, d	111.6, d	H7, H4	112.0, d
7	24.0, t	23.6, t	H6, H8	23.0, t
8	27.2, t	27.8, t	H7, H9	28.0, t
9	50.7, t	50.9, t	H11	51.0, t
11	47.2, t	48.0, t	H9	48.0, t
12	20.8, t	21.1, t	H11, H13	21.0, t
13	25.1, t	18.0, t	H12	18.0, t
14	36.2, s	113.0, s	H16, H13, H12	108.0, s
15	101.7, s	168.0, s	H16, H13, H11, H9	168.0, s
16	149.7, d	30.9, t	H19	37.0, t
17	131.5, s	141.0, s	H16, H19	80.0, s
18	173.0, s	188.0, s	H16, H19	188.0, s
19		110.8, t	H16	23.0, q
20	10.6, q	10.6, q	H4	10.4, q
21	10.4, q			52.0, q

[a]100 MHz; solution in CDCl₃ referenced at δ 77.0.

Pandanus species from other regions of Southeast Asia. Plant samples from Jember, East Java and from Jambi, Sumatra, Indonesia were both found to contain alkaloids.

Acid–base extraction and chromatographic separations yielded two new alkaloids named as pandamarilactam-3x (**5**) and -3y (**6**). These alkaloids were related structurally to the pandamarilactones in that they also contained the γ-alkylidene-α,β-unsaturated lactone moiety, but they differed in the lower part where there was a pyrrolidinone ring in each alkaloid instead of the lactone ring that had previously been found. Both pyrrolidinones were isolated from the chloroform fraction of an aqueous ethanolic extract from *P. amaryllifolius* leaves collected near Jambi, and purified by reverse-phase flash chromatography then by reverse-phase HPLC using 80% CH_3CN in H_2O (35).

The molecular ions of pandamarilactam-3x (**5**) and pandamarilactam-3y (**6**) both corresponded to the molecular formula $C_{13}H_{17}NO_3$ with six double bond equivalents. DQFCOSY and HMQC data afforded assignments for the γ-alkylidene-α,β-unsaturated lactone moiety of each compound, while HMBC cross-peaks together with the DQFCOSY data enabled the formation of the fragment containing C-6 to C-9. The pyrrolidinone ring containing the carbonyl at δ 175.0 and three methylenes at δ 47.2, 18.0, and 31.0 was also evident from the COSY and HMBC data. The connectivity of C-9 to the amide carbonyl of the pyrrolidinone ring and other long-range correlations shown in Table III were established through the HMBC data.

Table III ^1H and ^{13}C NMR assignments for pandamarilactam-3x (**5**) and -3y (**6**)

#	5		6		HMBC
	$^1H^{a,b}$	$^{13}C^c$	$^1H^{a,b}$	$^{13}C^c$	
2	–	171.0	–	171.0	H4, H15
3	–	131.0	–	129.4	H15
4	7.30 (1.0, 0.5)	135.6	6.98 (1.0, 0.5)	137.7	H6, H15
5	–	148.5	–	148.8	H4, H6, H7
6	5.58 (8.5, 1.0)	112.3	5.18 (7.6, 1.0)	112.8	H7, H8
7	2.25 (8.5, 7.6)	24.0	2.30 (7.6, 7.0)	23.4	H8, H9
8	1.69 (7.6, 7.1)	27.4	1.72 (7.4, 7.0)	26.7	H6, H7, H9
9	3.31 (7.1)	42.0	3.30 (7.4)	41.8	H7, H8
11	3.37 (7.0)	47.1	3.37 (7.1)	47.2	H9, H12, H13
12	2.04 (8.0, 7.0)	17.4	2.02 (7.9, 7.1)	18.0	H11, H13
13	2.37 (8.0)	31.0	2.35 (7.8)	31.0	H11, H12
14	–	175.0	–	175.0	H9, H11, H12, H13
15	2.03 (0.5)	10.8	1.98 (0.5)	10.5	H4

^{a}500 MHz; solution in $CDCl_3$ referenced at δ 7.25.
^{b}*J* values (Hz) in brackets.
^{c}Inverse detection at 500 MHz; solution in $CDCl_3$ referenced at δ 77.0.

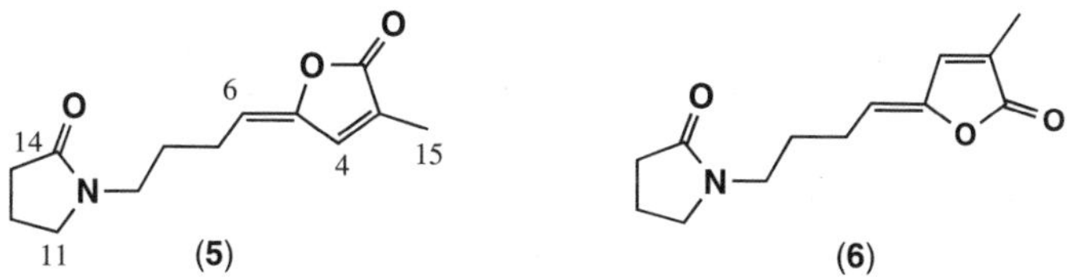

Pandamarilactam-3x (*E*-isomer) (**5**)
Pandamarilactam-3y (*Z*-isomer) (**6**)

The minor alkaloid pandamarilactam-3x (**5**) showed ^{1}H and ^{13}C spectra, which were closely similar to those of pandamarilactam-3y (**6**), except for the alkene signals at H-4 and H-6 which resonated at δ 7.30 and 5.58, respectively, instead of at δ 6.98 and 5.18 in **6**. The H-6 signal of pandamarilactone-1 (**2**), for which Z stereochemistry has been determined by a ROESY experiment, resonated at δ 5.04, which was only 0.14 ppm upfield of the chemical shift value for H-6 in (**6**), but was 0.54 ppm upfield of the H-6 signal in (**5**). In nuclear Overhauser effect (NOE) experiment on a mixture of pandamarilactams-3x and -3y, irradiation of the δ 5·18 signal of alkaloid **6** resulted in enhancement of H-4 at δ 6.98, but there was no NOE enhancement of H-4 at δ 7.30 when the δ 5.58 signal of alkaloid **5** was irradiated. Thus, alkaloids **5** and **6** were the *E*- and *Z*-isomers, respectively. The carbon signal for C-4 of **5** was at δ 135.6 compared to δ 137.7 in **6**.

D. Pandamarilactonines (36–43)

Fresh young leaves of *P. amaryllifolius*, purchased at the flower market in Bangkok, Thailand, afforded two new alkaloids, pandamarilactonine-A (**7**) and pandamarilactonine-B (**8**), both possessing a pyrrolidinyl α,β-unsaturated-γ-lactone residue and a γ-alkylidene-α,β-unsaturated-γ-lactone residue. The alkaloids were isolated, together with the known alkaloid pandamarilactone-1 (**2**), by a series of SiO$_2$ column chromatography procedures on the alkaloidal fraction obtained by acid–base extraction of an EtOH extract of the sample (**36**).

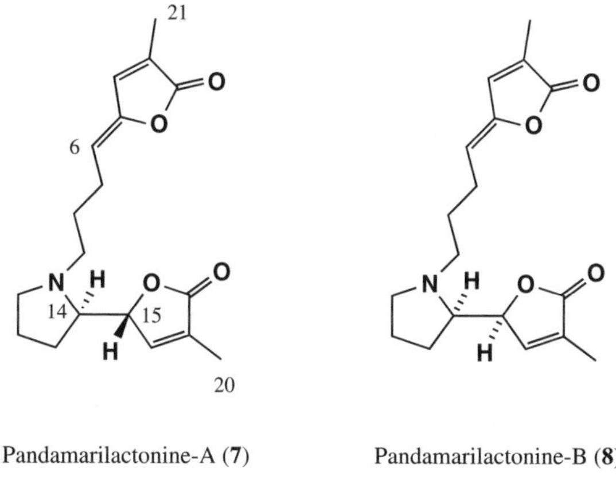

Pandamarilactonine-A (**7**)   Pandamarilactonine-B (**8**)

Pandamarilactonine-A (**7**) was obtained as an amorphous powder and was optically active with $[\alpha]_D^{23}$=+35 (c. 4.37, CHCl$_3$). The high-resolution FAB-MS gave an m/z 318.1721 [M+H]$^+$ for C$_{18}$H$_{23}$NO$_4$ suggesting that **7** was an isomer of the co-occurring alkaloid pandamarilactone-1 (**2**). The γ-alkylidene-α-methyl-α,β-unsaturated-γ-lactone moiety of **7** was suggested by UV absorption at 275 nm and was constructed from the diagnostic ^{1}H and ^{13}C NMR signals that were closely similar to those of **1–4**. The three-carbon methylene chain C-7 through C-9 was connected to C-6 of the γ-alkylidene-α-methyl-α,β-unsaturated-γ-lactone moiety through the analyses of the COSY, HMQC, and HMBC data. Likewise, the presence of the α-methyl-α,β-unsaturated-γ-lactone residue was recognized from the ^{1}H and ^{13}C NMR signals that were similar to those found in pandamarilactone-1 (**2**). The residual four carbons comprising three methylenes (C-11 to C-13), one methine (C-14), and the sole nitrogen atom were suggested to comprise a pyrrolidine ring. The downfield proton signals at δ 2.83 (m) and at δ 4.80 (ddd, J=1.8, 1.8, 5.5 Hz) were assigned to H-14 and H-15, respectively. The HMBC spectrum showed correlations from H-14 of the pyrrolidine ring to the methylene carbon at δ 55.0 assigned to C-9 and to the signal at δ 147.0 assigned to C-16 of the α,β-unsaturated-γ-lactone ring. In addition, H-15 of the γ-lactone ring showed connectivity with C-14 and C-13 (δ 23.8) in the pyrrolidine ring. The Z-configuration of the γ-alkylidene-α,β-unsaturated-γ-lactone moiety was apparent from the correlation between H-4 and H-6 in the NOESY spectrum. All the observed spectral data pointed to the structure **7** for the new alkaloid, having a novel pyrrolidinyl α,β-unsaturated-γ-lactone skeleton, except that the relative configuration of the vicinal asymmetric centers (C-14/C-15) remained to be determined.

The isomer pandamarilactonine-B (**8**) was also obtained as an amorphous powder but was found to be optically inactive with $[\alpha]_D$ ^{23}=0 (c. 0.20, CHCl$_3$). Analyses of the UV, mass, and ^{1}H and ^{13}C NMR spectra suggested that **8** had an identical carbon skeleton to **7**. When an NOE difference experiment revealed a Z-configuration for the γ-alkylidene-α,β-unsaturated-γ-lactone moiety of **8**, it became apparent that **7** and **8** were diastereomers that differed in configuration at either C-14 or at C-15. Thus, **7/8** represented a *threo/erythro* pair of stereoisomers. Complete ^{1}H and ^{13}C assignments for the diastereomers **7** and **8** are presented in Table IV. In particular the downfield proton signals at δ 2.70 (m) and at δ 4.71 (ddd, J=1.7, 2.0, 5.9 Hz) were assigned to H-14 and H-15, respectively, of **8**.

In synthetic work on alkaloids of the Stemonaceae family, Martin et al. (37) assigned the relative configuration of *threo/erythro* diastereomers with a pyrrolidinyl α,β-unsaturated-γ-lactone skeleton by comparison of their ^{1}H and ^{13}C and ^{1}H chemical shifts at positions equivalent to C-14 and C-15 of the pandamarilactonines. However, the chemical shifts values for H-14/H-15 and for C-14/C-15 of **7** and **8** were closely similar, and so could not be used to define the stereochemistry of **7** and **8**. Instead, the use of J-resolved configurational analysis was applied. Three possible staggered conformations for each diastereomer (*erythro*: E1, E2, and E3; *threo* T1, T2, and T3) are shown in Figure 2. The NMR evidence required to distinguish these stereochemical possibilities relies on measurement of $^3J_{H-H}$ values, together with $^3J_{C-H}$ values from PFG J-HMBC 2D

Table IV ¹H and ¹³C assignments for pandamarilactonine-A **(7)** and -B **(8)**

#	7		8	
	¹H $\delta^{a,b}$	¹³C^c	¹H $\delta^{a,b}$	¹³C^c
2		171.1		171.1
3		129.1		129.1
4	6.99 (1H, d, 1.5)	137.7	7.00 (1H, d, 1.5)	137.7
5		148.6		148.5
6	5.18 (1H, dd, 7.9, 7.9)	114.1	5.18 (1H, dd, 7.8, 8.0)	114.1
7	2.43 (2H, dd, 7.3, 15.0)	24.0	2.42–2.48 (2H, m) 2.36 (1H, m)	24.0
8	1.59–1.70 (2H, m)	28.3	1.59–1.67 (2H, m).	28.4
9	2.88 (1H, ddd, 4.0, 7.9, 12.9) 2.45 (1H, m)	55.0	2.73 (1H, m) 2.42–2.48 (1H, m)	55.8
11	3.12 (1H, dd, 6.7, 7.6) 2.21 (1H, m)	54.2	3.12 (1H, m) 2.25 (1H, m)	54.2
12	1.70–1.80 (1H, m) 1.42 (1H, m)	25.7	1.73–1.87 (2H, m)	27.1
13	1.70–1.80 (1H, m) 1.59–1.70 (1H, m)	23.8	1.73–1.87 (2H, m)	24.0
14	2.83 (1H, m)	65.3	2.70 (1H, m)	66.3
15	4.80 (1H, ddd, 1.8, 1.8, 5.5)	83.4	4.71 (1H, ddd, 1.7, 2.0, 5.9)	83.4
16	7.09 (1H, dd, 1.5, 1.8)	147.0	7.05 (1H, dd, 1.5, 1.7)	147.5
17		131.2		130.8
18		174.3		174.3
20	1.93 (3H, dd, 1.5, 1.8)	10.7	1.93 (3H, dd, 1.7, 1.7)	10.8
21	1.99 (3H, d, 0.9)	10.5	1.99 (3H, d, 0.7)	10.5

^{a}500 MHz, CDCl$_3$.
^{b}*J* values (Hz) in brackets.
^{c}125 MHz, CDCl$_3$.

spectroscopy (38), which enables measurement of the torsion angle between two heteroatoms, such as *H–C–C–C* (39), in combination with NOE difference measurements that confirm spatial relationships in individual rotamers. In acetone, medium-sized coupling constants were observed between H-14 and H-15 in each diastereomer (**7**, *J*=4.9 Hz; **8**, *J*=3.9 Hz). A large coupling constant (*J*=5.6 Hz) between H-14 and C-16 and a small coupling constant (*J*=3.7 Hz) between H-15 and C-13 were measured for pandamarilactonine-A (**7**) by PFG J-HMBC 2D spectroscopy, indicating their *anti* and *gauche* orientations, respectively (40–42). For pandamarilactonine-B (**8**) small coupling constants of 2.7 Hz ($^3J_{CH}$ H-14/C-16) and 3.3 Hz ($^3J_{CH}$ H-15/C-13), supported the *gauche* relationships of both H-14/C-16 and H-15/C-13. In alkaloid **7**, NOEs were observed between H-13 and H-16, while in alkaloid **8** there were NOEs between H-14 and H-16. Initially, these data were interpreted in favour of **7** and **8** as the *erythro* and *threo* isomers, respectively, with rotamers E1 and T1 as the

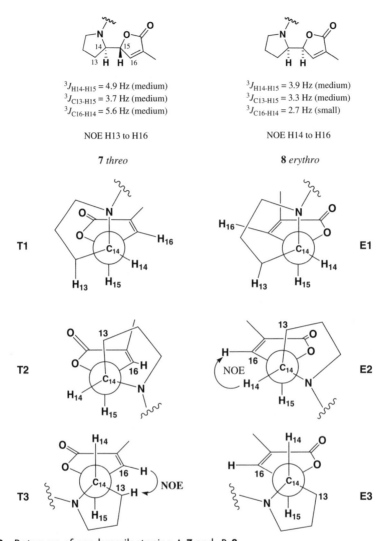

Figure 2 Rotamers of pandamarilactonine-A **7** and -B **8**.

predominant conformers, respectively (36). However, the NMR data measured in CDCl$_3$ hinted that the conformational situation was more complex since the $^3J_{H-H}$ values (**7**, J=5.5 Hz; **8**, J=5.9 Hz) were larger than those measured in acetone. This suggested that the rotamers E3 and T3, in both of which H-14 and H-15 are diaxial, contributed to the overall conformational equilibrium. A synthetic study of racemic pandamarilactonines-A and -B was undertaken as described in detail later (in Section III, A) and showed that pandamarilactonine-A was in fact the *threo* isomer, as shown above for **7**, while pandamarilactonine-B was the *erythro* isomer **8** (43). With this knowledge in hand, the conformational picture for pandamarilactonine-A could be reinterpreted as a mixture of rotamers T2 and T3, and for pandamarilactonine-B as a mixture of rotamers E2 and E3.

By use of these conformational preferences, the NOEs between H-13 and H-16 in 7 (rotamer T3) and between H-14 and H-16 in 8 (rotamer E2) were explained as shown in Figure 2.

Pandamarilactonine-A (7) exhibited $[\alpha]_D^{23}$ +35.0, in contrast, the specific rotation of pandamarilactonine-B (8) was almost zero. The optical purity of 7 and 8 was later studied using racemic samples prepared in a synthetic study (Section III, A). Chiral HPLC analysis of synthetic pandamarilactonine-A (7) yielded two peaks at retention times of 43.2 and 51.7 min with Chiralcel OB while synthetic pandamarilactonine-B (2) showed two peaks at 18.7 and 20.4 min using Chiralcel OD. The natural pandamarilactonine-A sample contained predominantly the (+) enantiomer in a ratio of 63:37 while the natural 8 was a racemate (36). Finally, the absolute configuration of (+)-pandamarilactonine-A has been established as (14*R*, 15*R*) by chiral synthesis as described in Section III A (44). The absolute configuration of (+)-pandamarilactonine-B is as yet unknown; consequently the structure 8 shown implies relative stereochemistry alone.

The synthetic work of Takayama *et al.* (Section III, A) (36) implied that there might be an acid-catalyzed interconversion of 7 and 8 during the isolation process so that 8 might be an isolation artifact of 7. The C-15 position of the α-methyl-α,β-unsaturated-γ-lactone fragment would be sensitive to epimerization under either acidic or basic conditions (45), which suggested that the two alkaloids differed in configuration at C-15 rather than at C-14. However, when the natural alkaloids 7 and 8 were treated with 5% H_2SO_4 under the same conditions as used for the acid–base partitioning of the alkaloids, NMR of the recovered products did not reveal any interconversion of the two alkaloids. It was therefore concluded that 8 was not an isolation artifact since the optical purity of the recovered 7 was unchanged (36).

Subsequent work by Takayama *et al.* (43) provided two additional pandamarilactonine metabolites pandamarilactonine-C (9) and pandamarilactonine-D (10) from the sample of *P. amaryllifolius* that had earlier yielded 7 and 8. As before (36), the crude alkaloid fraction was separated by SiO_2 column chromatography using a $CHCl_3$/MeOH gradient, with the 2–5% MeOH/$CHCl_3$ eluate then subjected to SiO_2 medium pressure liquid chromatography using 2% EtOH/$CHCl_3$. Pandamarilactonine-C (9) was obtained as an amorphous powder with $[\alpha]_D^{23}$ +26.2 (*c*. 0.99, $CHCl_3$) while Pandamarilactonine-D (10) was an amorphous powder, $[\alpha]_D^{25}$ 0 (*c*. 0.21, $CHCl_3$).

These two new alkaloids gave respectively m/z 318.1691 [M+H]$^+$ (calcd 318.1704) and m/z 318.1704 [M+H]$^+$ (calcd 318.1704) by HRFAB-MS (NBA). This established the molecular formulae of the two alkaloids as $C_{18}H_{23}NO_4$, and indicated that they were isomers of the co-occurring alkaloids, pandamarilactonine-A (7) and -B (8) (36). The spectroscopic data of 9 and 10 were similar to 7 and 8, revealing an identical carbon skeleton, except that there were upfield shifts in the signals for C-4 in 9 (δ 130.3) and 10 (δ 133.9) compared to those of 7 (δ 137.7) and 8 (δ 137·7). These data were rationalized in terms of the γ-gauche effect of C-7 on C-4 arising from the *E*-configuration of the γ-alkylidene-γ-lactone moiety, an interpretation that was supported by the NOE observed between H-4 and H-7 in both 9 and 10. Using COSY, HMQC, and HMBC

Table V ^{1}H and ^{13}C assignments for pandamarilactonine-C (**9**) and -D (**10**)

#	9		10	
	^{1}H^{a,b}	^{13}C^c	^{1}H^{a,b}	^{13}C^c
2		171.0		171.0
3		129.1		130.3
4	7.31 (1H, dd, 0.9, 1.5)	130.3	7.35 (1H, m)	133.9
5		148.7		149.0
6	5.64 (1H, dd, 8.2, 8.5)	113.4	5.60 (1H, dd, 8.3, 8.8)	113.3
7	2.35 (1H, ddd, 7.0, 8.2, 14.7)	24.2	2.29 (2H,m)	24.0
	2.25 (1H, m)			
8	1.62–1.68 (2H, m)	29.1	1.63 (2H, m).	28.7
9	2.91 (1H, ddd, 7.9, 8.2, 11.9)	55.2	2.77 (1H, m)	55.1
	2.45 (1H, ddd, 5.8, 7.0, 11.9)		2.42 (1H, m)	
11	3.12 (1H, d-like, 6.7, 7.3)	54.2	3.08 (1H, m)	54.0
	2.25 (1H, m)		2.25 (1H, m)	
12	1.71–1.82 (1H, m)	26.1	1.78–1.92 (2H, m)	26.3
	1.45 (1H, m)			
13	1.71–1.82 (1H, m)	23.8	1.78–1.92 (2H, m)	23.9
	1.62–1.68 (1H, m)			
14	2.78 (1H, m)	65.6	2.72 (1H, m)	65.9
15	4.80 (1H, ddd, 1.8, 1.8, 5.5)	83.8	4.74 (1H, m)	83.0
16	7.04 (1H, d, 1.5)	146.7	6.99 (1H, dd, 1.5, 1.7)	147.1
17		131.3		131.0
18		174.2		174.3
20	1.94 (3H, d, 1.8)	10.8	1.94 (3H, dd, 1.5, 2.0)	10.8
21	2.02 (3H, d, 0.9)	10.8	2.01 (3H, m)	10.7

^{a}500 MHz, CDCl$_3$.
^{b}J values (brackets) in Hz.
^{c}125 MHz, CDCl$_3$.

experiments, the proton and carbon signals of **9** and **10** were unambiguously assigned, as shown in Table V.

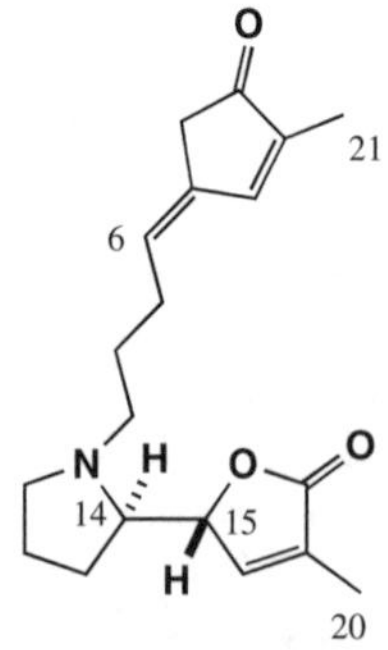

Pandamarilactonine-C (**9**)   Pandamarilactonine-D (**10**)

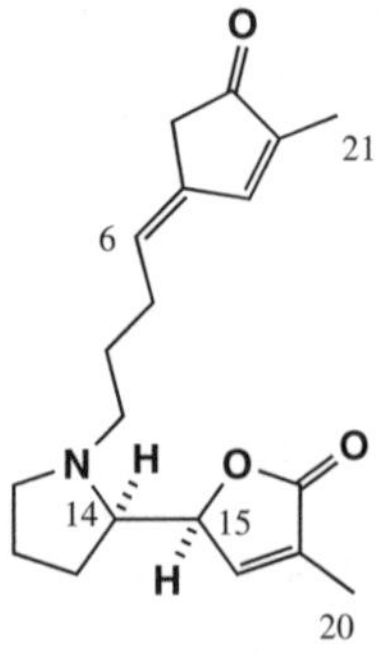

In CDCl₃, alkaloid **9** showed a coupling of 5.5 Hz between H-14 and H-15, suggesting a multi-conformer equilibrium, while the multiplet appearance of the H-15 signal in alkaloid **10** prevented measurement of the equivalent $^3J_{H-H}$ value. Confirmation of the relative configuration of the vicinal asymmetric centers at the C-14 and C-15 positions of **9** and **10** therefore involved a total synthesis that is detailed in the next section. Working in the *threo* series, the final product of the synthesis was found to be a mixture of *E*- and *Z*-pandamarilactonines those were identical to pandamarilactonine-C (**9**) and pandamarilactonine-A (**7**), respectively. Thus, pandamarilactonine-C and pandamarilactonine-D were *threo* and *erythro* isomers, respectively. The optical activities of the *E*-isomers showed a similar trend to the *Z* isomers; the *threo* compound was optically active and enriched in the (+)-isomer, while the *erythro* compound was optically inactive. The structures shown for **9** and **10** imply relative configuration only.

E. Norpandamarilactonines (46)

In addition to the four pandamarilactonine metabolites, acid–base extraction of the EtOH extract of fresh young leaves of *P. amaryllifolius* obtained from a market in Bangkok, Thailand followed by chromatography gave two minor alkaloids named norpandamarilactonine-A (**11**) and norpandamarilactonine-B (**12**). The alkaloids were characterized by PFG J-HMBC 2D spectroscopy, and the structures then confirmed by total synthesis. Both norpandamarilactonines lacked optical activity from which it could be concluded that they were racemates. Thus, the structures shown for natural **11** and **12** imply relative configuration alone.

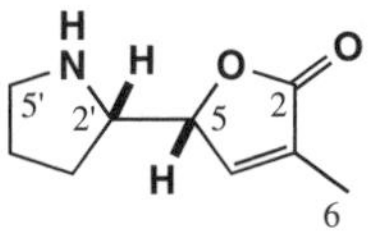

Norpandamarilactonine-A (**11**) 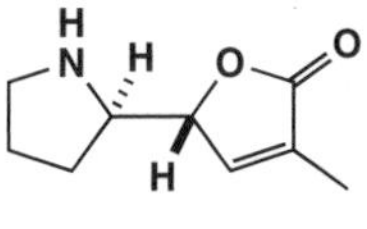   Norpandamarilactonine-B (**12**)

Norpandamarilactonine-A (**11**), obtained as an amorphous powder, gave *m/z* 168.1039 by HRFAB-MS (NBA) analysis to fit the molecular formula $C_9H_{13}NO_2$. The UV ((MeOH) λ_{max} (log ε) 274 (0.44), 252 (0.35), 207 (2.29) nm) and IR signal at 1750 cm^{-1} both suggested an α,β-unsaturated-γ-lactone. Characteristic signals in the ^1H and ^{13}C NMR spectra for the presence of an α-methyl-α,β-unsaturated-γ-lactone residue were observed while the pyrrolidine ring was constructed from the residual four carbons (three methylenes and one methine) and the sole nitrogen atom. The HMBC spectrum provided the connectivity of the methine proton at δ 3.18 for H-2′ with C-4 (δ 147.7) of the α,β-unsaturated-γ-lactone ring, while the methine proton (δ 4.73) for H-5 of the γ-lactone ring showed correlations to both C-2′ (δ 60·4) and C-3′ (δ 27·9) of the pyrrolidine ring. These data provided the complete structure except for the relative configuration of the

Table VI ^{1}H and ^{13}C assignments for norpandamarilactonine-A (**11**) and -B (**12**)

#	11		12	
	^{1}H^{a,b}	^{13}C^c	^{1}H^{a,b}	^{13}C^c
2		174.3		174.1
3		130.7		131.2
4	7.13 (1H, ddd, 0.8, 1.6, 1.6)	147.7	7.02 (1H, ddd, 1.4, 1.7, 3.0)	146.6
5	4.73 (1H, ddd, 1.6, 1.9, 6.6)	83.8	4.79 (1H, dddd, 1.6, 1.9, 3.0, 6.6)	84.3
2′	3.18 (1H, ddd, 6.6, 6.6, 7.4)	60.4	3.20 (1H, ddd, 6.6, 7.1, 7.4)	60.2
5′	2.96 (1H, ddd, 6.3, 6.3, 10.4)	47.1	2.98 (1H, ddd, 5.8, 7.1, 12.9)	46.5
	2.93 (1H, ddd, 6.8, 6.8, 10.4)		2.91 (1H, ddd, 6.6, 7.7, 14.3)	
6	1.93 (3H, s)	10.7	1.93 (3H, s).	10.7
3′	1.84–1.92 (1H, m)	27.9	1.87 (1H, dddd, 3.0, 7.4, 10.7, 15.4)	26.8
	1.63 (1H, dddd, 6.3, 6.3, 6.6, 12.9)		1.56 (1H, dddd, 5.2, 6.9, 7.1, 15.4)	
4′	1.72–1.90 (2H, m)	25.6	1.81 (1H, m)	25.1
			1.74 (1H, m)	

^{a}500 MHz, CDCl$_3$.
^{b}J values (brackets) in Hz.
^{c}125 MHz, CDCl$_3$.

vicinal asymmetric centers at C-2′ and C-5. Since the γ-alkylidene-α,β-unsaturated-γ-lactone unit that was present in the pandamarilactonines (**7–10**) was missing, the new alkaloid was named norpandamarilactonine-A (**11**) (46).

Norpandamarilactonine-B (**12**) was also obtained as an amorphous powder with UV and MS data almost identical to those of **11**. Analysis of the NMR data indicated that **11** and **12** were diastereomeric at the C-5 and C-2′ positions. In CDCl$_3$, both **11** and **12** showed a coupling of 6.6 Hz between H-2′ and H-5, and the chemical shift values of C-2′ and C-5′ were closely similar; the *erythro* and *threo* compounds could not be safely distinguished from these values. Thus, confirmation of the structures and relative configuration at C-5 and C-2′ of **11** and **12** was carried out by a total synthesis that is described in detail in Section III. In this way, norpandamarilactonine-A and -B were assigned *erythro* and *threo* stereochemistry, respectively. Table VI presents the ^{1}H and ^{13}C NMR assignments for **11** and **12**.

F. Pandanamine (47)

An alkaloidal extract prepared in the same way from fresh leaves of *P. amaryllifolius* was subjected to silica gel column chromatography. The polar fraction, obtained after elution of the fractions containing pandamarilactonines, was further purified using reverse-phase column chromatography to give the secondary amine pandanamine **13** as an amorphous powder in 0.21% yield from the crude alkaloid fraction.

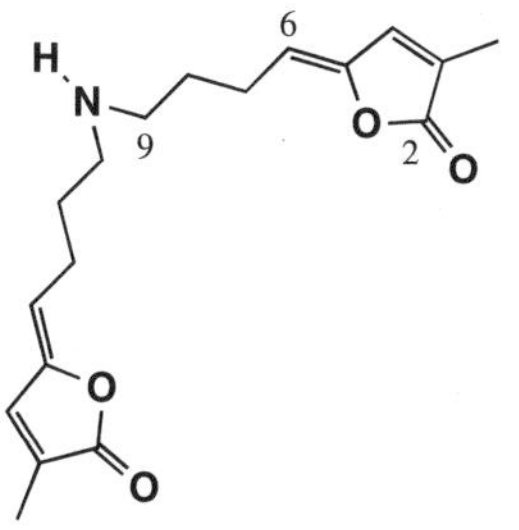

Pandanamine (**13**)

The new alkaloid **13** gave an m/z 318.1721 $[M+H]^+$ by HRFAB-MS for $C_{18}H_{23}NO_4$, and showed a UV maxima at 273 nm. The ^{13}C spectrum showed only nine signals indicative of a symmetrical structure. The characteristic ^1H and ^{13}C NMR signals for the γ-alkylidene-α-methyl-α,β-unsaturated-γ-lactone moiety [δ 7.03 (2H, d, J=1.5 Hz, H-4), 5.14 (2H, dd, J=7.9, 7.9 Hz, H-6), 1.99 (6H, s); δ 170.9 (C-2), 129.8 (C-3), 137.8 (C-4), 149.3 (C-5), 111.3 (C-6), 10.5 (C-21)] were observed and accounted for six of the nine carbons; the remaining carbons (δ 47.7, 254, and 23.1) were all methylenes with the signal at δ 47.7 assigned to C-9 on chemical shift grounds. Analysis of 2D spectra established that the signal at δ 23.1 (C-7) was connected to C-6 of the γ-alkylidene-γ-lactone moiety. The observation of an NOE between H-4 and H-6 demonstrated the Z-configuration in the γ-alkylidene-α,β-unsaturated-γ-lactone moiety. All of the above findings allowed the construction of the symmetrical molecular structure pandanamine **13** (47). This lactone compound was identical with a synthetic compound previously prepared as an intermediate for the synthesis of pandamarilactonines (36).

The isolation of pandanamine was of significant interest since its lactam equivalent had earlier been proposed as a biogenetic intermediate in the biogenesis of pandamarine **1** by Byrne *et al.* (29). Pandanamine (13) itself was also postulated as the common biogenetic precursor of pandamarilactonines-A and -B (**7** and **8**) and of pandamarilactone-1 (**2**). Thus, the isolation of this compound from Nature strongly supported the proposed biogenetic pathways to the alkaloids (**2**, **7**, and **8**). This is discussed further in Section IV.

G. Additional Pyrrolidine Alkaloids (48)

Salim *et al.* (48) isolated three new alkaloids, the two pyrrolidine-type alkaloids (**14** and **15**) together with the 6E-isomer of pandanamine (**16**), and five known alkaloids (**7–10** and **13**) from *P. amaryllifolius* leaves collected from West Java, Indonesia. Evidence that the three new alkaloids all possessed two α-methyl-α,β-unsaturated-γ-lactone functionality was apparent from the spectroscopic studies.

Alkaloid **15** also contained a seven-membered ring, a structural feature which had not been encountered previously in *Pandanus* alkaloids, but which is reminiscent of alkaloids of the Stemonaceae (37).

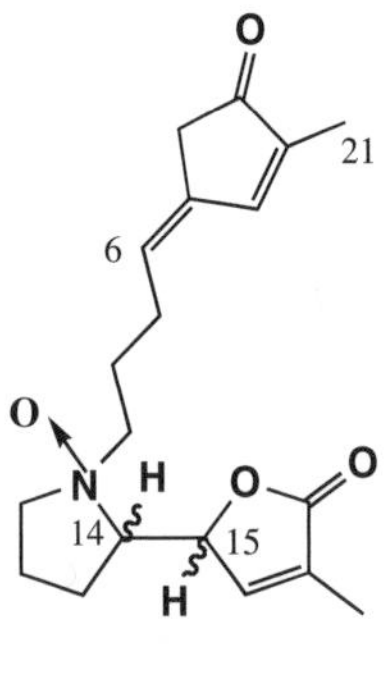

Alkaloids **14** and **15** were isolated from dried *P. amaryllifolius* leaves using conventional acid–base extraction methods and purification by reverse-phase HPLC using a gradient of H_2O/CH_3CN that contained 0.1% TFA to improve peak resolution.

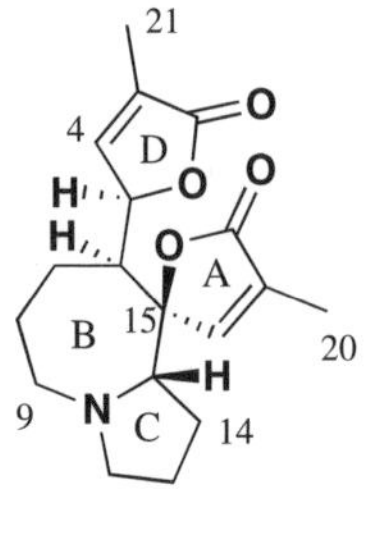

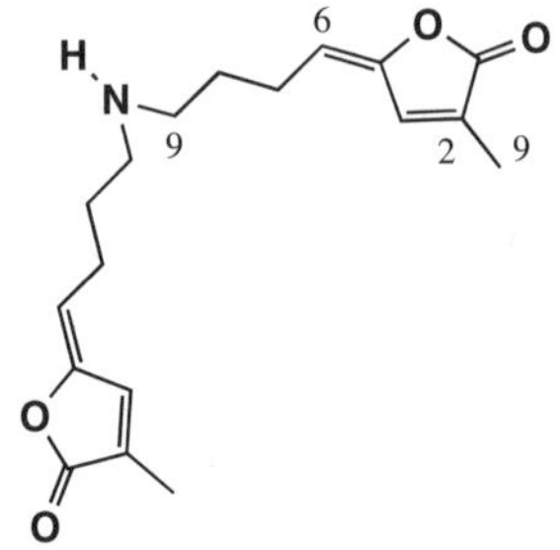

Alkaloid **14** was isolated as a colorless amorphous solid; no optical rotation was obtained. The compound had a molecular formula of $C_{18}H_{23}NO_5$ (HRESIMS), which corresponded to that of the pandamarilactonines except for an additional oxygen atom. Characteristic ^1H and ^{13}C NMR signals revealed the presence of both γ-alkylidene-α-methyl-α,β-unsaturated-γ-lactone moiety and α-methyl-α,β-unsaturated-γ-lactone ring, while the pyrrolidine ring was constructed from the remainder of the NMR signals. Overall, the NMR data of **14** in CDCl$_3$ were closely similar to those of pandamarilactonines-A and -B (**7** and **8**), except that there were significant differences at δ 3.78 (H-9a), 2.93 (H-9b), 4.01 (H-11a), 3.03 (H-11b), 3.08 (H-14), and 5.86 (H-15), where the chemical shifts were approximately 1 ppm lower than the published values for **7** and **8**. These NMR data indicated that the nitrogen in **1** was positively charged, and thus **14** was determined to be an N_b-oxide. In their paper, the authors proposed *erythro* stereochemistry at C-14/C-15 of **14**, but they did not provide any evidence that directly supported this stereochemical conclusion. Complete NMR data are given in Table VII.

Alkaloid **15** was isolated as a colorless amorphous solid of low optical activity, $[\alpha]_D^{23}=-4.35$ (*c.* 0.16, CHCl$_3$), and with the molecular formula $C_{18}H_{23}NO_4$ by HRESIMS. Characteristic ^1H and ^{13}C NMR signals again indicated the presence of an α-methyl-α,β-unsaturated lactone moiety but the C-5/C-6 alkene functionality was absent; in particular, the HMBC data showed a correlation between a proton at δ 4.67 assigned to H-5 and C-4 at δ 147.0. The signal at δ 79.2 for C-5 suggested a methine carbon adjacent to an oxygen atom, confirming the presence of a γ-lactone residue. The DQFCOSY data linked H-5 to another methine proton

Table VII   NMR assignments of alkaloid **14**

#	$\delta^{13}C^a$	$\delta^1H\ (J)^{b,c}$	HMBC^{d,e}	COSY
2	170.9	–	4, 21	
3	130.0	–	4, 21	
4	137.7	7.04, d (1.3)	6, 21	21
5	149.5	–	4, 6, 7	–
6	110.5	5.20, dd (7.8, 7.8)	7, 21	7
7	23.2	2.51–2.45, m	6, 8, 9	6, 8a/b
8a	24.6	2.21, m	6, 7, 9	7, 8b, 9a/b
8b		1.96, m	–	7, 8b, 9a/b
9a	54.7	3.78, ddd (12.1, 12.1, 4.6)	7, 8, 11, 14	8a/b, 9b
9b		2.93, m	–	8a/b, 9a
11a	54.1	4.01 ddd (7.3, 4.6)	9	11b, 12a/b
11b		3.03, m	–	11a, 12a/b
12a	21.7	2.33, m	13	11a/b, 12b, 13a/b
12b		2.10, m		11a/b, 12a, 13a/b
13a	26.5	2.23, m	11, 14	12a/b, 13b, 14
13b		2.21, m		12a/b, 13a, 14
14	70.5	3.08, ddd (7.8)	15, 20	13a/b
15	78.5	5.86, dd (7.8)	14, 20	20
16	144.4	7.07, br s	15, 20	20
17	132.9	–	15, 16, 20	–
18	172.1	–	16, 20	–
20	10.8	1.95, br s	16	16
21	10.5	2.00, br s	4	4

[a]125 MHz, CDCl$_3$, referenced to ^{13}C at δ 77.0 ppm.
[b]500 MHz, referenced to ^{1}H at 7.26 ppm.
[c]Coupling constants in Hz.
[d]HMBC connectivity from C to H.
[e]Correlations observed for one bond J_{C-H} of 135 Hz and long range J_{C-H} of 7 Hz.

at δ 3.27 (H-6), which in turn, was coupled to signals at δ 1.64 (H-7a) and 1.59 (H-7b). The C-7 to C-9 and C-11 to C-14 portions were then established by DQFCOSY and HSQC data. The ^{13}C shifts of C-11 (δ 59.0) and C-14 (δ 69.7) suggested that these carbons were each adjacent to a nitrogen atom. In this way the pyrrolidine ring was inferred. The remainder of the signals provided a second α-methyl-α,β-unsaturated-γ-lactone moiety; there were HMBC correlations between a methine proton at δ 8.06 (H-16) and both C-15 (δ 87.0) and C-17 (δ 134.3). The signal at δ 87.0 for C-15 was consistent with a quaternary carbon next to an oxygen atom. In the HMBC spectrum, the signal at δ 58.2 (C-9) showed HMBC correlations to both H-11 at δ 4.06 and H-14 at δ 3.92, while C-15 was correlated to H-5, H-6, H-7, H-13, H-14, H-16, and H-20. These data generated the carbon skeleton of **15** as shown.

The relative configuration of **15**, as shown in Figure 3, was assigned from coupling constant and NOESY data together with modeling studies.

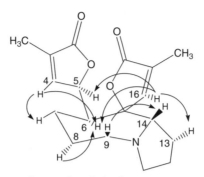

Figure 3   Selected NOESY correlations for alkaloid **15**.

Since alkaloid **15** had four chiral centers, the 16 possible stereoisomers were modeled and individual stereostructures excluded if their NOESY-correlated protons were separated by >4 Å. The model with 5*R*,6*S*,14*S*,15*R* stereochemistry and a chair conformation in the seven-membered ring, was found to fit the experimental NOESY data, with all the inter-proton distances that corresponded to observed NOESY correlations in the range 2.6–3.2 Å. The lowest energy conformation indicated a dihedral angle of 70° between H-5 and H-6, which was consistent with the 1.7 Hz coupling constant measured from the ^{1}H spectrum. Complete HMBC, DQFCOSY, and NOESY data are given in Table VIII.

Diagnostic signals at δ 8.06, 4.67, 4.06, 3.92, and 3.75 that were markers for alkaloid **15** were not observed in the spectrum of the extract before HPLC. Similarly for alkaloid **14**, there was also no evidence for this alkaloid prior to HPLC. Therefore, the isolated alkaloids **14** and **15** were not the "natural products" of *P. amaryllifolius*, and were assumed to be artifacts isolated as a result of the acidic conditions used during HPLC. This evidence suggested that the "natural" *Pandanus* alkaloids were sensitive to the acidic conditions used in the HPLC purification.

A comparative study of the alkaloids isolated from two different extraction methods was then carried out, involving a solvent partitioned method (method A; sequential partitioning between 50% aq. EtOH with hexane followed by chloroform, then normal phase flash chromatography) and an acid–base extraction method (method B; HCl/ether and NH$_4$OH/CHCl$_3$ followed by normal phase flash chromatography). An ethanolic extract prepared from a second batch of dried *Pandanus* leaves was divided in half; one portion was further worked up with method A, and the other portion was purified using method B. In method A, acid and base were excluded in order to minimize the formation of artifacts. The two extraction methods gave different alkaloid products.

The solvent partitioned method (method A) produced a mixture of two alkaloids. The major component was pandanamine (**13**) (47) while the second alkaloid was identified as the 6*E*-isomer **16**. In **16**, the alkene signals at δ 7.43 (H-4) and 5.51 (H-6) appeared at lower field [δ 7.00 (H-4) and 5.14 (H-6)] than for pandanamine **13**. Earlier studies on *P. amaryllifolius* alkaloids had showed that an *E*-configuration of the γ-alkylidenebutenolide moiety results in lower field proton

Table VIII NMR assignments of alkaloid **15**

#	$\delta^{13}C^{a,b}$	δ^1H $(J)^{c,d}$	HMBC^{e,f}	COSY	NOESY
2	173.2 (s)	–	4, 21	–	–
3	132.6 (s)	–	4, 5, 21	–	–
4	147.0 (d)	7.03, dd (3.3,1.6)	5, 6, 21	5, 21	5, 7b, 21
5	79.2 (d)	4.67, dd (3.3, 1.7)	6, 7, 21	4, 6, 21	4, 6, 7b, 16
6	44.6 (d)	3.27, ddd (10.2, 3.0, 1.7)	5, 7, 20, 21	5, 7a/b	5, 7a/b, 8a, 16
7a	21.6 (t)	1.64 m	5, 6, 9	6, 7b, 8a/b	–
7b		1.59, m		6, 7a, 8a/b	4, 5, 6, 8b
8a	26.5 (t)	2.24, m	6, 7, 9	7a/b, 8b, 9a/b	6, 7a, 8a
8b		2.08, m		7a/b, 8a, 9a/b	8b
9a	58.2 (t)	3.75, m	7, 8, 11, 14	8a/b, 9b	9a
9b		2.95, m		8a/b, 9a	9b, 14
11a	59.0 (t)	4.06 ddd (10.7, 6.3, 4.4)	12, 13	11a, 12a/b	11a, 12a
11b		2.92, m		11a, 12a/b	11b
12a	23.3 (t)	2.03, m	11, 13, 14	11a/b, 12b	
12b		1.93, m		11a/b, 12a	11b
13a	26.5 (t)	2.14, m	11, 12, 14	12a/b, 13b, 14	14
13b		1.64, m		12a/b, 13a, 14	16
14	69.7 (d)	3.92, dd (9.5, 5.1)	9, 11, 12	13a/b	9b, 13a
15	87.0 (s)	–	5, 6, 7, 13, 14, 16, 20	–	–
16	148.2 (d)	8.06, d (1.3)	14, 20	20	5, 6, 13b, 20
17	134.3 (s)	–	16, 20	–	–
18	171.3 (s)	–	16, 20	–	–
20	10.6 (q)	2.04, d (1.3)	16	16	16
21	10.7 (q)	1.91 d (1.6)	4	4, 5	4

[a]125 MHz, CDCl$_3$, referenced to ^{13}C at δ 77.0 ppm.
[b]Multiplicity from DEPT.
[c]500 MHz, referenced to ^{1}H at 7.26 ppm.
[d]Coupling constants in Hz.
[e]HMBC connectivity from C to H.
[f]Correlations observed for one bond J_{C-H} of 135 Hz and long range J_{C-H} of 7 Hz.

chemical shifts for H-4 and H-6 (35,43). Additionally, the C-4 signal (δ 134.0) for **16** appeared at higher field than in pandanamine. In contrast, four alkaloids were isolated from the acid–base extraction (method B) including pandamarilactonines-A and -B (**7** and **8**) (major components) and pandamarilactonines-C and -D (**9** and **10**) (minor components) (36,43). Changing the acid used in the extraction (HCl vs. H_2SO_4) did not affect the type of alkaloids obtained from method B.

The absence of pandamarilactonine products from the solvent partitioned method suggested that they were artifacts formed during the acid–base treatment. The most likely precursors of these alkaloids were the pandanamines **13** and **16**, which were isolated from the solvent partitioned method (47,48). In a biomimetic synthesis, Takayama *et al.* had previously shown that **13** cyclized to **7** and **8** on treatment with a catalytic amount of TFA in CH_3CN (36). When the pandanamine products **13** and **16** isolated using the method A procedure were subjected to acid–base treatment similar to that in the acid–base extraction procedure (method B), the products were found to be the pandamarilactonines **7–10**. Conversion of the pandanamines to the pandamarilactonines was proposed to involve Michael addition of the nitrogen onto the double bond at C-6 followed by protonation at C-15.

The pandamarilactonines-A and -B (**7** and **8**) isolated from the crude alkaloidal fraction under acid–base conditions could themselves be the precursors to alkaloid **15**. Under acidic conditions (such as in the presence of TFA), the lactone ring of **7** or **8** may enolize, followed by attack of C-15 on C-6, ultimately leading to the formation of **15** after reprotonation at C-5 (48).

The authors commented that the use of conventional acid–base extraction to isolate *Pandanus* alkaloids should be avoided since it can lead to the formation of artifacts, and speculated whether the alkaloids isolated in earlier studies (29,30,35,36,43,46) were "natural product" or artifacts formed during the isolation process.

The mixed pandamarilactonine sample isolated from the method B procedure using HCl was anticipated to be a racemate, but the sample in fact had low optical activity, $[\alpha]_D$ +4.0 (*c.* 2.56, $CHCl_3$) that at the time was attributed to the difficulty in removing trace impurities during the chromatography (48). The configurational instability of pandamarilactonines is further explored below (Section III).

In summary, it appears that *P. amaryllifolius* has several subspecies or chemotypes, which may produce different types of alkaloids. Since the very first isolation of pandamarine **1** from *P. amaryllifolius* (29), a number of structural variations in the alkaloids associated with *P. amaryllifolius* has been observed from plants collected in the Philippines, Thailand, and Indonesia. Piperidinyl alkaloids with lactam (29) or lactone (30) moieties have been isolated from Philippine specimens of *P. amaryllifolius*. Pyrrolidinone (35) and pyrrolidine (36,43,46,48) rings feature in alkaloids from *P. amaryllifolius* collected in Indonesia and Thailand, respectively. The majority of *P. amaryllifolius* alkaloids have at least one α,β-unsaturated-γ-lactone ring, and are suggested to be derived from the same common precursor (36), a symmetrical secondary amine called pandanamine (**13**), which has also been isolated from Nature (47). Some of the interesting "natural" product structures may in fact be artifacts of isolation or epimerization products since some compounds show sensitivity towards acidic and/or basic conditions.

III. SYNTHESIS

A. Pandamarilactonines

The pandamarilactonine metabolites (**7–10**) (36,43) consist of two moieties, a γ-butylidene-α-methyl-α,β-unsaturated-γ-lactone and a structurally unique pyrrolidinyl-α-methyl-α,β-unsaturated-γ-lactone, the latter of which corresponds to norpandamarilactonines (**11 and 12**) (46). Five independent methods for the construction of the norpandamarilactonine unit have been developed, and these can be classified as biomimetic, racemic, or enantiomeric syntheses.

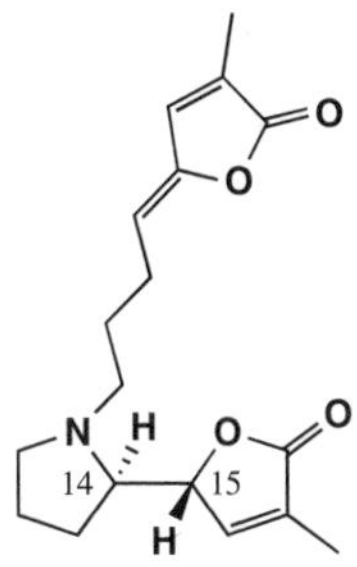

Pandamarilactonine-A (**7**) 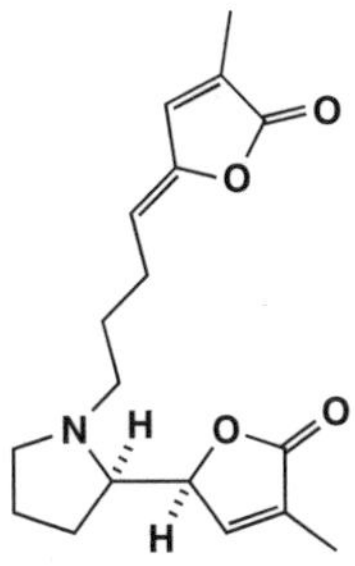   Pandamarilactonine-B (**8**)

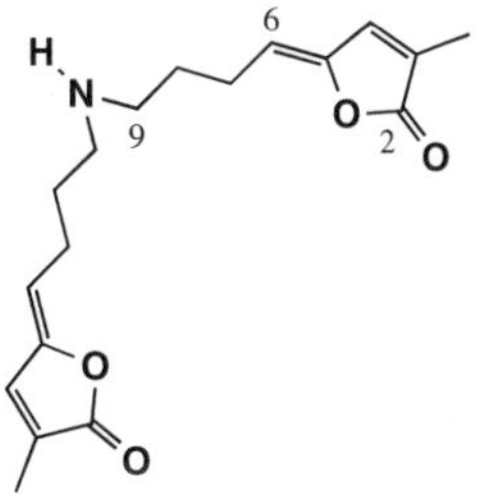

Pandanamine (**13**)

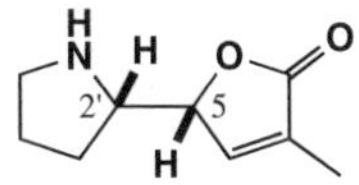

Norpandamarilactonine-A (**11**) 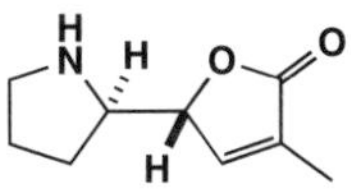   Norpandamarilactonine-B (**12**)

1. Biomimetic Synthesis

The total synthesis of **7** and **8** *via* a route that mimics the final step of the proposed biosynthesis (from **13** to **7/8**, see Section IV), was achieved as shown in Scheme 1 (36). The key secondary amine derivative **13**, which corresponds to the hypothetical biogenetic intermediate, and was isolated from *Pandanus* plants (45), was prepared as follows. N-Dialkylation of benzylamine (**17**) with O-tetrahy-dropyranyl (THP) 4-chlorobutanol (**18**) was carried out using potassium carbonate in the presence of a catalytic amount of sodium iodide to obtain tertiary amine **19** in 61% yield. Then, the benzyl group was converted into a β,β,β-trichloroethoxycarbonyl group to give carbamate **20**. After removal of the THP ether in **20**, the free alcohol groups in **21** were converted into the dialdehyde **22** in 72% yield *via* Swern oxidation. Aldol reaction of dialdehyde **22** with siloxyfuran **23** using boron trifluoride etherate (BF₃·Et₂O) gave a mixture of stereoisomeric adducts **24** in quantitative yield. Next, installation of the *exo*-double bond was accomplished by treating the mixed adducts **24** with a combination of TMSCl and DBU to give the γ-alkylidenebutenolide **25** in 41% yield, together with the Z, E-isomer in 25% yield. Finally, the protecting group on

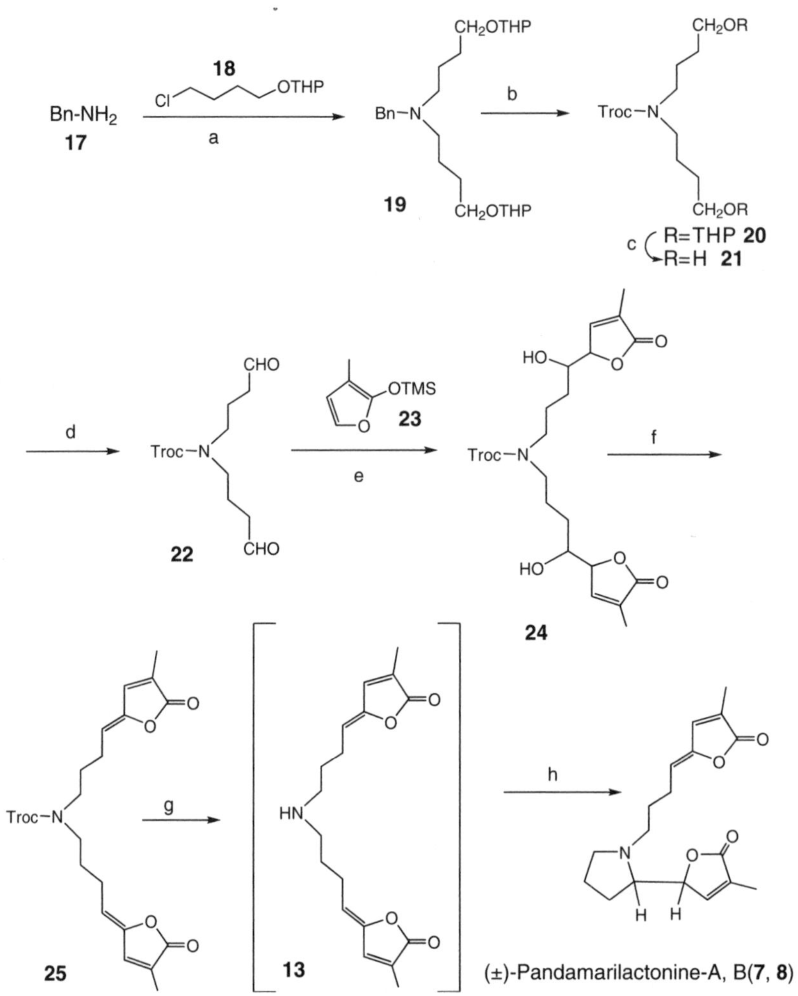

Scheme 1 Reagents and conditions: a, K$_2$CO$_3$, NaI, CH$_3$CN, 61%; b, β,β,β-trichloroethoxycarbonylchloride, CH$_3$CN, 76%; c, p-TsOH·H$_2$O, aq. CH$_3$CN, 80%; d, Swern oxidation, 72%; e, BF$_3$·Et$_2$O, CH$_2$Cl$_2$, quant; f, TMSCl, DBU, CH$_2$Cl$_2$, **25**, 41%, Z, E-isomer, 25%; g, Zn, AcOH; h, cat. TFA, CH$_3$CN, **7**, 9%, **8**, 9%.

nitrogen in **25** was removed by reacting with Zn in AcOH. Crude amine **13** was directly treated with a catalytic amount of trifluoroacetic acid in CH$_3$CN. By careful separation of the crude products, ($\pm$)-pandamarilactonine-A (**7**) and -B (**8**) were obtained in 9% yield each. Direct comparison revealed that they were identical with their respective natural products. Although the chemical yield at the final step was low, the first biomimetic total synthesis of the alkaloids was accomplished, providing chemical support for the suggested biogenetic route of **7** and **8**, as well as proof of the chemical structures deduced by spectroscopic analysis.

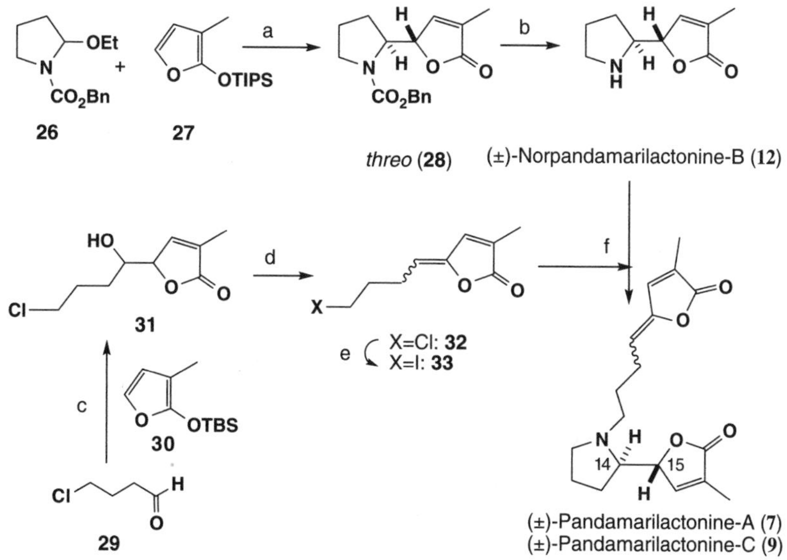

Scheme 2   Reagents and conditions: a, BF₃ · Et₂O, CH₂Cl₂, −78°C, 91%(*threo:erythro* = 4:1); b, TMSI, CH₃CN, r.t., 94%; c, BF₃ · Et₂O, CH₂Cl₂; d, Tf₂O, Py, CH₂Cl₂, 48%(2 steps); e, NaI, acetone, 47%; f, K₂CO₃, CH₃CN, **7**, 33%, **9**, 7%.

2. Syntheses of the Racemic Form

The lower part of the alkaloids, i.e., the pyrrolidinyl-α-methyl-α,β-unsaturated-γ-lactone residue, was synthesized according to the procedure of Martin *et al.* (37) as follows (Scheme 2) (43). Compound **28**, whose stereochemistry at the vicinal positions was established to be *threo* by X-ray analysis, was prepared by vinylogous Mannich coupling reaction of **26** and **27**. The protecting group on the nitrogen in **28** was removed by treatment with TMSI in CH₃CN to give (±)-norpandamarilactonine-B (**12**) in 94% yield. Iodide **33** corresponding to the upper part of the pandamarilactonines was prepared in three steps: (i) condensation of aldehyde **29** with siloxyfuran **30** in the presence of BF₃ · Et₂O; (ii) dehydration of resultant aldol adduct **31** with trifluoromethanesulfonic anhydride and pyridine (48% overall yield); and (iii) halogen exchange of **32** with NaI in acetone (47% yield). The resulting nine-carbon unit **33**, which consisted of the Z- and E-isomers in the ratio of 4.6:1, was condensed with the secondary amine **12** in acetonitrile in the presence of K₂CO₃ to give two pandamarilactonines in 33% and 7% yields, both of which possessed the *threo* form at the C-14 and C-15 positions.

Based on the results of this synthetic study, the relative configuration at the vicinal positions of pandamarilactonines was unambiguously established. Thus, pandamarilactonine-A and -B were respectively the *threo* and *erythro* isomers.

A synthesis by Figueredo's group (45,49) starting from (S)-prolinol also provided racemic samples of these alkaloids and consequently provided a plausible mechanistic explanation for the configurational instability of the pandamarilactonines. The work involved the transformation of an oxirane to an α-methylbutenolide using the dianion of 2-phenylselenopropionic acid (Scheme 3). Initially, the γ-butylidene-α-methyl-α,β-unsaturated-γ-lactone moiety was prepared in the following manner. Starting from hydroxydithiane **34**, aldehyde **35** was prepared in 88% yield. The vinylogous Mukaiyama reaction of **35** with silyloxyfuran **27** afforded a 7:1 mixture of *threo* and *erythro* alcohols **36** and **37** in 82% yield. Treatment of the diastereomeric alcohols with TMSCl/DBU in chloroform gave a mixture of olefins **38** and **39** in 94% yield after purification by silica gel chromatography in an approximately 3:1 ratio, and the olefins were respectively assigned as the *Z*- and *E*-isomers. Desilylation was accomplished in 86% yield to produce the free alcohols, and these were converted into sulfonates without intermediate purification. Isomeric sulfonates **40** and **41** were chromatographically separated (74% total yield). Synthesis of the pyrrolidine fragment was accomplished starting from carbamate **42**, which was prepared from (S)-prolinol. Oxidation of **42** with MCPBA furnished two oxiranes **43** and **44** in 77% total yield and an *erythro*/*threo* ratio of 1.5:1.

The oxiranes were separated and major isomer **43** was converted into *erythro*-methyl butenolide **45** *via* a three-step protocol consisting of addition of the dianion of 2-phenylselenopropionic acid to **43**, followed by acid-induced lactonization and oxidation of the selenide function with subsequent thermal elimination, giving an overall yield of 61%. Treatment of *erythro*-methyl butenolide **45** with TMSI in chloroform at reflux resulted in cleavage of the carbamate with concomitant epimerization of the stereogenic center of the

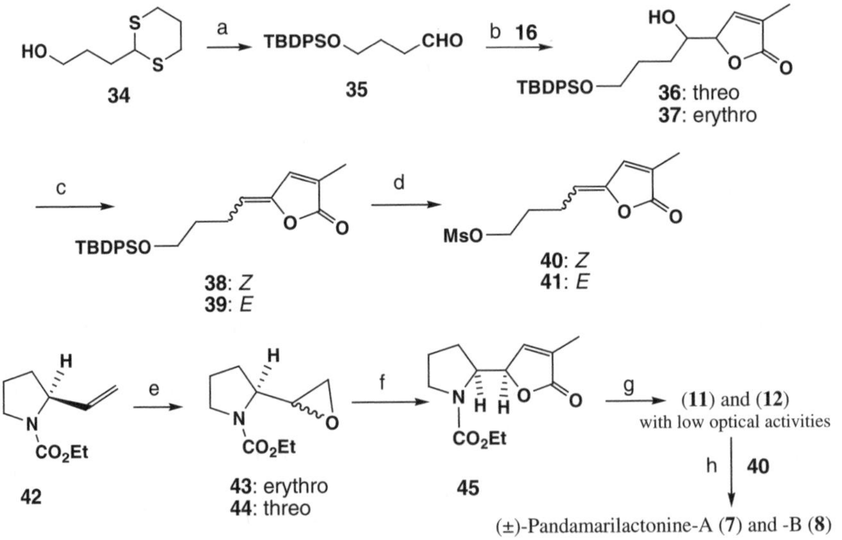

Scheme 3 Reagents and conditions: a, TBDPSCl, Im, DMF, 96%, then CaCO₃, MeI, aq. CH₃CN, 92%; b, BF₃·Et₂O, CH₂Cl₂, −78°C, 82% (**36:37** = 7:1); c, TMSCl, DBU, CHCl₃, reflux, 94%, (**38:39** = 3:1); d, Bu₄NF, THF, 86%, then MsCl, Py, CH₂Cl₂, 74%; e, MCPBA, CHCl₃, 77% (**43:44** = 1.5:1); f, (i) separation of diastereomers, (ii) PhSeCH(CH₃)CO₂H, LDA (2 equiv.), THF, (iii) AcOH, THF, reflux, (iv) H₂O₂, AcOH, 0°C, 61% (4 steps); g, TMSI, CHCl₃, reflux, 84%; h, Py, DMF, 60°C, 44%.

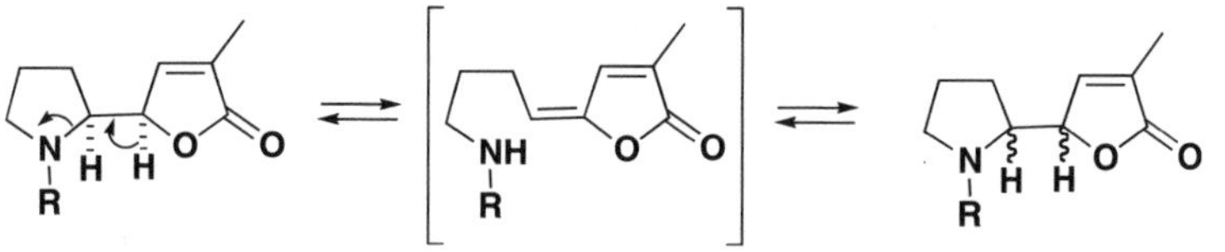

Figure 4 A mechanism that explains the configurational instability of the *Pandanus* alkaloids (45).

lactone moiety, furnishing an approximately 1:1 mixture of two norpandamarilactonines (**11** and **12**), which were separated by silica gel chromatography. The specific rotation measured for **11** was $[\alpha]_D^{20}$ −7 (*c.* 1.5, CHCl$_3$) and that for **12** was $[\alpha]_D^{20}$ −3 (*c.* 2.6, CHCl$_3$). The authors suspected that the reason for such low optical activity values could be that, besides epimerization, racemization also occurred to some extent. They proposed ring opening of the pyrrolidinyl ring under neutral or basic conditions, followed by ring closure, as shown in Figure 4. This sequence explains the loss of stereochemistry in the pandamarilactonines at both C-14 and at C-15. The facile epimerization of *erythro* intermediate (**45**) under reflux in CHCl$_3$ contrasts with the configurational stability of *threo*-(**28**) in CH$_3$CN at room temperature when each were exposed to TMSI.

When the synthetic norpandamarilactonine mixture (**11/12**) was treated with an equimolar amount of freshly prepared mesylate **40** in DMF in the presence of pyridine at 60°C, pandamarilactonines **7** and **8** were slowly formed in an approximate 1:1 ratio. After 3 days, the reaction mixture was purified by normal phase flash chromatography over silica gel to give pandamarilactonine-A (**7**) and pandamarilactonine-B (**8**) in an overall isolated yield of 44%. Neither diastereomer, when purified, showed any optical activity. However, it should be noted that the starting materials for this reaction themselves showed low optical activity.

Figueredo *et al.* (45) made stereochemical assignments of oxiranes **43** and **44** and butenolide **45** on the basis of the NMR trends of pyrrolidinyl-α,β-unsaturated-γ-lactones that had been noted earlier by Martin *et al.* (37). With issues of relative configuration resolved, it is now apparent the same trends apply to both the norpandamarilactonines and pandamarilactonines series. The proton signal corresponding to H-14 of the pandamarilactonines (Tables IV and V) or to H-2' of the norpandamarilactonines (Table VI) is upfield in the *erythro* series of compounds, while the corresponding carbon signal is downfield.

3. Syntheses of the Chiral Form

Two papers have described the chiral syntheses of pandamarilactonine alkaloids. In 2005, Takayama *et al.* described an approach to pandamarilactonine-A (**7**) which established the absolute configuration of the (+)-isomer (**44**), while in 2006, Honda *et al.* (50) reported the synthesis of a norpandamarilactonine precursor using an alternative synthetic approach, and converted this to a mixture of pandamarilactonines-A and -B (**7/8**).

In order to construct the chiral norpandamarilactonine moiety, a Reformatsky-type condensation using l-prolinal and 2-(bromomethyl)acrylate was employed by Takayama *et al.* (44). l-Prolinol (46) was converted into aldehyde (47, $[\alpha]_D^{26}$ −63.1 (*c.* 1.14, MeOH)) in two steps, and then the Reformatsky-type condensation with ethyl 2-(bromomethyl)acrylate was carried out (Scheme 4). When Zn metal was used, the adducts were obtained in 52% yield, which contained the *erythro* (more polar, 48) and the *threo* (less polar, 49) isomers in the ratio of 4:1, the stereochemistry of which was determined in the later stage of the synthesis, as described below. On the other hand, indium-mediated coupling gave the same adducts in 80% yield, the diastereoselectivity of which was 2:1. The *erythro* isomer 48, which had an undesirable stereochemistry, was transformed into the *threo* isomer 49 by inversion of the secondary alcohol. Initially, the intramolecular Mitsunobu reaction was applied to carboxylic acid 50 that was prepared by the alkaline hydrolysis of 48. Treatment of 50 with di-*tert*-butyl azodicarboxylate (DTAD) and PPh₃ in THF at room temperature gave the lactone derivative in quantitative yield, which contained the *threo* isomer 51 and its *erythro* isomer in the ratio of 7.3:1. The unexpected minor *erythro* lactone, which showed retention of the configuration of the alcohol, was formed *via* an acyloxyphosphonium ion intermediate. Application of the conventional inter-molecular Mitsunobu reaction to ester 48 resulted in the recovery of the starting material.

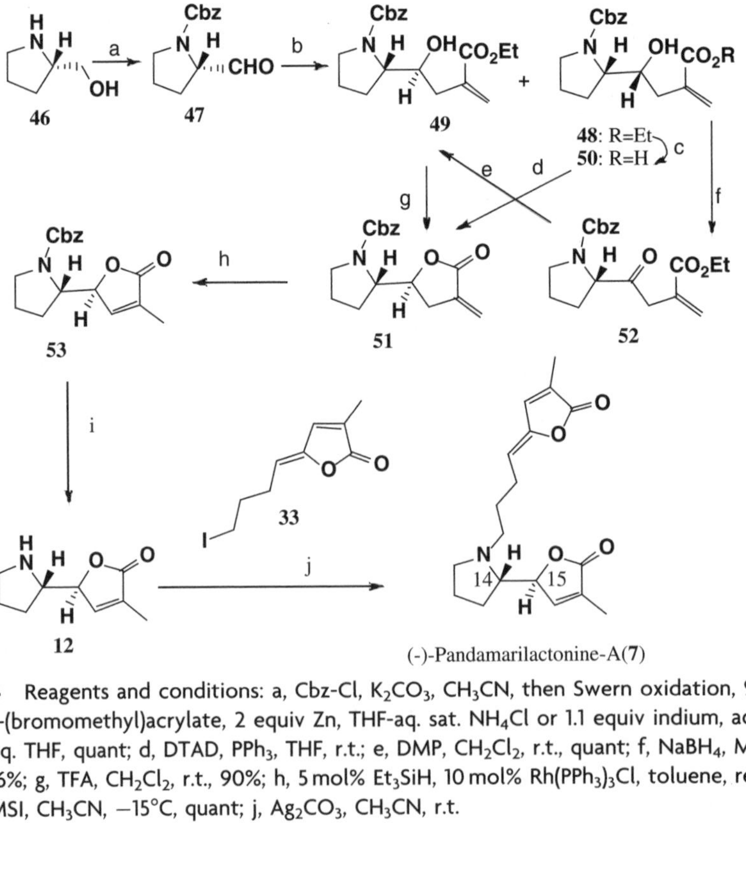

Scheme 4 Reagents and conditions: a, Cbz-Cl, K₂CO₃, CH₃CN, then Swern oxidation, 91%; b, ethyl 2-(bromomethyl)acrylate, 2 equiv Zn, THF-aq. sat. NH₄Cl or 1.1 equiv indium, aq. EtOH; c, LiOH, aq. THF, quant; d, DTAD, PPh₃, THF, r.t.; e, DMP, CH₂Cl₂, r.t., quant; f, NaBH₄, MeOH, −20°C, 86%; g, TFA, CH₂Cl₂, r.t., 90%; h, 5 mol% Et₃SiH, 10 mol% Rh(PPh₃)₃Cl, toluene, reflux, 86%; i, TMSI, CH₃CN, −15°C, quant; j, Ag₂CO₃, CH₃CN, r.t.

For the inversion of the alcohol in **48**, an oxidation–reduction sequence was attempted. Ketone derivative **52**, prepared from **48** by oxidation with Dess–Martin periodinane, was reduced with $NaBH_4$ in MeOH at $-20°C$ to afford *threo* **49** and *erythro* **48** alcohols in the ratio of 2.6:1. The major alcohol **49** thus obtained was treated with trifluoroacetic acid in CH_2Cl_2 to give lactone **51** in 90% yield, which was identical with the major product obtained *via* the intramolecular Mitsunobu reaction described above. Next, the isomerization of the double bond in **51** from the *exo* to the *endo* position was performed using Et_3SiH (5 mol%) and tris(triphenylphosphine)rhodium chloride (10 mol%) in refluxing toluene to afford the α-methyl butenolide **53** in 86% yield. The ^1H- and ^{13}C NMR spectra of **53** ($[\alpha]_D^{25}$ −183 (*c.* 0.49, $CHCl_3$), 100% *ee* based on chiral HPLC analysis) were identical with those of racemic material prepared previously by Martin *et al.* (37), the relative stereochemistry at C-14 and C-15 of which was established to be *threo* by X-ray analysis. Careful removal of the Cbz group (TMSI in CH_3CN at $-15°C$ for 30 min) gave the secondary amine, norpandamarilactonine-B (**12**) ($[\alpha]_D^{24}$ −49.4 (*c.* 1.22, $CHCl_3$)) in quantitative yield.

The final stage of the total synthesis of **7** was the coupling of optically active amine **12** with the nine-carbon unit containing a γ-alkylidene butenolide moiety. Iodide **22**, which consisted of the *Z*- and *E*-isomers in the ratio of 5:1, was condensed with freshly prepared **12** in CH_3CN in the presence of Ag_2CO_3 at room temperature to furnish the adducts in 66% yield. After repeated column chromatography, pure pandamarilactonine-A (**7**) was obtained in 48% yield. The synthetic compound, with 14*S* and 15*S* configurations, displayed spectroscopic data, including ^1H- and ^{13}C NMR, UV, IR, MS, and HR-MS, completely identical with those of the natural product, and exhibited $[\alpha]_D^{23}$ −94.0 (*c.* 0.12, $CHCl_3$). This demonstrated that the absolute configuration of the major enantiomer in "natural" pandamarilactonine-A (**7**) was 14*R* and 15*R*.

Honda *et al.* have prepared enantiopure *N*-Boc-norpandamarilactonine-A by a double ring-closing metathesis (RCM) reaction of tetraene derivative **64** that yields a dehydro analogue (Scheme 5) (50). Initially, L-serine methyl ester hydrochloride was converted into the known compound **54** which, on allylation with allyl iodide and NaH in DMF, gave the *N*-allyl compound **55**. Removal of the acetonide in **55** on treatment with *p*-toluenesulfonic acid gave diol **56**. After selective protection of the primary alcohol of **56** with *tert*-butyldimethylsilyl chloride, the resulting silyl ether **57** was protected as its MOM ether **58**, which was further converted into primary alcohol **59** by treatment with TBAF. Oxidation of **59** with Dess–Martin periodinane proceeded smoothly to give aldehyde **60** which, on methylenation with methyltriphenylphosphonium bromide and *n*-BuLi in the usual manner, afforded the desired triene **61** in good yield. After the two-step manipulation of the protecting groups in **61**, which involved removal of the Boc and MOM groups by acid hydrolysis, followed by protection of the resulting amino group of **62** with Boc_2O, the resulting secondary alcohol **63** was esterified with methacrylic acid in the presence of DCC and DMAP to provide the desired tetraene derivative **64**. With this tetraene available, a study was conducted to determine the best conditions for the double RCM reaction.

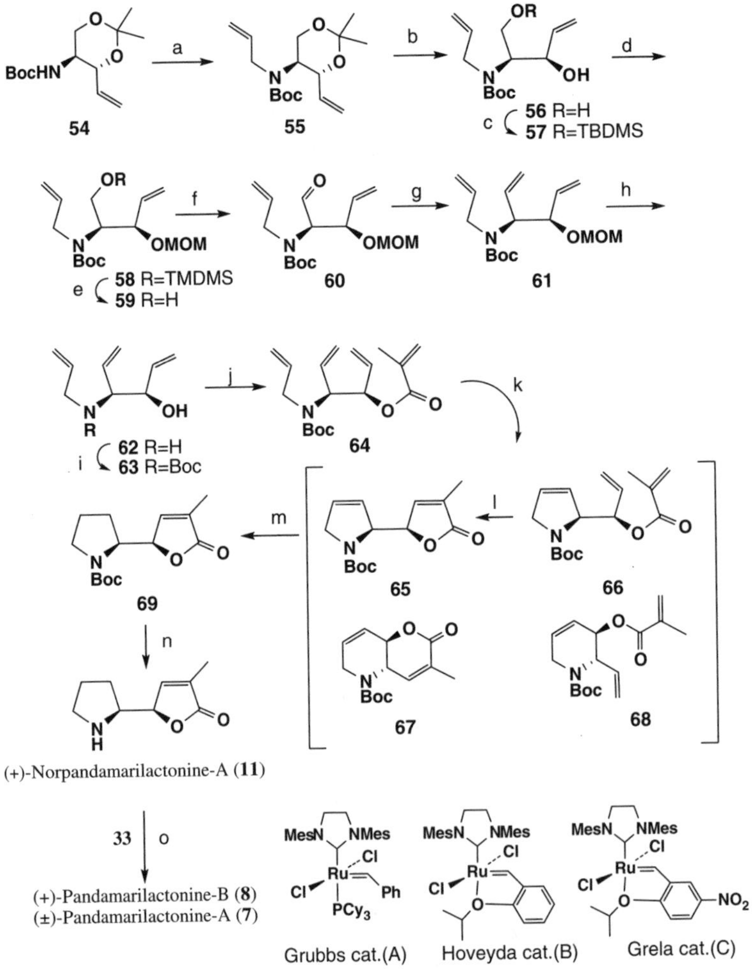

Scheme 5 Reagents and conditions: a, allyl iodide, NaH, DMF, 63%; b, p-TsOH, MeOH, 92%; c, TBDMSCl, Im, DMF, 93%; d, MOMCl, iPr₂NEt, DMAP, CH₂Cl₂, 77%; e, TBAF, THF, 95%; f, Dess–Martin periodinane, CH₂Cl₂, 96%; g, CH₃P⁺Ph₃Br⁻, n-BuLi, THF, 85%; h, 10% HCl, MeOH, 73%; i, Boc₂O, Et₃N, THF, 76%; j, methacrylic acid, DCC, DMAP, CH₂Cl₂, 86%; k, (i) 10% 2nd Grubbs cat., benzene, 80°C, **65** 2%, **66** 25%, **67** 7%, **68** 25%, (ii) 10% Hoveyda cat., benzene, 80°C, **65** 6%, **66** 2%, **67** 28%, **68** 24%, (iii) 10% Hoveyda cat., benzene, 60°C, **65** 73%, **66** 3%, **67** 0%, **68** 23%, (iv) 10% Grela cat., benzene, 60°C, **65** 76%, **66** 2%, **67** 0%, **68** 22%; l, 10% Hoveyda cat., benzene, 60°C, 76%; m, Wilkinson cat., H₂ (5 atm), CH₂Cl₂, 95%; n, TMSOTf, 2,6-lutidine, CH₂Cl₂, 43%; o, Ag₂CO₃, CH₃CN, (+)-(**8**) 62%, (±)-(**7**) 15%.

First, the authors attempted the double RCM of **64** using 10 mol% Grubbs's 2nd-generation ruthenium catalyst (A) in benzene at 80°C for 20 h; however, desired compound **65** was isolated in only 2% yield. The major products were mono-cyclized **66** and **68** in 25% yield each. When this reaction was carried out using 10 mol% Hoveyda catalyst (B) in benzene at 80°C for 20 h, bicyclic tetrahydropyridine derivative **67** was isolated as the major product in 28% yield, together with the desired compound **65** (6%), pyrrolidine derivative **66** (2%), and tetrahydropyridine derivative **68** (24%). Interestingly, a similar reaction of **64** with 10 mol% Hoveyda catalyst at 60°C gave the desired compound **65** in 73% yield, together with **68** (23%). On screening a variety of reaction conditions for RCM of **64**, the authors found that the use of 10 mol% Grela catalyst (C) in benzene at 60°C for 20 h afforded the desired compound **65** in 76% yield. Mono-cyclized pyrrolidine **66** could also be converted into **65** in 84% yield by further reaction with 10 mol% Hoveyda catalyst in benzene at 60°C for 20 h. The structure of **65** was unambiguously determined by X-ray crystallography. The fact that the cyclization product ratios depended on the reaction conditions, particularly the reaction temperature, implied that the bis-five-membered compound **65** is a kinetically controlled product.

Selective reduction of the double bond in the pyrrolidine ring of **65** was successfully achieved by catalytic hydrogenation with the Wilkinson catalyst under 5 atm hydrogen to afford enantiopure *N*-Boc-norpandamarilactonine-A (**69**) (m.p. 75–77°C, $[\alpha]_D$ −51.1 (*c*. 1.0, CHCl$_3$)) in 95% yield. Treatment of **69** with trimethylsilyl triflate provided (+)-norpandamarilactonine-A (**11**), $[\alpha]_D$ +55.0 (*c*. 0.82, CHCl$_3$) [lit. (50), +80.2 (*c*. 0.79, CHCl$_3$)], as the sole product. Compound **11**, however, gradually became a mixture with norpandamarilactonine-B (**12**), due to rapid epimerization. The conversion of (+)-**11** into pandamarilactonine-B **2** by coupling with iodide **22** was attempted according to Takayama's procedure, and gave racemic pandamarilactonine-A (**7**) and optically active pandamar-ilactonine-B (**8**) in 15% and 62% yields, respectively (50).

B. Pandamarilactams

Pandamarilactam-3y (**6**) and -3x (**5**) (35) are the members with the simplest structure among the known *Pandanus* alkaloids. Their synthesis was accom-plished by sequential coupling of three components: 2-pyrrolidinone (**70**), the propane fragment **71**, and siloxyfuran **30** (Scheme 6) (51). *N*-Alkylation of 2-pyrrolidinone (**70**) with THP 4-chlorobutanol (**71**) was efficiently carried out using potassium fluoride on alumina in the presence of a catalytic amount of sodium iodide to produce tertiary amide **72** in 76% yield. After removal of the THP ether in **72**, the resultant free alcohol was converted into aldehyde **73** in 85% yield by Swern oxidation. The aldol reaction of aldehyde **73** with siloxyfuran **30** using tetrabutylammonium fluoride (TBAF) in the presence of TBSOTf gave a mixture of adducts in 37% yield [*syn* (**74**): *anti* (**75**)=3:7] along with desilylated alcohols (**76** and **77**) (3:7) in 11% yield.

Condensation of aldehyde **73** with 2-trimethylsiloxy-3-methylfuran (**23**) in the presence of BF$_3$·Et$_2$O gave the adducts as a mixture of *syn* (**76**) and *anti* (**77**) isomers (ratio 88:12) in 80% yield. Treatment of a mixture of the *O*-silylated adduct [(**74** and **75**) (3:7 ratio)] with DBU gave *γ*-alkylidenebutenolides in 92% yield in a 9:1 ratio of the *Z*- and *E*-mixture. This result was ascribable to the E1cb mechanism that favored the formation of the thermodynamically more stable isomer. Dehydration of the major aldol adduct **76** with excess DEAD/PPh$_3$ resulted in *anti*-elimination to form exclusively the *Z*-isomer **6** in 82% yield.

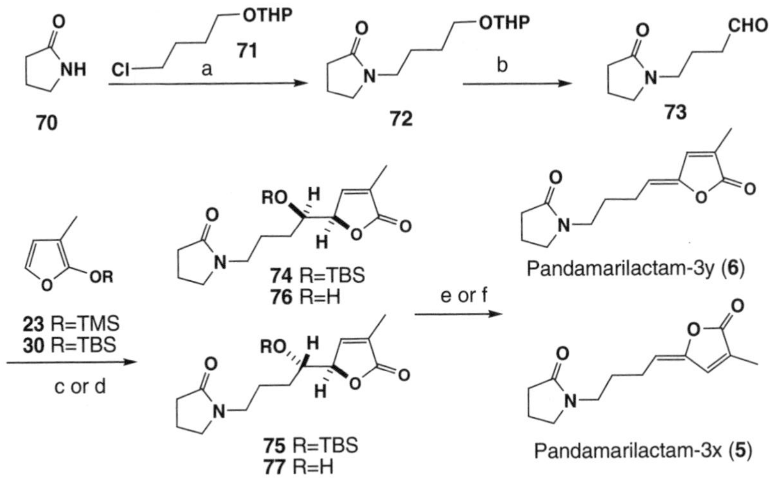

Scheme 6 Reagents and conditions: a, KF-alumina, cat. NaI, DMF, 76%; b, (i) p-TsOH, MeOH, (ii) Swern oxidation, 85%; c, TBAF, TBSOTf, CH$_2$Cl$_2$, 48%; d, BF$_3$·Et$_2$O, CH$_2$Cl$_2$, 80%; e, DBU, CHCl$_3$, 70°C, 92%; f, DEAD, PPh$_3$, THF, −40°C, 82%.

The geometric mixture obtained above by the elimination reaction with DBU was purified by HPLC with a C18-silica column using MeCN/H$_2$O as eluant. The major and minor isomers were identical with the natural products, pandamarilactam-3y (**6**) and -3x (**5**), respectively (35).

IV. BIOGENESIS

The *Pandanus* alkaloids represent a new structural class of alkaloids with a C9-N-C9 skeletal structure possessing a γ-alkylidene-α,β-unsaturated-γ-lactone or a γ-alkylidene-α,β-unsaturated-γ-lactam moiety. When considering how these alkaloids might be formed in the plant, the existence of a symmetrical secondary amine as a precursor intermediate can be postulated, which then undergoes intramolecular cyclization to give the isolated alkaloids.

It is likely that the five-membered lactone ring of these alkaloids is derived from 4-hydroxy-4-methylglutamic acid (**78**), which has been identified in some higher plants, and was found to occur in the related species *P. veitchii* (33). A biosynthetic study conducted by Peterson and Fowden suggested that leucine could be the biogenetic origin of 4-methylglutamic acid, which was then converted to 4-hydroxy-4-methylglutamic acid (34). 4-Hydroxy-4-methylglutamic acid, which is a suitable precursor for the five-membered ring structure can cyclize to either a lactone or a lactam ring product. The carbon chain C-6/C-9 and the nitrogen of the *Pandanus* alkaloids may derive from glutamic acid. The pandamarilactams (**5**, **6**) could then derive from combination of one unit of **78** derived from leucine with the C$_4$-N-C$_4$ unit **79** while the norpandamarilactonines (**11**, **12**) could derive from **78** and glutamate (Figure 5).

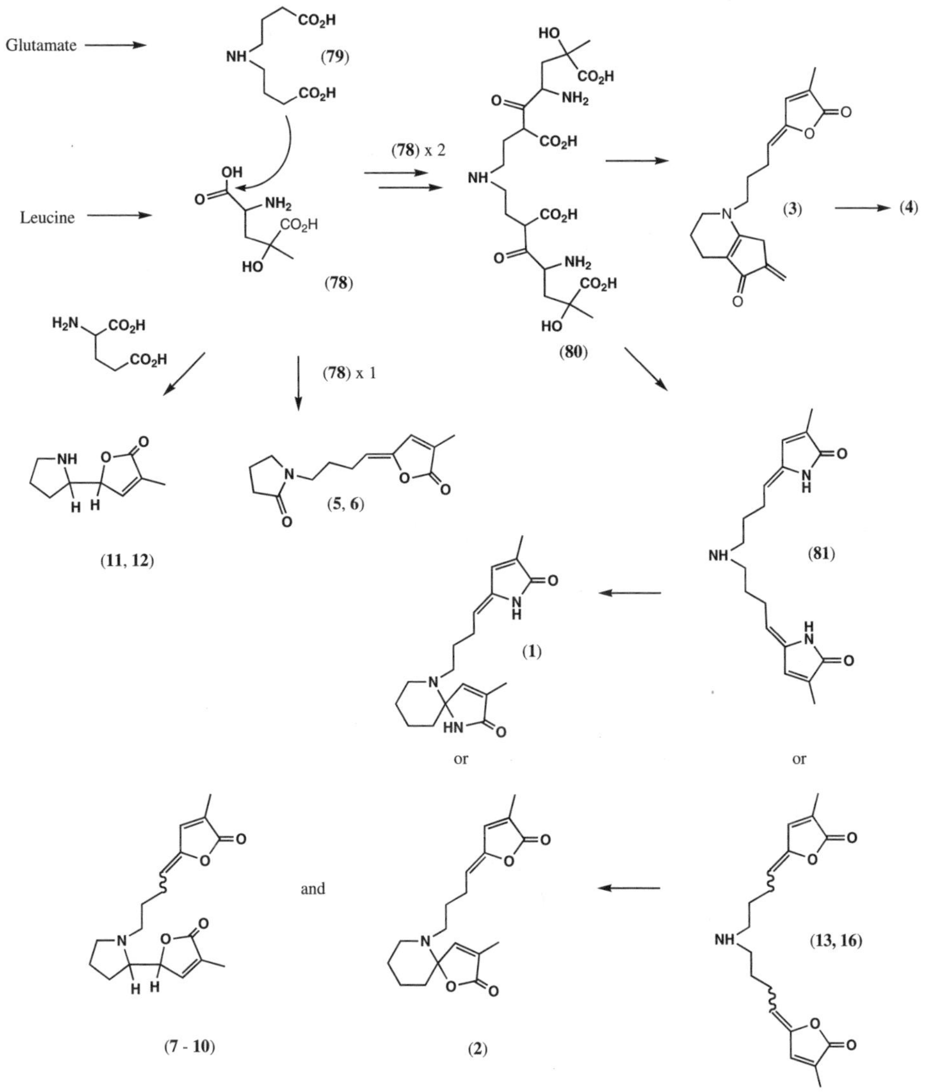

Figure 5 Biogenetic pathway of the *Pandanus* alkaloids.

The majority of *Pandanus* alkaloids derived from combination of two 4-hydroxy-4-methylglutamic acid units with the glutamate-derived C_4-N-C_4 unit *via* an intermediate equivalent to **80** that may undergo decarboxylation, dehydration, and transamination to generate the requisite γ-alkylidene-α,β-unsaturated lactone moieties. The pathway from precursor **80** to alkaloids **3** and **4** could involve the formation of a double bond between C-14 and C-15 followed by a "domino"-like cyclization triggered by attack of an imine or amine nitrogen on this double bond, with eventual C–C bond formation between C-14 and the carboxyl group.

The isolation of the symmetrical pandanamine (**13**) in Nature (47) strongly supports the final stages of the proposed biogenetic pathways (Figure 5) leading to the diverse *Pandanus* alkaloid structures (**2, 7–10, 14, 15**) described earlier in this Chapter. The involvement of the lactam equivalent **81** of pandanamine (**13**) in *Pandanus* chemistry was first discussed by Byrne *et al.* (29) when this compound was proposed as the precursor to pandamarine (**1**), and later by Takayama *et al.* (36). It is intriguing that **1**, whose structure was elucidated using X-ray diffraction (29), and **2** (30) differ in structure only by the replacement of a NH by an O in the two heterocyclic rings. This is especially interesting since the extract from which **2** was isolated was supposedly obtained from the same *Pandanus* species, but collected at a different time and place. No pandamarine (**1**) was observed in the plant extract from which **2** was obtained.

V. PHARMACOLOGY

The spiropiperidine structural unit present in pandamarine (**1**) and pandamar-ilactone-1 (**2**) has received considerable attention from several synthetic organic chemists (52,53). Natural alkaloids with spiropiperidine units are said to display interesting biological properties, such as ion transport inhibition at the cholinergic receptor and phospholipase A2 (PLA$_2$) activity inhibition, and thus are potential anti-inflammatory agents (54,55). To date, there are no reports of any biological activity of the *Pandanus* plants that is directly associated with the alkaloids.

ACKNOWLEDGMENTS

The authors wish to thank the various funding agencies, students, and colleagues who participated in the studies on the *Pandanus* plants described here. We also acknowledge the assistance provided by colleagues towards the completion of this manuscript.

REFERENCES

[1] J. C. Willis, "A Dictionary of the Flowering Plants and Ferns," 8th Edition, University Press, Cambridge, 1973.
[2] B. C. Stone, *Phil. J. Biol.* **5**, 1 (1976).
[3] M. W. Callmander, *Webbia* **55**, 317 (2000).
[4] M. W. Callmander, *Bot. J. Linn. Soc.* **137**, 353 (2001).
[5] M. W. Callmander, P. Chassot, P. Kupfer, and P. P. Lowry, *Taxon* **52**, 747 (2003).
[6] A. J. Macleod and N. M. Pieris, *Phytochemistry* **21**, 1653 (1982).
[7] H. Takayama, T. Kuwajima, M. Kitajima, M. G. Nonato, and N. Aimi, *Nat. Med.* **53**, 335 (1999).
[8] I. Vahirua-Lechat, C. Menut, B. Roig, J. M. Bessiere, and G. Lamaty, *Phytochemistry* **43**, 1277 (1996).
[9] T. T. Jong and S. W. Chau, *Phytochemistry* **49**, 2145 (1998).
[10] P. Peungvicha, S. S. Rhirawarapan, and H. Watanabe, *Jpn. J. Pharmacol.* **78**, 395 (1998).
[11] L. Wu, J. Tan, H. Chen, C. Xie, and Q. Pu, *Zhongcaoyao* **18**, 391 (1987).
[12] C. Ualat, MSc. Thesis, University of Santo Tomas, Manila, Philippines, 1989.
[13] R. G. Buttery, B. O. Juliano, and L. C. Ling, *Chem. Ind.* **12**, 478 (1983).
[14] T. Toyoda, (Ed.), *Flora of Bonin Islands*, Aboc Publishing Company, Kamakura, Japan, pp. 27–29 (1983).

[15] A. Inada, Y. Ikeda, H. Murada, Y. Inatomi, T. Nakanishi, K. Bhattacharyya, T. Kar, G. Bocelli, and A. Cantoni, *Phytochemistry* **66**, 2729 (2005).

[16] R. Braithwaite, *Aust. Nat. Hist.* **25**, 26 (1994–5).

[17] N. Ramarathan and P. R. Kulkarni, *Naturawissenshaften* **71**, 215 (1984).

[18] V. Laksanalamai and S. Ilangantileke, *Cer. Chem.* **70**, 381 (1993).

[19] N. Cheeptham and G. H. N. Towers, *Fitoterapia* **73**, 651 (2002).

[20] E. Quisumbing, "Medicinal Plants of the Philippines," Bureau of Printing, Manila, Philippines, 1978.

[21] M. Tan, "Philippine Medicinal Plants in Common Use: Their Phytochemistry and Pharmacology," AKAP, Quezon City, Philippines, 1980.

[22] E. A. M. Zuhud and Haryanto (eds.), Conservation and Utilisation of Indonesian Tropical Medicinal Plant Diversity, Bogor Agricultural Institute & the Indonesian Tropical Institute, Bogor, Indonesia, 1994.

[23] P. Peungvicha, S. S. Thirawarapan, and H. Watanabe, *Biol. Pharm. Bull.* **19**, 364 (1996).

[24] P. Peungvicha, R. Temsiririrkkul, J. K. Prasain, Y. Tezuka, S. Kadota, S. S. Thirawarapan, and H. Watanabe, *J. Ethnopharmacol.* **62**, 79 (1998).

[25] L. B. Lee, J. Su, and C. N. Ong, *J. Chromatogr.* **1048**, 263 (2004).

[26] K. H. Miean and S. Mohamed, *J. Agric. Food Chem.* **49**, 3106 (2001).

[27] L. S. M. Ooi, E. Y. L. Wong, S. S. M. Sun, and V. E. C. Ooi, *Peptides* **27**, 626 (2006).

[28] L. S. M. Ooi, S. S. M. Sun, and V. E. C. Ooi, *Int. J. Biochem. Cell Biol.* **36**, 1440 (2004).

[29] L. T. Byrne, B. Q. Guevara, W. C. Patalinghug, B. V. Recio, C. R. Ualat, and A. H. White, *Aust. J. Chem.* **45**, 1903 (1992).

[30] M. G. Nonato, M. J. Garson, R. W. Truscott, and J. A. Carver, *Phytochemistry* **34**, 1159 (1993).

[31] R. M. Silverstein, G. C. Bassler, and T. C. Morill, "Spectrometric Identification of Organic Compounds," 5th Edition, Wiley, Singapore, 1991.

[32] R. Zeisberg and F. Bohlman, *Chem. Ber.* **107**, 3800 (1974).

[33] E. A. Bell, L. K. Meier, and H. Sorensen, *Phytochemistry* **20**, 2213 (1981).

[34] J. P. Peterson and L. Fowden, *Phytochemistry* **11**, 663 (1972).

[35] A. Sjaifullah and M. J. Garson, *ACGC Chem. Res. Commun.* **5**, 24 (1996).

[36] H. Takayama, T. Ichikawa, T. Kuwajima, M. Kitajima, H. Seki, N. Aimi, and M. G. Nonato, *J. Am. Chem. Soc.* **122**, 8635 (2000).

[37] S. F. Martin, K. J. Barr, D. W. Smith, and S. K. Bur, *J. Am. Chem. Soc.* **121**, 6990 (1999).

[38] W. Willker and D. Leibfritz, *Magn. Reson. Chem.* **33**, 632 (1995).

[39] P. E. Hansen, *Prog. NMR Spectrosc.* **14**, 175 (1981).

[40] N. Matsumori, M. Murata, and K. Tachibana, *Tetrahedron* **51**, 12229 (1995).

[41] N. Matsumori, T. Nonomura, M. Sasaki, M. Murata, K. Tachibana, M. Satake, and T. Yasumoto, *Tetrahedron Lett.* **37**, 1269 (1996).

[42] K. Furihata and H. Seto, *Tetrahedron Lett.* **40**, 6271 (1999).

[43] H. Takayama, T. Ichikawa, M. Kitajima, M. G. Nonato, and N. Aimi, *Chem. Pharm. Bull.* **50**, 1303 (2002).

[44] H. Takayama, R. Sudo, and M. Kitajima, *Tetrahedron Lett.* **46**, 5795 (2005).

[45] P. Blanco, F. Busque, P. March, M. Figueredo, J. Font, and E. Sanfeliu, *Eur. J. Org. Chem.*, 48 (2004).

[46] H. Takayama, T. Ichikawa, M. Kitajima, M. G. Nonato, and N. Aimi, *J. Nat. Prod.* **64**, 1224 (2001).

[47] H. Takayama, T. Ichikawa, M. Kitajima, N. Aimi, D. Lopez, and M. G. Nonato, *Tetrahedron Lett.* **42**, 2995 (2001).

[48] A. Salim, M. J. Garson, and D. J. Craik, *J. Nat. Prod.* **67**, 54 (2004).

[49] F. Busque, P. March, M. Figueredo, J. Font, and E. Sanfeliu, *Tetrahedron Lett.* **43**, 5583 (2002).

[50] T. Honda, M. Ushiwata, and H. Mizutania, *Tetrahedron Lett.* **47**, 6251 (2006).

[51] H. Takayama, T. Kuwajima, M. Kitajima, M. G. Nonato, and N. Aimi, *Heterocycles* **50**, 75 (1999).

[52] M. J. Martin and F. Bermejo, *Tetrahedron Lett.* **36**, 7705 (1995).

[53] M. J. Martin-Lopez and F. Bermejo, *Tetrahedron* **54**, 12379 (1998).

[54] T. Tokuyama, K. Uenoyama, G. Brown, J. W. Daly, and B. Witkop, *Helv. Chim. Acta* **57**, 2597 (1974).

[55] Y. Inubushi and T. Ibuka, *Heterocycles* **8**, 633 (1977).